Equinoctial Regions of
America, Volume 3
By Alexander von Humboldt

The longitudes mentioned in the text refer always to the meridian of the Observatory of Paris.

The real is about 6 1/2 English pence.

The agrarian measure, called caballeria, is eighteen cordels, (each cordel includes twenty-four varas) or 432 square varas; consequently, as 1 vara = 0.835m., according to Rodriguez, a caballeria is 186,624 square varas, or 130,118 square metres, or thirty-two and two-tenths English acres.

20 leagues to a degree.

5000 varas = 4150 metres.

3403 square toises = 1.29 hectare.

An acre = 4044 square metres.

Five hundred acres = fifteen and a half caballerias.

Sugar-houses are thought to be very considerable that yield 2000 cases annually, or 32,000 arrobas (nearly 368,000 kilogrammes.)

An arroba of 25 Spanish pounds = 11.49 kilogrammes.

A quintal = 45.97 kilogrammes.

A tarea of wood = one hundred and sixty cubic feet.

VOLUME 3.
CHAPTER 3.25.
SPANISH GUIANA. ANGOSTURA. PALM-INHABITING TRIBES. MISSIONS OF THE CAPUCHINS. THE LAGUNA PARIME. EL DORADO. LEGENDARY TALES OF THE EARLY VOYAGERS.

I shall commence this chapter by a description of Spanish Guiana (Provincia de la Guyana), which is a part of the ancient Capitania general of Caracas. Since the end of the sixteenth century three towns have successively borne the name of St. Thomas of Guiana. The first was situated opposite to the island of Faxardo, at the confluence of the Carony and the Orinoco, and was destroyed* by the Dutch, under the command of Captain Adrian Janson, in 1579. (* The first of the voyages undertaken at Raleigh's expense was in 1595; the second, that of Laurence Keymis, in 1596; the third, described by Thomas Masham, in 1597; and the fourth, in 1617. The first and last only were performed by Raleigh in person. This celebrated man was beheaded on October the 29th, 1618. It is therefore the second town of Santo Tomas, now called Vieja Guyana, which existed in the time of Raleigh.) The second, founded by Antonio de Berrio in 1591, near twelve leagues east of the mouth of the Carony, made a courageous resistance to Sir Walter Raleigh, whom the Spanish writers of the conquest know only by the name of the pirate Reali. The third town, now the capital of the province, is fifty leagues west of the confluence of the Carony. It was begun in 1764, under the Governor Don Joaquin Moreno de Mendoza, and is distinguished in the public documents from the second town, vulgarly called the fortress (el castillo, las fortalezas), or Old Guayana (Vieja Guayana), by the name of Santo Thome de la Nueva Guayana. This name being very long, that of Angostura* (the strait) has been commonly substituted for it. (* Europe has learnt the existence of the town of Angostura by the trade carried on by the Catalonians in the Carony bark, which is the beneficial bark of the Bonplanda trifoliata. This bark, coming from Nueva Guiana, was called corteza or cascarilla del Angostura (Cortex Angosturae). Botanists so little guessed the origin of this geographical denomination that they began by writing Augustura, and then Augusta.)

Angostura, the longitude and latitude of which I have already indicated from astronomical observations, stands at the foot of a hill of amphibolic schist* bare of vegetation. (* Hornblendschiefer.) The streets are regular, and for the most part parallel with the course of the river. Several of the houses are built on the bare rock; and here, as at Carichana, and in many other parts of the missions, the action of black and strong strata, when strongly heated by the rays of the sun upon the atmosphere, is considered injurious to health. I think the small pools of stagnant water (lagunas y anegadizos), which extend behind the town in the direction of south-east, are more to be feared. The houses of Angostura are lofty and convenient; they are for the most part built of stone; which proves that the inhabitants have but little dread of earthquakes. But unhappily this security is not founded on induction from any precise data. It is true that the shore of Nueva Andalusia sometimes undergoes very violent shocks, without the commotion being propagated across the Llanos. The fatal catastrophe of Cumana, on the 4th of February, 1797, was not felt at Angostura; but in the great earthquake of 1766, which destroyed the same city, the granitic soil of the two banks of the Orinoco was agitated as far as the Raudales of Atures and Maypures. South of these Raudales shocks are sometimes felt, which are confined to the basin of the Upper Orinoco and the Rio Negro. They appear to depend on a volcanic focus distant from that of the Caribbee Islands. We were told by the missionaries at Javita and San Fernando de Atabapo that in 1798 violent earthquakes took place between the Guaviare and the Rio Negro, which were not propagated on the north towards Maypures. We cannot be sufficiently attentive to whatever relates to the simultaneity of the oscillations, and to the independence of the movements in contiguous ground. Everything seems to prove that the propagation of the commotion is not superficial, but depends on very deep crevices that terminate in different centres of action.

The scenery around the town of Angostura is little varied; but the view of the river, which forms a vast canal, stretching from south-west to north-east, is singularly majestic.

When the waters are high, the river inundates the quays; and it sometimes happens that, even in the town, imprudent persons become the prey of crocodiles. I shall transcribe from my journal a fact that took place during M. Bonpland's illness. A Guaykeri Indian, from the island of La Margareta, was anchoring his canoe in a cove where there were not three feet of water. A very fierce crocodile, which habitually haunted that spot, seized him by the leg, and withdrew from the shore, remaining on the surface of the water. The cries of the Indian drew together a crowd of spectators. This unfortunate man was first seen seeking, with astonishing presence of mind, for a knife which he had in his pocket. Not being able to find it, he seized the head of the crocodile and thrust his fingers into its eyes. No man in the hot regions of America is ignorant that this carnivorous reptile, covered with a buckler of hard and dry scales, is extremely sensitive in the only parts of his body which are soft and unprotected, such as the eyes, the hollow underneath the shoulders, the nostrils, and beneath the lower jaw, where there are two glands of musk. The Guaykeri Indian was less fortunate than the negro of Mungo Park, and the girl of Uritucu, whom I mentioned in a former part of this work, for the crocodile did not open its jaws and lose hold of its prey. The animal, overcome by pain, plunged to the bottom of the river, and, after having drowned the Indian, came up to the surface of the water, dragging the dead body to an island opposite the port. A great number of the inhabitants of Angostura witnessed this melancholy spectacle.

The crocodile, owing to the structure of its larynx, of the hyoidal bone, and of the folds of its tongue, can seize, though not swallow, its prey under water; thus when a man disappears, the animal is usually perceived some hours after devouring its prey on a neighbouring beach. The number of individuals who perish annually, the victims of their own imprudence and of the ferocity of these reptiles, is much greater than is believed in Europe. It is particularly so in villages where the neighbouring grounds are often inundated. The same crocodiles remain long in the same places. They become from year to year more daring, especially, as the Indians assert, if they have once tasted of human flesh. These animals are so wary, that they are killed with difficulty. A ball does not pierce their skin; and the shot is only mortal when it penetrates the throat or a part beneath the shoulder. The Indians, who know little of the use of fire-arms, attack the crocodile with lances, after the animal has been caught with large pointed iron hooks, baited with pieces of meat, and fastened by a chain to the trunk of a tree. They do not approach the animal till it has struggled a long time to disengage itself from the iron fixed in the upper jaw. There is little probability that a country in which a labyrinth of rivers without number brings every day new bands of crocodiles from the eastern back of the Andes, by the Meta and the Apure, toward the coast of Spanish Guiana, should ever be delivered from these reptiles. All that will be gained by civilization will be to render them more timid and more easily put to flight.

Affecting instances are related of African slaves, who have exposed their lives to save those of their masters, who had fallen into the jaws of a crocodile. A few years ago, between Uritucu and the Mission de Abaxo, a negro, hearing the cries of his master, flew to the spot, armed with a long knife (machete), and plunged into the river. He forced the crocodile, by putting out his eyes, to let go his prey and to plunge under the water. The slave bore his expiring master to the shore; but all succour was unavailing to restore him to life. He had died of suffocation, for his wounds were not deep. The crocodile, like the dog, appears not to close its jaws firmly while swimming.

The inhabitants of the banks of the Orinoco and its tributary streams discourse continually on the dangers to which they are exposed. They have marked the manners of the crocodile, as the torero has studied the manners of the bull. When they are assailed, they put in practice, with that presence of mind and that resignation which characterize the Indians, the Zamboes, and copper-coloured men in general, the counsels they have heard from their infancy. In countries where nature is so powerful and so terrible, man is constantly prepared for danger. We have mentioned before the answer of the young Indian girl, who delivered herself from the jaws of the crocodile: "I knew he would let me go if I thrust my fingers into his eyes." This girl belonged to the indigent class of the people, in whom the habits of physical want augment energy of character; but how can we avoid being surprised to observe in the countries convulsed by terrible earthquakes, on the table-land of the province of Quito, women belonging to the highest classes of society display in the moment of peril, the same calm, the same reflecting intrepidity? I shall mention one example only in support of this assertion. On the 4th of February, 1797, when 35,000 Indians perished in the space of a few minutes, a young mother saved herself and her children, crying out to them to extend their arms at the moment when the cracked ground was ready to swallow them up. When this courageous woman heard the astonishment that was expressed at a presence of mind so extraordinary, she answered, with great simplicity, "I had been told in my infancy: if the earthquake surprise you in a house, place yourself under a doorway that communicates from one apartment to another; if you be in the open air and feel the ground opening beneath you, extend both your arms, and try to support yourself on the edge of the crevice." Thus, in savage regions or in countries exposed to frequent convulsions, man is prepared to struggle with the beasts of the forest, to deliver himself from the jaws of the crocodile, and to escape from the conflict of the elements.

The town of Angostura, in the early years of its foundation, had no direct communication with the mother-country. The inhabitants were contented with carrying on a trifling contraband trade in dried meat and tobacco with the West India Islands, and with the Dutch colony of Essequibo, by the Rio Carony. Neither wine, oil, nor flour, three articles of importation the most sought after, was received directly from Spain. Some merchants, in 1771, sent the first schooner to Cadiz; and since that period a direct exchange of commodities with the ports of Andalusia and Catalonia has become extremely active. The population of Angostura,* after having been a long time languishing, has much increased since 1785. (* Angostura, or Santo Thome de la Nueva Guayana, in 1768, had only 500 inhabitants. Caulin page 63. They were numbered in 1780 and the result was 1513 (455 Whites, 449 Blacks, 363 Mulattoes and Zamboes, and 246 Indians). The population in the year 1789 rose to 4590; and in 1800 to 6600 souls. Official Lists manuscript. The capital of the English colony of Demerara, the town of Stabroek, the name of which is scarcely known in Europe, is only fifty leagues distant, south-east of the mouths of the Orinoco. It contains, according to Bolingbroke, nearly 10,000 inhabitants.) At the time of my abode in Guiana, however, it was far from being equal to that of Stabroek, the nearest English town. The mouths of the Orinoco have an advantage over every other part in Terra Firma. They afford the most prompt communications with the Peninsula. The voyage from Cadiz to Punta Barima is performed sometimes in eighteen or twenty days. The return to Europe takes from thirty to thirty-five days. These mouths being placed to windward of all the islands, the vessels of Angostura can maintain a more advantageous commerce with the West Indies than La Guayra and Porto Cabello. The merchants of Caracas, therefore, have been always jealous of the progress of industry in Spanish Guiana; and Caracas having been hitherto the seat of the supreme government, the port of Angostura has been treated with still less favour than the ports of Cumana and Nueva Barcelona. With respect to the inland trade, the most active is that of the province of Varinas, which sends mules, cacao, indigo, cotton, and sugar to Angostura; and in return receives generos, that is, the products of the manufacturing industry of Europe. I have seen long boats (lanchas) set off, the cargoes of which were valued at eight or ten thousand piastres. These boats went first up the Orinoco to Cabruta;

3

then along the Apure to San Vicente; and finally, on the Rio Santo Domingo, as far as Torunos, which is the port of Varinas Nuevas. The little town of San Fernando de Apure, of which I have already given a description, is the magazine of this river-trade, which might become more considerable by the introduction of steamboats.

I have now described the country through which we passed during a voyage of five hundred leagues; it remains for me to make known the small space of three degrees fifty-two minutes of longitude, that separates the present capital from the mouth of the Orinoco. Exact knowledge of the delta and the course of the Rio Carony is at once interesting to hydrography and to European commerce.

When a vessel coming from sea would enter the principal mouth of the Orinoco, the Boca de Navios, it should make the land at the Punta Barima. The right or southern bank is the highest: the granitic rock pierces the marshy soil at a small distance in the interior, between the Cano Barima, the Aquire, and the Cuyuni. The left, or northern bank of the Orinoco, which stretches along the delta towards the Boca de Mariusas and the Punta Baxa, is very low, and is distinguishable at a distance only by the clumps of moriche palm-trees which embellish the passage. This is the sago-tree* of the country (* The nutritious fecula or medullary flour of the sago-trees is found principally in a group of palms which M. Kunth has distinguished by the name of calameae. It is collected, however, in the Indian Archipelago, as an article of trade, from the trunks of the Cycas revoluta, the Phoenix farinifera, the Corypha umbraculifera, and the Caryota urens. (Ainslie, Materia Medica of Hindostan, Madras 1813.)) The quantity of nutritious matter which the real sago-tree of Asia affords (Sagus Rumphii, or Metroxylon sagu, Roxb.) exceeds that which is furnished by any other plant useful to man. One trunk of a tree in its fifteenth year sometimes yields six hundred pounds weight of sago, or meal (for the word sago signifies meal in the dialect of Amboyna). Mr. Crawfurd, who resided a long time in the Indian Archipelago, calculates that an English acre could contain four hundred and thirty-five sago-trees, which would yield one hundred and twenty thousand five hundred pounds avoirdupois of fecula, or more than eight thousand pounds yearly. History of the Indian Archipelago volume 1 pages 387 and 393. This produce is triple that of corn, and double that of potatoes in France. But the plantain produces, on the same surface of land, still more alimentary substance than the sago-tree.); it yields the flour of which the yuruma bread is made; and far from being a palm-tree of the shore, like the Chamaerops humilis, the common cocoa-tree, and the lodoicea of Commerson, is found as a palm-tree of the marshes as far as the sources of the Orinoco.* (* I dwell much on these divisions of the great and fine families of palms according to the distribution of the species: first, in dry places, or inland plains, Corypha tectorum; second, on the sea-coast, Chamaerops humilis, Cocos nucifera, Corypha maritima, Lodoicea seychellarum, Labill.; third, in the fresh-water marshes, Sagus Rumphii, Mauritia flexuosa; and 4th, in the alpine regions, between seven and fifteen hundred toises high, Ceroxylon andicola, Oreodoxa frigida, Kunthia montana. This last group of palmae montanae, which rises in the Andes of Guanacas nearly to the limit of perpetual snow, was, I believe, entirely unknown before our travels in America. (Nov. Gen. volume 1 page 317; Semanario de Santa Fe de Bogota 1819 Number 21 page 163.) In the season of inundations these clumps of mauritia, with their leaves in the form of a fan, have the appearance of a forest rising from the bosom of the waters. The navigator, in proceeding along the channels of the delta of the Orinoco at night, sees with surprise the summit of the palm-trees illumined by large fires. These are the habitations of the Guaraons (Tivitivas and Waraweties of Raleigh* (* The Indian name of the tribe of Uaraus (Guaraunos of the Spaniards) may be recognized in the Warawety (Ouarauoty) of Raleigh, one of the branches of the Tivitivas. See Discovery of Guiana, 1576 page 90 and the sketch of the habitations of the Guaraons, in Raleghi brevis Descrip. Guianae, 1594 tab 4.)), which are suspended from the trunks of trees. These tribes hang up mats in the air, which they fill with earth, and kindle, on a layer of moist clay, the fire necessary for their household wants. They have owed their liberty and their political independence for ages to the quaking and swampy soil, which they pass over in the time of drought, and on which they alone know how to walk in security to their solitude in the delta of the Orinoco; to their abode on the trees where religious enthusiasm will probably never lead any American stylites.* (* This sect was founded by Simeon Sisanites, a native of Syria.

He passed thirty-seven years in mystic contemplation, on five pillars, the last of which was thirty-six cubits high. The sancti columnares attempted to establish their aerial cloisters in the country of Treves, in Germany; but the bishops opposed these extravagant and perilous enterprises. Mosheim, Instit. Hist. Eccles page 192. See Humboldt's Views of Nature (Bohn) pages 13 and 136.) I have already mentioned in another place that the mauritia palm-tree, the tree of life of the missionaries, not only affords the Guaraons a safe dwelling during the risings of the Orinoco, but that its shelly fruit, its farinaceous pith, its juice, abounding in saccharine matter, and the fibres of its petioles, furnish them with food, wine,* and thread proper for making cords and weaving hammocks. (* The use of this moriche wine however is not very common. The Guaraons prefer in general a beverage of fermented honey.) These customs of the Indians of the delta of the Orinoco were found formerly in the Gulf of Darien (Uraba), and in the greater part of the inundated lands between the Guarapiche and the mouths of the Amazon. It is curious to observe in the lowest degree of human civilization the existence of a whole tribe depending on one single species of palm-tree, similar to those insects which feed on one and the same flower, or on one and the same part of a plant.

The navigation of the river, whether vessels arrive by the Boca de Navios, or risk entering the labyrinth of the bocas chicas, requires various precautions, according as the waters are high or low. The regularity of these periodical risings of the Orinoco has been long an object of admiration to travellers, as the overflowings of the Nile furnished the philosophers of antiquity with a problem difficult to solve. The Orinoco and the Nile, contrary to the direction of the Ganges, the Indus, the Rio de la Plata, and the Euphrates, flow alike from the south toward the north; but the sources of the Orinoco are five or six degrees nearer to the equator than those of the Nile. Observing every day the accidental variations of the atmosphere, we find it difficult to persuade ourselves that in a great space of time the effects of these variations mutually compensate each other: that in a long succession of years the averages of the temperature of the humidity, and of the barometric pressure, differ so little from month to month; and that nature, notwithstanding the multitude of partial perturbations, follows a constant type in the series of meteorological phenomena. Great rivers unite in one receptacle the waters which a surface of several thousand square leagues receives. However unequal may be the quantity of rain that falls during several successive years, in such or such a valley, the swellings of rivers that have a very long course are little affected by these local variations. The swellings represent the average of the humidity that reigns in the whole basin; they follow annually the same progression because their commencement and their duration depend also on the mean of the periods, apparently extremely variable, of the beginning and end of the rains in the different latitudes through which the principal trunk and its various tributary streams flow. Hence it follows that the periodical oscillations of rivers are, like the equality of temperature of caverns and springs, a sensible indication of the regular distribution of humidity and heat, which takes place from year to year on a considerable extent of land. They strike the imagination of the vulgar; as order everywhere astonishes, when we cannot easily ascend to first causes. Rivers that belong entirely to the torrid zone display in their periodical movements that wonderful regularity which is peculiar to a region where the same wind brings almost always strata of air of the same temperature; and where the change of the sun in its declination causes every year at the same period a rupture of equilibrium in the electric intensity, in the cessation of the breezes, and the commencement of the season of rains. The Orinoco, the Rio Magdalena, and the Congo or Zaire are the only great rivers of the equinoctial region of the globe, which, rising near the equator, have their mouths in a much higher latitude, though still within the tropics. The Nile and the Rio de la Plata direct their course, in the two opposite hemispheres, from the torrid zone towards the temperate.* (* In Asia, the Ganges, the Burrampooter, and the majestic rivers of Indo-China direct their course towards the equator. The former flow from the temperate to the torrid zone. This circumstance of courses pursuing opposite directions (towards the equator, and towards the temperate climates) has an influence on the period and the height of the risings, on the nature and variety of the productions on the banks of the rivers, on the less or greater activity of trade; and, I may add, from what we know of the nations of Egypt, Merce, and India, on the progress of civilization along the valleys of the rivers.)

As long as, confounding the Rio Paragua of Esmeralda with the Rio Guaviare, the sources of the Orinoco were sought towards the south-west, on the eastern back of the Andes, the risings of this river were attributed to a periodical melting of the snows. This reasoning was as far from the truth as that in which the Nile was formerly supposed to be swelled by the waters of the snows of Abyssinia. The Cordilleras of New Grenada, near which the western tributary streams of the Orinoco, the Guaviare, the Meta, and the Apure take their rise, enter no more into the limit of perpetual snows, with the sole exception of the Paramos of Chita and Mucuchies, than the Alps of Abyssinia. Snowy mountains are much more rare in the torrid zone than is generally admitted; and the melting of the snows, which is not copious there at any season, does not at all increase at the time of the inundations of the Orinoco.

The cause of the periodical swellings of the Orinoco acts equally on all the rivers that take rise in the torrid zone. After the vernal equinox, the cessation of the breezes announces the season of rains. The increase of the rivers (which may be considered as natural pluviometers) is in proportion to the quantity of water that falls in the different regions. This quantity, in the centre of the forests of the Upper Orinoco and the Rio Negro, appeared to me to exceed 90 or 100 inches annually. Such of the natives, therefore, as have lived beneath the misty sky of the Esmeralda and the Atabapo, know, without the smallest notion of natural philosophy, what Eudoxus and Eratosthenes knew heretofore,* that the inundations of the great rivers are owing solely to the equatorial rains. (* Strabo lib. 17 page 789. Diod. Sic. lib. 1 c. 5.) The following is the usual progress of the oscillations of the Orinoco. Immediately after the vernal equinox (the people say on the 25th of March) the commencement of the rising is perceived. It is at first only an inch in twenty-four hours; sometimes the river again sinks in April; it attains its maximum in July; remains at the same level from the end of July till the 25th of August; and then decreases progressively, but more slowly than it increased. It is at its minimum in January and February. In both worlds the rivers of the northern torrid zone attain the greatest height nearly at the same period. The Ganges, the Niger, and the Gambia reach the maximum, like the Orinoco, in the month of August.* (* Nearly forty or fifty days after the summer solstice.) The Nile is two months later, either on account of some local circumstances in the climate of Abyssinia, or of the length of its course, from the country of Berber, or 17.5 degrees of latitude, to the bifurcation of the delta. The Arabian geographers assert that in Sennaar and in Abyssinia the Nile begins to swell in the month of April (nearly as the Orinoco); the rise, however, does not become sensible at Cairo till toward the summer solstice; and the water attains its greatest height at the end of the month of September.* (* Nearly eighty or ninety days after the summer solstice.) The river keeps at the same level till the middle of October; and is at its minimum in April and May, a period when the rivers of Guiana begin to swell anew. It may be seen from this rapid statement, that, notwithstanding the retardation caused by the form of the natural channels, and by local climatic circumstances, the great phenomenon of the oscillations of the rivers of the torrid zone is everywhere the same. In the two zodiacs vulgarly called the Tartar and Chaldean, or Egyptian (in the zodiac which contains the sign of the Rat, an in that which contains those of the Fishes and Aquarius), particular constellations are consecrated to the periodical overflowings of the rivers. Real cycles, divisions of time, have been gradually transformed into divisions of space; but the generality of the physical phenomena of the risings seems to prove that the zodiac which has been transmitted to us by the Greeks, and which, by the precession of the equinoxes, becomes an historical monument of high antiquity, may have taken birth far from Thebes, and from the sacred valley of the Nile. In the zodiacs of the New World—in the Mexican, for instance, of which we discover the vestiges in the signs of the days, and the periodical series which they compose—there are also signs of rain and of inundation corresponding to the Chou (Rat) of the Chinese* and Thibetan cycle of Tse, and to the Fishes and Aquarius of the dodecatemorion. (* The figure of water itself is often substituted for that of the Rat (Arvicola) in the Tartar zodiac. The Rat takes the place of Aquarius. Gaubil, Obs. Mathem. volume 3 page 33.) These two Mexican signs are Water (Atl) and Cipactli, the sea-monster furnished with a horn. This animal is at once the Antelope-fish of the Hindoos, the Capricorn of our zodiac, the Deucalion of the Greeks, and the Noah (Coxcox) of the Azteks.* (* Coxcox bears also the denomination of Teo-Cipactli, in which the root god or divine is added to the name of the sign Cipactli. It is the man of the Fourth Age; who, at the fourth destruction of the world (the last renovation of nature), saved himself with his wife, and reached the mountain of Colhuacan. According to the commentator Germanicus, Deucalion was placed in Aquarius; but the three signs of the Fishes, Aquarius and Capricorn (the Antelope-fish) were heretofore intimately linked together. The animal, which, after having long inhabited the waters, takes the form of an antelope, and climbs the mountains, reminds people, whose restless imagination seizes the most remote similitudes, of the ancient traditions of Menou, of Noah, and of those Deucalions celebrated among the Scythians and the Thessalians. As the Tartarian and Mexican zodiacs contain the signs of the Monkey and the Tiger, they, no doubt, originated in the torrid zone. With the Muyscas, inhabitants of New Grenada, the first sign, as in eastern Asia, was that of water, figured by a Frog. It is also remarkable that the astrological worship of the Muyscas came to the table-land of Bogota from the eastern side, from the plains of San Juan, which extend toward the Guaviare and the Orinoco.) Thus we find the general results of comparative hydrography in the astrological monuments, the divisions of time and the religious traditions of nations the most remote from each other in their situation and in their degree of intellectual advancement.

As the equatorial rains take place in the flat country when the sun passes through the zenith of the place, that is, when its declination becomes homonymous with the zone comprised between the equator and one of the tropics, the waters of the Amazon sink, while those of the Orinoco rise perceptibly. In a very judicious discussion on the origin of the Rio Congo,* (* Voyage to the Zaire page 17.) the attention of philosophers has been already called to the modifications which the periods of the risings must undergo in the course of a river, the sources and the mouth of which are not on the same side of the equinoctial line.* (* Among the rivers of America this is the case with the Rio Negro, the Rio Branco, and the Jupura.) The hydraulic systems of the Orinoco and the Amazon furnish a combination of circumstances still more extraordinary. They are united by the Rio Negro and the Cassiquiare, a branch of the Orinoco; it is a navigable line, between two great basins of rivers, that is crossed by the equator. The river Amazon, according to the information which I obtained on its banks, is much less regular in the periods of its oscillations than the Orinoco; it generally begins, however, to increase in December, and attains its maximum of height in March.* (* Nearly seventy or eighty days after our winter solstice, which is the summer solstice of the southern hemisphere.) It sinks from the month of May, and is at its minimum of height in the months of July and August, at the time when the Lower Orinoco inundates all the surrounding land. As no river of America can cross the equator from south to north, on account of the general configuration of the ground, the risings of the Orinoco have an influence on the Amazon; but those of the Amazon do not alter the progress of the oscillations of the Orinoco. It results from these data, that in the two basins of the Amazon and the Orinoco, the concave and convex summits of the curve of progressive increase and decrease correspond very regularly with each other, since they exhibit the difference of six months, which results from the situation of the rivers in opposite hemispheres. The commencement of the risings only is less tardy in the Orinoco. This river increases sensibly as soon as the sun has crossed the equator; in the Amazon, on the contrary, the risings do not commence till two months after the equinox. It is known that in the forests north of the line the rains are earlier than in the less woody plains of the southern torrid zone. To this local cause is joined another, which acts perhaps equally on the tardy swellings of the Nile. The Amazon receives a great part of its waters from the Cordillera of the Andes, where the seasons, as everywhere among mountains, follow a peculiar type, most frequently opposite to that of the low regions.

The law of the increase and decrease of the Orinoco is more difficult to determine with respect to space, or to the magnitude of the oscillations, than with regard to time, or the period of the maxima and minima. Having been able to measure but imperfectly the risings of the river, I report, not without hesitation, estimates that differ much from each other.* (*

Tuckey, Maritime Geogr. volume 4 page 309. Hippisley, Expedition to the Orinoco page 38. Gumilla volume 1 pages 56 to 59. Depons volume 3 page 301. The greatest height of the rise of the Mississippi is, at Natchez, fifty-five English feet. This river (the largest perhaps of the whole temperate zone) is at its maximum from February to May; at its minimum in August and September. Ellicott, Journal of an Expedition to the Ohio.) Foreign pilots admit ninety feet for the ordinary rise in the Lower Orinoco. M. Depons, who has in general collected very accurate notions during his stay at Caracas, fixes it at thirteen fathoms. The heights naturally vary according to the breadth of the bed and the number of tributary streams which the principal trunk receives.

The people believe that every five years the Orinoco rises three feet higher than common; but the idea of this cycle does not rest on any precise measures. We know by the testimony of antiquity, that the oscillations of the Nile have been sensibly the same with respect to their height and duration for thousands of years; which is a proof, well worthy of attention, that the mean state of the humidity and the temperature does not vary in that vast basin. Will this constancy in physical phenomena, this equilibrium of the elements, be preserved in the New World also after some ages of cultivation? I think we may reply in the affirmative; for the united efforts of man cannot fail to have an influence on the general causes on which the climate of Guiana depends.

According to the barometric height of San Fernando de Apure, I find from that town to the Boca de Navios the slope of the Apure and the Lower Orinoco to be three inches and a quarter to a nautical mile of nine hundred and fifty toises.* (* The Apure itself has a slope of thirteen inches to the mile.) We may be surprised at the strength of the current in a slope so little perceptible; but I shall remind the reader on this occasion, that, according to measurements made by order of Mr. Hastings, the Ganges was found, in a course of sixty miles (comprising the windings,) to have also only four inches fall to a mile; that the mean swiftness of this river is, in the seasons of drought, three miles an hour, and in those of rains six or eight miles. The strength of the current, therefore, in the Ganges as in the Orinoco, depends less on the slope of the bed, than on the accumulation of the higher waters, caused by the abundance of the rains, and the number of tributary streams. European colonists have already been settled for two hundred and fifty years on the banks of the Orinoco; and during this long period of time, according to a tradition which has been propagated from generation to generation, the periodical oscillations of the river (the time of the beginning of the rising, and that when it attains its maximum) have never been retarded more than twelve or fifteen days.

When vessels that draw a good deal of water sail up toward Angostura in the months of January and February, by favour of the sea-breeze and the tide, they run the risk of taking the ground. The navigable channel often changes its breadth and direction; no buoy, however, has yet been laid down, to indicate any deposit of earth formed in the bed of the river, where the waters have lost their original velocity. There exists on the south of Cape Barima, as well by the river of this name as by the Rio Moroca and several estuaries (esteres) a communication with the English colony of Essequibo. Small vessels can penetrate into the interior as far as the Rio Poumaron, on which are the ancient settlements of Zealand and Middleburg. Heretofore this communication interested the government of Caracas only on account of the facility it furnished to an illicit trade; but since Berbice, Demerara, and Essequibo have fallen into the hands of a more powerful neighbour, it fixes the attention of the Spanish Americans as being connected with the security of their frontiers. Rivers which have a course parallel to the coast, and are nowhere farther distant from it than five or six nautical miles, characterize the whole of the shore between the Orinoco and the Amazon.

Ten leagues distant from Cape Barima, the great bed of the Orinoco is divided for the first time into two branches of two thousand toises in breadth. They are known by the Indian names of Zacupana and Imataca. The first, which is the northernmost, communicates on the west of the islands Congrejos and del Burro with the bocas chicas of Lauran, Nuina, and Mariusas. As the Isla del Burro disappears in the time of great inundations, it is unhappily not suited to fortifications. The southern bank of the brazo Imataca is cut by a labyrinth of little channels, into which the Rio Imataca and the Rio Aquire flow. A long series of little granitic hills rises in the fertile savannahs between the Imataca and the Cuyuni;

8

it is a prolongation of the Cordilleras of Parima, which, bounding the horizon south of Angostura, forms the celebrated cataracts of the Rio Caroni, and approaches the Orinoco like a projecting cape near the little fort of Vieja Guyana. The populous missions of the Caribbee and Guiana Indians, governed by the Catalonian Capuchins, lie near the sources of the Imataca and the Aquire. The easternmost of these missions are those of Miamu, Camamu, and Palmar, situate in a hilly country, which extends towards Tupuquen, Santa Maria, and the Villa de Upata. Going up the Rio Aquire, and directing your course across the pastures towards the south, you reach the mission of Belem de Tumeremo, and thence the confluence of the Curumu with the Rio Cuyuni, where the Spanish post or destacamento de Cuyuni was formerly established. I enter into this topographical detail because the Rio Cuyuni, or Cuduvini, runs parallel to the Orinoco from west to east, through an extent of 2.5 or 3 degrees of longitude,* and furnishes an excellent natural boundary between the territory of Caracas and that of English Guiana. (* Including the Rio Juruam, one of the principal branches of the Cuyuni. The Dutch military post is five leagues west of the union of Cuyuni with the Essequibo, where the former river receives the Mazuruni.)

The two great branches of the Orinoco, the Zacupana and the Imataca, remain separate for fourteen leagues: on going up farther, the waters of the river are found united* in a single channel extremely broad. (* At this point of union are found two villages of Guaraons. They also bear the names of Imataca and Zacupana.) This channel is near eight leagues long; at its western extremity a second bifurcation appears; and as the summit of the delta is in the northern branch of the bifurcated river, this part of the Orinoco is highly important for the military defence of the country. All the channels* that terminate in the bocas chicas, rise from the same point of the trunk of the Orinoco. (* Cano de Manamo grande, Cano de Manamo chico, Cano Pedernales, Cano Macareo, Cano Cutupiti, Cano Macuona, Cano grande de Mariusas, etc. The last three branches form by their union the sinuous channel called the Vuelta del Torno.) The branch (Cano Manamo) that separates from it near the village of San Rafael has no ramification till after a course of three or four leagues; and by placing a small fort above the island of Chaguanes, Angostura might be defended against an enemy that should attempt to penetrate by one of the bocas chicas. In my time the station of the gun-boats was east of San Rafael, near the northern bank of the Orinoco. This is the point which vessels must pass in sailing up toward Angostura by the northern channel, that of San Rafael, which is the broadest but the most shallow.

Six leagues above the point where the Orinoco sends off a branch to the bocas chicas is placed an ancient fort (los Castillos de la Vieja or Antigua Guayana,) the first construction of which goes back to the sixteenth century. In this spot the bed of the river is studded with rocky islands; and it is asserted that its breadth is nearly six hundred and fifty toises. The town is almost destroyed, but the fortifications subsist, and are well worthy the attention of the government of Terra Firma. There is a magnificent view from the battery established on a bluff north-west of the ancient town, which, at the period of great inundations, is entirely surrounded with water. Pools that communicate with the Orinoco form natural basins, adapted for the reception of vessels that want repairs.

After having passed the little forts of Vieja Guayana, the bed of the Orinoco again widens. The state of cultivation of the country on the two banks affords a striking contrast. On the north is seen the desert part of the province of Cumana, steppes (Llanos) destitute of habitations, and extending beyond the sources of the Rio Mamo, toward the tableland or mesa of Guanipa. On the south we find three populous villages belonging to the missions of Carony, namely, San Miguel de Uriala, San Felix and San Joaquin. The last of these villages, situate on the banks of the Carony, immediately below the great cataract, is considered as the embarcadero of the Catalonian missions. On navigating more to the east, between the mouth of the Carony and Angostura, the pilot should avoid the rocks of Guarampo, the sandbank of Mamo, and the Piedra del Rosario. From the numerous materials which I brought home, and from astronomical discussions, the principal results of which I have indicated above, I have constructed a map of the country bounded by the delta of the Orinoco, the Carony, and the Cuyuni. This part of Guiana, from its proximity to the coast, will some day offer the greatest attraction to European settlers.

The whole population of this vast province in its present state is, with the exception of a few Spanish parishes, scattered on the banks of the Lower Orinoco, and subject to two monastic governments. Estimating the number of the inhabitants of Guiana, who do not live in savage independence, at thirty-five thousand, we find nearly twenty-four thousand settled in the missions, and thus withdrawn as it were from the direct influence of the secular arm. At the period of my voyage, the territory of the Observantin monks of St. Francis contained seven thousand three hundred inhabitants, and that of the Capuchinos Catalanes seventeen thousand; an astonishing disproportion, when we reflect on the smallness of the latter territory compared to the vast banks of the Upper Orinoco, the Atabapo, the Cassiquiare and the Rio Negro. It results from these statements that nearly two-thirds of the population of a province of sixteen thousand eight hundred square leagues are found concentrated between the Rio Imataca and the town of Santo Thome del Angostura, on a space of ground only fifty-five leagues in length, and thirty in breadth. Both of these monastic governments are equally inaccessible to Whites, and form status in statu. The first, that of the Observantins, I have described from my own observations; it remains for me to record here the notions I could procure respecting the second of these governments, that of the Catalonian Capuchins. Fatal civil dissensions and epidemic fevers have of late years diminished the long-increasing prosperity of the missions of the Carony; but, notwithstanding these losses, the region which we are going to examine is still highly interesting with respect to political economy.

The missions of the Catalonian Capuchins, which in 1804 contained at least sixty thousand head of cattle grazing in the savannahs, extend from the eastern banks of the Carony and the Paragua as far as the banks of the Imataca, the Curumu, and the Cuyuni; at the south-east they border on English Guiana, or the colony of Essequibo; and toward the south, in going up the desert banks of the Paragua and the Paraguamasi, and crossing the Cordillera of Pacaraimo, they touch the Portuguese settlements on the Rio Branco. The whole of this country is open, full of fine savannahs, and no way resembling that through which we passed on the Upper Orinoco. The forests become impenetrable only on advancing toward the south; on the north are meadows intersected with woody hills. The most picturesque scenes lie near the falls of the Carony, and in that chain of mountains, two hundred and fifty toises high, which separates the tributary streams of the Orinoco from those of the Cuyuni. There are situate the Villa de Upata,* the capital of these missions, Santa Maria, and Cupapui. (* Founded in 1762. Population in 1797, 657 souls; in 1803, 769 souls. The most populous villages of these missions, Alta Gracia, Cupapui, Santa Rosa de Cura, and Guri, had between 600 and 900 inhabitants in 1797; but in 1818 epidemic fevers diminished the population more than a third. In some missions these diseases have swept away nearly half of the inhabitants.) Small table-lands afford a healthy and temperate climate. Cacao, rice, cotton, indigo, and sugar grow in abundance wherever a virgin soil, covered with a thick coat of grasses, is subjected to cultivation. The first Christian settlements in those countries are not, I believe, of an earlier date than 1721. The elements of which the present population is composed are the three Indian races of the Guayanos, the Caribs and the Guaycas. The last are a people of mountaineers and are far from being so diminutive in size as the Guaycas whom we found at Esmeralda. It is difficult to fix them to the soil; and the three most modern missions in which they have been collected, those of Cura, Curucuy, and Arechica, are already destroyed. The Guayanos, who early in the sixteenth century gave their name to the whole of that vast province, are less intelligent but milder; and more easy, if not to civilize, at least to subjugate, than the Caribs. Their language appears to belong to the great branch of the Caribbee and Tamanac tongues. It displays the same analogies of roots and grammatical forms, which are observed between the Sanscrit, the Persian, the Greek, and the German. It is not easy to fix the forms of what is indefinite by its nature; and to agree on the differences which should be admitted between dialects, derivative languages and mother-tongues. The Jesuits of Paraguay have made known to us another tribe of Guayanos* in the southern hemisphere, living in the thick forests of Parana. (* They are also called Guananas, or Gualachas.) Though it cannot be denied in general that in consequence of distant migrations,* (* Like the celebrated migrations of the Omaguas, or Omeguas.) the nations that are settled north and south of the Amazon have had communications with each

10

other, I will not decide whether the Guayanos of Parana and of Uruguay exhibit any other relation to those of Carony, than that of an homonomy, which is perhaps only accidental.

The most considerable Christian settlements are now concentrated between the mountains of Santa Maria, the mission of San Miguel and the eastern bank of the Carony, from San Buenaventura as far as Guri and the embarcadero of San Joaquin; a space of ground which has not more than four hundred and sixty square leagues of surface. The savannahs to the east and south are almost uninhabited; we find there only the solitary missions of Belem, Tumuremo, Tupuquen, Puedpa, and Santa Clara. It were to be wished that the spots preferred for cultivation were distant from the rivers where the land is higher and the air more favourable to health. The Rio Carony, the waters of which, of an admirable clearness, are not well stocked with fish, is free from shoals from the Villa de Barceloneta, a little above the confluence of the Paragua, as far as the village of Guri. Farther north it winds between innumerable islands and rocks; and only the small boats of the Caribs venture to navigate amid these raudales, or rapids of the Carony. Happily the river is often divided into several branches; and consequently that can be chosen which, according to the height of the waters, presents the fewest whirlpools and shoals. The great fall, celebrated for the picturesque beauty of its situation, is a little above the village of Aguacaqua, or Carony, which in my time had a population of seven hundred Indians. This cascade is said to be from fifteen to twenty feet high; but the bar does not cross the whole bed of the river, which is more than three hundred feet broad. When the population is more extended toward the east, it will avail itself of the course of the small rivers Imataca and Aquire, the navigation of which is pretty free from danger. The monks, who like to keep themselves isolated, in order to withdraw from the eye of the secular power, have been hitherto unwilling to settle on the banks of the Orinoco. It is, however, by this river only, or by the Cuyuni and the Essequibo, that the missions of Carony can export their productions. The latter way has not yet been tried, though several Christian settlements* are formed on one of the principal tributary streams of the Cuyuni, the Rio Juruario. (* Guacipati, Tupuquen, Angel de la Custodia, and Cura, where the military post of the frontiers was stationed in 1800, which had been anciently placed at the confluence of the Cuyuni and the Curumu.) This stream furnishes, at the period of the great swellings, the remarkable phenomenon of a bifurcation. It communicates by the Juraricuima and the Aurapa with the Rio Carony; so that the land comprised between the Orinoco, the sea, the Cuyuni, and the Carony, becomes a real island. Formidable rapids impede the navigation of the Upper Cuyuni; and hence of late an attempt has been made to open a road to the colony of Essequibo much more to the south-east, in order to fall in with the Cuyuni much below the mouth of the Curumu.

The whole of this southern territory is traversed by hordes of independent Caribs; the feeble remains of that warlike people who were so formidable to the missionaries till 1733 and 1735, at which period the respectable bishop Gervais de Labrid,* (* Consecrated a bishop for the four parts of the world (obispo para las quatro partes del mundo) by pope Benedict XIII.) canon of the metropolitan chapter of Lyon, Father Lopez, and several other ecclesiastics, perished by the hands of the Caribs. These dangers, too frequent formerly, exist no longer, either in the missions of Carony, or in those of the Orinoco; but the independent Caribs continue, on account of their connection with the Dutch colonists of Essequibo, an object of mistrust and hatred to the government of Guiana. These tribes favour the contraband trade along the coast, and by the channels or estuaries that join the Rio Barima to the Rio Moroca; they carry off the cattle belonging to the missionaries, and excite the Indians recently converted, and living within the sound of the bell, to return to the forests. The free hordes have everywhere a powerful interest in opposing the progress of cultivation and the encroachments of the Whites. The Caribs and the Aruacas procure fire-arms at Essequibo and Demerara; and when the traffic of American slaves (poitos) was most active, adventurers of Dutch origin took part in these incursions on the Paragua, the Erevato, and the Ventuario. Man-hunting took place on these banks, as heretofore (and probably still) on those of the Senegal and the Gambia. In both worlds Europeans have employed the same artifices, and committed the same atrocities, to maintain a trade that dishonours humanity. The missionaries of the Carony and the Orinoco attribute all the evils they suffer from the independent Caribs to the hatred of their neighbours, the Calvinist preachers of Essequibo.

11

Their works are therefore filled with complaints of the secta diabolica de Calvino y de Lutero, and against the heretics of Dutch Guiana, who also think fit sometimes to go on missions, and spread the germs of social life among the savages.

Of all the vegetable productions of those countries, that which the industry of the Catalonian Capuchins has rendered the most celebrated is the tree that furnishes the Cortex angosturae, which is erroneously designated by the name of cinchona of Carony. We were fortunate enough to make it first known as a new genus distinct from the cinchona, and belonging to the family of meliaceae, or of zanthoxylus. This salutary drug of South America was formerly attributed to the Brucea ferruginea which grows in Abyssinia, to the Magnolia glauca, and to the Magnolia plumieri. During the dangerous disease of M. Bonpland, M. Ravago sent a confidential person to the missions of Carony, to procure for us, by favour of the Capuchins of Upata, branches of the tree in flower which we wished to be able to describe. We obtained very fine specimens, the leaves of which, eighteen inches long, diffused an agreeable aromatic smell. We soon perceived that the cuspare (the indigenous name of the cascarilla or corteza del Angostura) forms a new genus; and on sending the plants of the Orinoco to M. Willldenouw, I begged he would dedicate this plant to M. Bonpland. The tree, known at present by the name of Bonplandia trifoliata, grows at the distance of five or six leagues from the eastern bank of the Carony, at the foot of the hills that surround the missions Capapui, Upata and Alta Gracia. The Caribbee Indians make use of an infusion of the bark of the cuspare, which they consider as a strengthening remedy. M. Bonpland discovered the same tree west of Cumana, in the gulf of Santa Fe, where it may become one of the articles of exportation from New Andalusia.

The Catalonian monks prepare an extract of the Cortex angosturae which they send to the convents of their province, and which deserves to be better known in the north of Europe. It is to be hoped that the febrifuge and anti-dysenteric bark of the bonplandia will continue to be employed, notwithstanding the introduction of another, described by the name of False Angostura bark, and often confounded with the former. This false Angostura, or Angostura pseudo-ferruginea, comes, it is said, from the Brucea antidysenterica; it acts powerfully on the nerves, produces violent attacks of tetanus, and contains, according to the experiments of Pelletier and Caventon, a peculiar alkaline substance* analogous to morphine and strychnine. (* Brucine. M. Pelletier has wisely avoided using the word angosturine, because it might indicate a substance taken from the real Cortex angosturae, or Bonplandia trifoliata. (Annales de Chimie volume 12 page 117.) We saw at Peru the barks of two new species of weinmannia and wintera mixed with those of cinchona; a mixture less dangerous, but still injurious, on account of the superabundance of tannin and acrid matter contained in the false cascarilla.) As the tree which yields the real Cortex angosturae does not grow in great abundance, it is to be wished that plantations of it were formed. The Catalonian monks are well fitted to spread this kind of cultivation; they are more economical, industrious, and active than the other missionaries. They have already established tan-yards and cotton-spinning in a few villages; and if they suffer the Indians henceforth to enjoy the fruit of their labours, they will find great resources in the native population. Concentered on a small space of land, these monks have the consciousness of their political importance, and have from time to time resisted the civil authority, and that of their bishop. The governors who reside at Angostura have struggled against them with very unequal success, according as the ministry of Madrid showed a complaisant deference for the ecclesiastical hierarchy, or sought to limit its power. In 1768 Don Manuel Centurion carried off twenty thousand head of cattle from the missionaries, in order to distribute them among the indigent inhabitants. This liberality, exerted in a manner not very legal, produced very serious consequences. The governor was disgraced on the complaint of the Catalonian monks though he had considerably extended the territory of the missions toward the south, and founded the Villa de Barceloneta, above the confluence of the Carony with the Rio Paragua, and the Ciudad de Guirior, near the union of the Rio Paragua and the Paraguamusi. From that period the civil administration has carefully avoided all intervention in the affairs of the Capuchins, whose opulence has been exaggerated like that of the Jesuits of Paraguay.

The missions of the Carony, by the configuration of their soil* and the mixture of savannahs and arable lands, unite the advantages of the Llanos of Calabozo and the valleys of Aragua. (* It appears that the little table-lands between the mountains of Upata, Cumanu, and Tupuquen, are more than one hundred and fifty toises above the level of the sea.) The real wealth of this country is founded on the care of the herds and the cultivation of colonial produce. It were to be wished that here, as in the fine and fertile province of Venezuela, the inhabitants, faithful to the labours of the fields, would not addict themselves too hastily to the research of mines. The example of Germany and Mexico proves, no doubt, that the working of metals is not at all incompatible with a flourishing state of agriculture; but, according to popular traditions, the banks of the Carony lead to the lake Dorado and the palace of the gilded man* (* El Dorado, that is, el rey o hombre dorado. See volume 2.23.): and this lake, and this palace, being a local fable, it might be dangerous to awaken remembrances which begin gradually to be effaced. I was assured that in 1760, the independent Caribs went to Cerro de Pajarcima, a mountain to the south of Vieja Guayana, to submit the decomposed rock to the action of washing. The gold-dust collected by this labour was put into calabashes of the Crescentia cujete and sold to the Dutch at Essequibo. Still more recently, some Mexican miners, who abused the credulity of Don Jose Avalo, the intendant of Caracas, undertook a very considerable work in the centre of the missions of the Rio Carony, near the town of Upata, in the Cerros del Potrero and de Chirica. They declared that the whole rock was auriferous; stamping-mills, brocards, and smelting-furnaces were constructed. After having expended very large sums, it was discovered that the pyrites contained no trace whatever of gold. These essays, though fruitless, served to renew the ancient idea that every shining rock in Guiana is teeming with gold (una madre del oro). Not contented with taking the mica-slate to the furnace, strata of amphibolic slates were shown to me near Angostura, without any mixture of heterogeneous substances, which had been worked under the whimsical name of black ore of gold (oro negro).

This is the place to make known, in order to complete the description of the Orinoco, the principal results of my researches on El Dorado, the White Sea, or Laguna Parime, and the sources of the Orinoco, as they are marked in the most recent maps. The idea of an auriferous earth, eminently rich, has been connected, ever since the end of the sixteenth century, with that of a great inland lake, which furnishes at the same time waters to the Orinoco, the Rio Branco and the Rio Essequibo. I believe, from a more accurate knowledge of the country, a long and laborious study of the Spanish authors who treat of El Dorado, and, above all, from comparing a great number of ancient maps, arranged in chronological order, I have succeeded in discovering the source of these errors. All fables have some real foundation; that of El Dorado resembles those myths of antiquity, which, travelling from country to country, have been successively adapted to different localities. In the sciences, in order to distinguish truth from error, it often suffices to retrace the history of opinions, and to follow their successive developments. The discussion to which I shall devote the end of this chapter is important, not only because it throws light on the events of the Conquest, and that long series of disastrous expeditions made in search of El Dorado, the last of which was in the year 1775; it also furnishes, in addition to this simply historical interest, another, more substantial and more generally felt, that of rectifying the geography of South America, and of disembarrassing the maps published in our days of those great lakes, and that strange labyrinth of rivers, placed as if by chance between sixty and sixty-six degrees of longitude. No man in Europe believes any longer in the wealth of Guiana and the empire of the Grand Patiti. The town of Manoa and its palaces covered with plates of massy gold have long since disappeared; but the geographical apparatus serving to adorn the fable of El Dorado, the lake Parima, which, similar to the lake of Mexico, reflected the image of so many sumptuous edifices, has been religiously preserved by geographers. In the space of three centuries, the same traditions have been differently modified; from ignorance of the American languages, rivers have been taken for lakes, and portages for branches of rivers; one lake, the Cassipa, has been made to advance five degrees of latitude toward the south, while another, the Parima or Dorado, has been transported the distance of a hundred leagues from the western to the eastern bank of the Rio Branco. From these various changes, the problem we are going to solve has become much more complicated than is generally supposed. The number of geographers who discuss the basis of a map, with regard to the three points of measures, of the comparison of descriptive works, and of the etymological study* of names, is extremely small. (* I use this expression, perhaps an improper one, to mark a species of philological examination, to which the names of rivers, lakes, mountains, and tribes, must be subjected, in order to discover their identity in a great number of maps. The apparent diversity of names arises partly from the difference of the dialects spoken by one and the same family of people, partly from the imperfection of our European orthography, and from the extreme negligence with which geographers copy one another. We recognize with difficulty the Rio Uaupe in the Guaupe or Guape; the Xie, in the Guaicia; the Raudal de Atures, in Athule; the Caribbees, in the Calinas and Galibis; the Guaraunos or Uarau, in the Oaraw-its; etc. It is, however, by similar mutations of letters, that the Spaniards have made hijo of filius; hambre, of fames; and Felipo de Urre, and even Utre, of the Conquistador Philip von Huten; that the Tamanacs in America have substituted choraro for soldado; and the Jews in China, Ialemeiohang for Jeremiah. Analogy and a certain etymological tact must guide geographers in researches of this kind, in which they would be exposed to serious errors, if they were not to study at the same time the respective situations of the upper and lower tributary streams of the same river. Our maps of America are overloaded with names, for which rivers have been created. This desire of compiling, of filling up vacancies, and of employing, without investigation, heterogeneous materials, has given our maps of countries the least visited an appearance of exactness, the falsity of which is discovered when we arrive on the spot.) Almost all the maps of South America which have appeared since the year 1775 are, in what regards the interior of the country, comprised between the steppes of Venezuela and the river of the Amazons, between the eastern back of the Andes and the coast of Cayenne, a simple copy of the great Spanish map of La Cruz Olmedilla. A line, indicating the extent of country which Don Jose Solano boasted of having discovered and pacified by his troops and emissaries, was taken for the road followed by that officer, who never went beyond San Fernando de Atabapo, a village one hundred and sixty leagues distant from the pretended lake Parima. The study of the work of Father Caulin, who was the historiographer of the expedition of Solano, and who states very clearly, from the testimony of the Indians, how the name of the river Parima gave rise to the fable of El Dorado, and of an inland sea, has been neglected. No use either has been made of a map of the Orinoco, three years posterior to that of La Cruz, and traced by Surville from the collection of true or hypothetical materials preserved in the archives of the Despacho universal de Indias. The progress of geography, as manifested on our maps, is much slower than might be supposed from the number of useful results which are found scattered in the works of different nations. Astronomical observations and topographic information accumulate during a long lapse of years, without being made use of; and from a principle of stability and preservation, in other respects praiseworthy, those who construct maps often choose rather to add nothing, than to sacrifice a lake, a chain of mountains, or an interbranching of rivers, which have figured there during ages.

The fabulous traditions of El Dorado and the lake Parima having been diversely modified according to the aspect of the countries to which they were to be adapted, we must distinguish what they contain that is real from what is merely imaginary. To avoid entering here into minute particulars, I shall begin first to call the attention of the reader to those spots which have been, at various periods, the theatre of the expeditions undertaken for the discovery of El Dorado. When we have learnt to know the aspect of the country, and the local circumstances, such as they can now be described, it will be easy to conceive how the different hypotheses recorded on our maps have taken rise by degrees, and have modified each other. To oppose an error, it is sufficient to recall to mind the variable forms in which we have seen it appear at different periods.

Till the middle of the eighteenth century, all that vast space of land comprised between the mountains of French Guiana and the forests of the Upper Orinoco, between the sources of the Carony and the River Amazon (from 0 to 4 degrees of north latitude, and from 57 to 68 degrees of longitude), was so little known that geographers could place in it lakes where they pleased, create communications between rivers, and figure chains of mountains more or less lofty. They have made full use of this liberty; and the situation of lakes, as well as the course and branches of rivers, has been varied in so many ways that it would not be surprising if among the great number of maps some were found that trace the real state of things. The field of hypotheses is now singularly narrowed. I have determined the longitude of Esmeralda in the Upper Orinoco; more to the east amid the plains of Parima (a land as unknown as Wangara and Dar-Saley, in Africa), a band of twenty leagues broad has been travelled over from north to south along the banks of the Rio Carony and the Rio Branco in the longitude of sixty-three degrees. This is the perilous road which was taken by Don Antonio Santos in going from Santo Thome del Angostura to Rio Negro and the Amazon; by this road also the colonists of Surinam communicated very recently with the inhabitants of Grand Para. This road divides the terra incognita of Parima into two unequal portions; and fixes limits at the same time to the sources of the Orinoco, which it is no longer possible to carry back indefinitely toward the east, without supposing that the bed of the Rio Branco, which flows from north to south, is crossed by the bed of the Upper Orinoco, which flows from east to west. If we follow the course of the Rio Branco, or that strip of cultivated land which is dependent on the Capitania General of Grand Para, we see lakes, partly imaginary and partly enlarged by geographers, forming two distinct groups. The first of these groups includes the lakes which they place between the Esmeralda and the Rio Branco; and to the second belong those that are supposed to lie between the Rio Branco and the mountains of Dutch and French Guiana. It results from this sketch that the question whether there exists a lake Parima on the east of the Rio Branco is altogether foreign to the problem of the sources of the Orinoco.

Beside the country which we have just noticed (the Dorado de la Parime, traversed by the Rio Branco), another part of America is found, two hundred and sixty leagues toward the west, near the eastern back of the Cordillera of the Andes, equally celebrated in the expeditions to El Dorado. This is the Mesopotamia between the Caqueta, the Rio Negro, the Uaupes, and the Yurubesh, of which I have already given a particular account; it is the Dorado of the Omaguas which contains Lake Manoa of Father Acunha, the Laguna de oro of the Guanes and the auriferous land whence Father Fritz received plates of beaten gold in his mission on the Amazon, toward the end of the seventeenth century.

The first and above all the most celebrated enterprises attempted in search of El Dorado were directed toward the eastern back of the Andes of New Grenada. Fired with the ideas which an Indian of Tacunga had given of the wealth of the king or zaque of Cundirumarca, Sebastian de Belalcazar, in 1535, sent his captains Anasco and Ampudia, to discover the valley of El Dorado,* twelve days' journey from Guallabamba, consequently in the mountains between Pasto and Popayan. (* El valle del Dorado. Pineda relates: que mas adelante de la provincia de la Canela se hallan tierras muy ricas, adonde andaban los hombres armados de piecas y joyas de oro, y que no havia sierra, ni montana. [Beyond the province of Canela there are found very rich countries (though without mountains) in which the natives are adorned with trinkets and plates of gold.] Herrera dec. 5 lib. 10 cap. 14 and dec. 6 lib. 8 cap. 6 Geogr. Blaviana volume 11 page 261. Southey tome 1 pages 78 and 373.) The information which Pedro de Anasco had obtained from the natives, joined to that which was received subsequently (1536) by Diaz de Pineda, who had discovered the provinces of Quixos and Canela, between the Rio Napo and the Rio Pastaca, gave birth to the idea that on the east of the Nevados of Tunguragua, Cayambe, and Popayan, were vast plains, abounding in precious metals, and where the inhabitants were covered with armour of massy gold. Gonzales Pizarro, in searching for these treasures, discovered accidentally, in 1539, the cinnamon-trees of America (Laurus cinnamomoides, Mut.); and Francisco de Orellana went down the Napo, to reach the river Amazon. Since that period expeditions were undertaken at the same time from Venezuela, New Grenada, Quito, Peru, and even from Brazil and the Rio de la Plata,* for the conquest of El Dorado. (* Nuno de Chaves went from the Ciudad de la Asumpcion, situate on Rio Paraguay, to discover, in the latitude of 24 degrees south, the vast empire of El Dorado, which was everywhere supposed to lie on the eastern back of the Andes.) Those of which the remembrance have been best preserved, and which have most contributed to spread the fable of the riches of the Manaos, the Omaguas, and the Guaypes, as well as the existence of the lagunas de oro, and the town of the gilded king (Grand Patiti, Grand Moxo, Grand Paru, or Enim), are the incursions made to the south of the Guaviare, the Rio Fragua, and the Caqueta. Orellana, having found idols of massy gold, had fixed men's ideas on an auriferous land between the Papamene and the Guaviare. His narrative, and those of the voyages of Jorge de Espira (George von Speier), Hernan Perez de Quesada, and Felipe de Urre (Philip von Huten), undertaken in 1536, 1542, and 1545, furnish, amid much exaggeration, proofs of very exact local knowledge.* (* We may be surprised to see, that the expedition of Huten is passed over in absolute silence by Herrera (dec. 7 lib. 10 cap. 7 volume 4 238). Fray Pedro Simon gives the whole particulars of it, true or fabulous; but he composed his work from materials that were unknown to Herrera.) When these are examined merely in a geographical point of view, we perceive the constant desire of the first conquistadores to reach the land comprised between the sources of the Rio Negro, of the Uaupes (Guape), and of the Jupura or Caqueta. This is the land which, in order to distinguish it from El Dorado de la Parime, we have called El Dorado des Omaguas.* (* In 1560 Pedro de Ursua even took the title of Governador del Dorado y de Omagua. Fray Pedro Simon volume 6 chapter 10 page 430.) No doubt the whole country between the Amazon and the Orinoco was vaguely known by the name of las Provincias del Dorado; but in this vast extent of forests, savannahs, and mountains, the progress of those who sought the great lake with auriferous banks, and the town of the gilded king, was directed towards two points only, on the north-east and south-west of the Rio Negro; that is, to Parima (or the isthmus between the Carony, the Essequibo, and the Rio Branco), and to the ancient abode of the Manaos, the inhabitants of the banks of the Yurubesh. I have just mentioned the situation of the latter spot, which is celebrated in the history of the conquest from 1535 to 1560; and it remains for me to speak of the configuration of the country between the Spanish missions of the Rio Carony, and the Portuguese missions of the Rio Branco or Parima. This is the country lying near the Lower Orinoco, the Esmeralda, and French and Dutch Guiana, on which, since the end of the sixteenth century, the enterprises and exaggerated narratives of Raleigh have shed so bright a splendour.

From the general disposition of the course of the Orinoco, directed successively towards the west, the north, and the east, its mouth lies almost in the same meridian as its sources: so that by proceeding from Vieja Guyana to the south the traveller passes through the whole of the country in which geographers have successively placed an inland sea (Mar Blanco), and the different lakes which are connected with the El Dorado de la Parime. We find first the Rio Carony, which is formed by the union of two branches of almost equal magnitude, the Carony properly so called, and the Rio Paragua. The missionaries of Piritu call the latter river a lake (laguna): it is full of shoals, and little cascades; but, passing through a country entirely flat, it is subject at the same time to great inundations, and its real bed (su verdadera caxa) can scarcely be discovered. The natives have given it the name of Paragua or Parava, which means in the Caribbee language sea, or great lake. These local circumstances and this denomination no doubt have given rise to the idea of transforming the Rio Paragua, a tributary stream of the Carony, into a lake called Cassipa, on account of the Cassipagotos,* who lived in those countries. (* Raleigh pages 64 and 69. I always quote, when the contrary is not expressly said, the original edition of 1596. Have these tribes of Cassipagtos, Epuremei, and Orinoqueponi, so often mentioned by Raleigh, disappeared? or did some misapprehension give rise to these denominations? I am surprised to find the Indian words [of one of the different Carib dialects?] Ezrabeta cassipuna aquerewana, translated by Raleigh, the great princes or greatest commander. Since acarwana certainly signifies a chief, or any person who commands (Raleigh pages 6 and 7), cassipuna perhaps means great, and lake Cassipa is synonymous with great lake. In the same manner Cass-iquiare may be a great river, for iquiare, like veni, is, in the north of the Amazon, a termination common to all rivers. Goto, however, in Cassipa-goto, is a Caribbee term denoting a tribe.) Raleigh gives this basin forty miles in breadth; and, as all the lakes of Parima must have auriferous sands, he does not fail to assert that in summer, when the waters retire, pieces of gold of considerable weight are found there.

The sources of the tributary streams of the Carony, the Arui, and the Caura (Caroli, Arvi, and Caora,* of the ancient geographers (* D'Anville names the Rio Caura, Coari; and the Rio Arui, Aroay. I have not been able hitherto to guess what is meant by the Aloica (Atoca, Atoica of Raleigh), which issues from the lake Cassipa, between the Caura and the Arui.)) being very near each other, this suggested the idea of making all these rivers take their rise from the pretended lake Cassipa.* (* Raleigh makes only the Carony and the Arui issue from it (Hondius, Nieuwe Caerte van het wonderbare landt Guiana, besocht door Sir Walter Raleigh, 1594 to 1596): but in later maps, for instance that of Sanson, the Rio Caura issues also from Lake Cassipa.) Sanson has so much enlarged this lake, that he gives it forty-two leagues in length, and fifteen in breadth. The ancient geographers placed opposite to each other, with very little hesitation, the tributary streams of the two banks of a river; and they place the mouth of the Carony, and lake Cassipa, which communicates by the Carony with the Orinoco, sometimes* ABOVE the confluence of the Meta. (* Sanson, Map for the Voyage of Acunha, 1680. Id. South America, 1659. Coronelli, Indes occidentales, 1689.) Thus it is carried back by Hondius as far as the latitudes of 2 and 3 degrees, giving it the form of a rectangle, the longest sides of which run from north to south. This circumstance is worthy of remark, because, in assigning gradually a more southern latitude to lake Cassipa, it has been detached from the Carony and the Arui, and has taken the name of Parima. To follow this metamorphosis in its progressive development, we must compare the maps which have appeared since the voyage of Raleigh till now. La Cruz, who has been copied by all the modern geographers, has preserved the oblong form of the lake Cassipa for his lake Parima, although this form is entirely different from that of the ancient lake Parima, or Rupunuwini, of which the great axis was directed from east to west. The ancient lake (that of Hondius, Sanson, and Coronelli) was also surrounded by mountains, and gave birth to no river; while the lake Parima of La Cruz and the modern geographers communicates with the Upper Orinoco, as the Cassipa with the Lower Orinoco.

I have stated the origin of the fable of the lake Cassipa, and the influence it has had on the opinion that the lake Parima is the source of the Orinoco. Let us now examine what relates to this latter basin, this pretended interior sea, called Rupunuwini by the geographers of the sixteenth century. In the latitude of four degrees or four degrees and a half (in which direction unfortunately, south of Santo Thome del Angostura to the extent of eight degrees, no astronomical observation has been made) is a long and narrow Cordillera, that of Pacaraimo, Quimiropaca, and Ucucuamo; which, stretching from east to south-west, unites the group of mountains of Parima to the mountains of Dutch and French Guiana. It divides its waters between the Carony, the Rupunury or Rupunuwini, and the Rio Branco, and consequently between the valleys of the Lower Orinoco, the Essequibo, and the Rio Negro. On the north-west of the Cordillera de Pacaraimo, which has been traversed but by a small number of Europeans (by the German surgeon, Nicolas Hortsmann, in 1739; by a Spanish officer, Don Antonio Santos, in 1775; by the Portuguese colonel, Barata, in 1791; and by several English settlers, in 1811), descend the Noeapra, the Paraguamusi, and the Paragua, which fall into the Rio Carony; on the north-east, the Rupunuwini, a tributary stream of the Rio Essequibo. Toward the south, the Tacutu and the Uariquera form together the famous Rio Parima, or Rio Branco.

This isthmus, between the branches of the Rio Essequibo and the Rio Branco (that is, between the Rupunuwini on one side, and the Pirara, the Mahu, and the Uaricuera or Rio Parima on the other), may be considered as the classical soil of the Dorado of Parima. The rivers at the foot of the mountains of Pacaraimo are subject to frequent overflowings. Above Santa Rosa, the right bank of the Uariapara, a tributary stream of the Uaricuera, is called el Valle de la Inundacion. Great pools are also found between the Rio Parima and the Xurumu. These are marked on the maps recently constructed in Brazil, which furnish the most ample details of those countries. More to the west, the Cano Pirara, a tributary stream of the Mahu, issues from a lake covered with rushes. This is the lake Amucu described by Nicolas Hortsmann, and respecting which some Portuguese of Barcelos, who had visited the Rio Branco (Rio Parima or Rio Paravigiana), gave me precise notions during my stay at San Carlos del Rio Negro. The lake Amucu is several leagues broad, and contains two small islands, which Santos heard called Islas Ipomucena. The Rupunuwini (Rupunury), on the banks of which Hortsmann discovered rocks covered with hieroglyphical figures, approaches very near this lake, but does not communicate with it. The portage between the Rupunuwini and the Mahu is farther north, where the mountain of Ucucuamo* rises, the natives still call the mountain of gold. (* I follow the orthography of the manuscript journal of Rodriguez; it is the Cerro Acuquamo of Caulin, or rather of his commentator. Hist. corogr. page 176.) They advised Hortsmann to seek round the Rio Mahu for a mine of silver (no doubt mica with large plates), of diamonds, and emeralds. He found nothing but rocky crystals. His account seems to prove that the whole length of the mountains of the Upper Orinoco (Sierra Parima) toward the east, is composed of granitic rocks, full of druses and open veins, the Peak of Duida. Near these lands, which still enjoy a great celebrity for their riches, on the western limits of Dutch Guiana, live the Macusis, Aturajos, and Acuvajos. The traveller Santos found them stationed between the Rupunuwini, the Mahu, and the chain of Pacaraimo. It is the appearance of the micaceous rocks of the Ucucuamo, the name of the Rio Parima, the inundations of the rivers Urariapara, Parima, and Xurumu, and more especially the existence of the lake Amucu (near the Rio Rupunuwini, and regarded as the principal source of the Rio Parima), which have given rise to the fable of the White Sea and the Dorado of Parima. All these circumstances (which have served on this very account to corroborate the general opinion) are found united on a space of ground which is eight or nine leagues broad from north to south, and forty long from east to west. This direction, too, was always assigned to the White Sea, by lengthening it in the direction of the latitude, till the beginning of the sixteenth century. Now this White Sea is nothing but the Rio Parima, which is called the White River (Rio Branco, or Rio del Aguas blancas), and runs through and inundates the whole of this land. The name of Rupunuwini is given to the White Sea on the most ancient maps, which identifies the place of the fable, since of all the tributary streams of the Rio Essequibo the Rupunuwini is the nearest to the lake Amucu. Raleigh, in his first voyage (1595), had formed no precise idea of the situation of El Dorado and the lake Parima, which he believed to be salt, and which he calls another Caspian Sea. It was not till the second voyage (1596), performed equally at the expense of Raleigh, that Laurence Keymis fixed so well the localities of El Dorado, that he appears to me to have no doubt of the identity of the Parima de Manao with the lake Amucu, and with the isthmus between the Rupunuwini (a tributary stream of the Essequibo) and the Rio Parima or Rio Branco. "The Indians," says Keymis, "go up the Dessekebe [Essequibo] in twenty days, towards the south. To mark the greatness of this river, they call it the brother of the Orinoco. After twenty days' navigating they convey their canoes by a portage of one day, from the river Dessekebe to a lake, which the Jaos call Roponowini, and the Caribbees Parime. This lake is as large as a sea; it is covered with an infinite number of canoes; and I suppose" [the Indians then had told him nothing of this] "that this lake is no other than that which contains the town of Manoa."* (* Cayley's Life of Raleigh volume 1 pages 159, 236 and 283. Masham in the third voyage of Raleigh (1596) repeats these accounts of the Lake Rupunuwini.) Hondius has given a curious plate of this portage; and, as the mouth of the Carony was then supposed to be in latitude 4 degrees (instead of 8 degrees 8 minutes), the portage of Parima was placed close to the equator. At the same period the Viapoco (Oyapoc) and the Rio Cayenne (Maroni?) were made to issue from this lake Parima. The same name being given by the Caribs to the western branch of the Rio Branco has perhaps contributed as much to the imaginary enlargement of the lake Amucu, as the inundations of the various tributary streams of the Uraricuera, from the confluence of the Tacutu to the Valle de la Inundacion.

We have shown above that the Spaniards took the Rio Paragua, or Parava, which falls into the Carony, for a lake, because the word parava signifies sea, lake, river. Parima seems also to denote vaguely great water; for the root par is found in the Carib words that designate rivers, pools, lakes, and the ocean.* (* In Persian the root water (ab) is found also in lake (abdan). For other etymologies of the words Parima and Manoa see Gili volume 1 pages 81 and 141; and Gumilla volume 1 page 403.) In Arabic and in Persian, bahr and deria are also applied at the same time to the sea, to lakes, and to rivers; and this practice, common to many nations in both worlds, has, on our ancient maps, converted lakes into rivers and rivers into lakes. In support of what I here advance, I shall appeal to very respectable testimony, that of Father Caulin. "When I inquired of the Indians," says this missionary, who sojourned longer than I on the banks of the Lower Orinoco, "what Parima was, they answered that it was nothing more than a river that issued from a chain of mountains, the opposite side of which furnished waters to the Essequibo." Caulin, knowing nothing of lake Amucu, attributes the erroneous opinion of the existence of an inland sea solely to the inundations of the plains (a las inundaciones dilatadas por los bajos del pais).

According to him, the mistakes of geographers arise from the vexatious circumstance of all the rivers of Guiana having different names at their mouths and near their sources. "I have no doubt," he adds, "that one of the upper branches of the Rio Branco is that very Rio Parima which the Spaniards have taken for a lake (a quien suponian laguna)." Such are the opinions which the historiographer of the Expedition of the Boundaries had formed on the spot. He could not expect that La Cruz and Surville, mingling old hypotheses with accurate ideas, would reproduce on their maps the Mar Dorado or Mar Blanco. Thus, notwithstanding the numerous proofs which I have furnished since my return from America, of the non-existence of an inland sea the origin of the Orinoco, a map has been published in my name,* on which the Laguna Parima figures anew. (* Carte de l'Amerique, dressee sur les Observations de M. de Humboldt, par Fried. Vienna 1818.)

From the whole of these statements it follows, first, that the Laguna Rupunuwini, or Parima of the voyage of Raleigh and of the maps of Hondius, is an imaginary lake, formed by the lake Amucu* (* This is the lake Amaca of Surville and La Cruz. By a singular mistake, the name of this lake is transformed to a village on Arrowsmith's map.) and the tributary streams of the Uraricuera, which often overflow their banks; secondly, that the Laguna Parime of Surville's map is the lake Amucu, which gives rise to the Rio Pirara and (conjointly with the Mahu, the Tacutu, the Uraricuera, or Rio Parima, properly so called) to the Rio Branco; thirdly, that the Laguna Parime of La Cruz is an imaginary swelling of the Rio Parime (confounded with the Orinoco) below the junction of the Mahu with the Xurumu. The distance from the mouth of the Mahu to that of the Tacutu is scarcely 0 degrees 40 minutes; La Cruz enlarges it to 7 degrees of latitude. He calls the upper part of the Rio Branco (that which receives the Mahu) Orinoco or Purumu. There can be no doubt of its being the Xurumu, one of the tributary streams of the Tacutu, which is well known to the inhabitants of the neighbouring fort of San Joaquim. All the names that figure in the fable of El Dorado are found in the tributary streams of the Rio Branco. Slight local circumstances, joined to the remembrances of the salt lake of Mexico, more especially of the celebrated lake Manoa in the Dorado des Omaguas, have served to complete a picture created by the imagination of Raleigh and his two lieutenants, Keymis and Masham. The inundations of the Rio Branco, I conceive, may be compared at the utmost to those of the Red River of Louisiana, between Nachitoches and Cados, but not to the Laguna de los Xarayes, which is a temporary swelling of the Rio Paraguay.* (* Southey volume 1 page 130. These periodical overflowings of the Rio Paraguay have long acted the same part in the southern hemisphere, as lake Parima has been made to perform in the northern. Hondius and Sanson have made the Rio de la Plata, the Rio Topajos (a tributary stream of the Amazon), the Rio Tocantines, and the Rio de San Francisco, issue from the Laguna de los Xarayes.)

We have now examined a White Sea,* (* That of D'Anville and La Cruz, and of the greater part of the modern maps.) which the principal of the Rio Branco is made to traverse; and another,* (* The lake of Surville, which takes the place of lake Amucu.) which is placed on the east of this river, and communicates with it by the Cano Pirara. A third lake* (* The lake which Surville calls Laguna tenida hasta ahora or La una Parime.) is figured on the west of the Rio Branco, respecting which I found recently some curious details in the manuscript journal of the surgeon Hortsmann. "At the distance of two days' journey below the confluence of the Mahu (Tacutu) with the Rio Parima (Uraricuera) a lake is found on top of a mountain. This lake is stocked with the same fish as the Rio Parima; but the waters of the former are black, and those of the latter white." May not Surville, from a vague notion of this basin, have imagined, in his map prefixed to Father Caulin's work, an Alpine lake of ten leagues in length, near which, towards the east, rise at the same time the Orinoco, and the Rio Idapa, a tributary stream of the Rio Negro? However vague may be the account of the surgeon of Hildesheim, it is impossible to admit that the mountain, which has a lake at its summit, is to the north of the parallel of 2 degrees 30 minutes: and this latitude coincides nearly with that of the Cerro Unturan. Hence it follows that the Alpine lake of Hortsmann, which has escaped the attention of D'Anville, and which is perhaps situate amid a group of mountains, lies north-east of the portage from the Idapa to the Mavaca, and south-east of the Orinoco, where it goes up above Esmeralda.

Most of the historians who have treated of the first ages of the conquest seem persuaded that the name provincias or pais del Dorado denoted originally every region abounding in gold. Forgetting the precise etymology of the word El Dorado (the gilded), they have not perceived that this tradition is a local fable, as were almost all the ancient fables of the Greeks, the Hindoos, and the Persians. The history of the gilded man belongs originally to the Andes of New Grenada, and particularly to the plains in the vicinity of their eastern side: we see it progressively advance, as I observed above, three hundred leagues toward the east-north-east, from the sources of the Caqueta to those of the Rio Branco and the Essequibo. Gold was sought in different parts of South America before 1536, without the word El Dorado having been ever pronounced, and without the belief of the existence of any other centre of civilization and wealth, than the empire of the Inca of Cuzco. Countries which now do not furnish commerce with the smallest quantities of the precious metals, the coast of Paria, Terra Firma (Castillo del Oro), the mountains of Santa Marta, and the isthmus of Darien, then enjoyed the same celebrity which has been more recently acquired by the auriferous lands of Sonora, Choco, and Brazil.

Diego de Ordaz (1531) and Alonzo de Herrera (1535) directed their journeys of discovery along the banks of the Lower Orinoco. The former is the famous Conquistador of Mexico, who boasted that he had taken sulphur out of the crater of the Peak of Popocatepetl, and whom the emperor Charles V permitted to wear a burning volcano on his armorial bearings. Ordaz, named Adelantado of all the country which he could conquer between Brazil and the coast of Venezuela, which was then called the country of the German Company of Welsers (Belzares) of Augsburg, began his expedition by the mouth of the Maranon. He there saw, in the hands of the natives, "emeralds as big as a man's fist." They were, no doubt, pieces of that saussurite jade, or compact feldspar, which we brought home from the Orinoco, and which La Condamine found in abundance at the mouth of the Rio Topayos. The Indians related to Diego de Ordaz that on going up during a certain number of suns toward the west, he would find a large rock (pena) of green stone; but before they reached this pretended mountain of emerald (rocks of euphotide?) a shipwreck put an end to all farther discovery. The Spaniards saved themselves with difficulty in two small vessels. They hastened to get out of the mouth of the Amazon; and the currents, which in those parts run with violence to the north-west, led Ordaz to the coast of Paria where, in the territory of the cacique Yuripari (Uriapari, Viapari), Sedeno had constructed the Casa fuerte de Paria. This post being very near the mouth of the Orinoco, the Mexican Conquistador resolved to attempt an expedition on this great river. He sojourned first at Carao (Caroa, Carora), a large Indian village, which appears to me to have been a little to the east of the confluence of the Carony; he then went up the Cabruta (Cabuta, Cabritu), and to the mouth of the Meta (Metacuyu), where he found great difficulty in passing his boats through the Raudal of Cariven. The Aruacas, whom Ordaz employed as guides, advised him to go up the Meta; where, on advancing towards the west, they asserted he would find men clothed, and gold in abundance. Ordaz pursued in preference the navigation of the Orinoco, but the cataracts of Tabaje (perhaps even those of the Atures) compelled him to terminate his discoveries.

It is worthy of remark that in this voyage, far anterior to that of Orellana, and consequently the greatest which the Spaniards had then performed on a river of the New World, the name of the Orinoco was for the first time heard. Ordaz, the leader of the expedition, affirms that the river, from its mouth as far as the confluence of the Meta, is called Uriaparia, but that above this confluence it bears the name of Orinucu. This word (formed analogously with the words Tamanacu, Otomacu, Sinarucu) is, in fact, of the Tamanac tongue; and, as the Tamanacs dwell south-east of Encaramada, it is natural that the conquistadores heard the actual name of the river only on drawing near the Rio Meta.* (* Gili volume 3 page 381. The following are the most ancient names of the Orinoco, known to the natives near its mouth, and which historians give us altered by the double fault of pronunciation and orthography; Yuyapari, Yjupari, Huriaparia, Urapari, Viapari, Rio de Paria. The Tamanac word Orinucu was disfigured by the Dutch pilots into Worinoque. The Otomacs say Joga-apurura (great river); the Cabres and Guaypunabis, Paragua, Bazagua Parava, three words signifying great water, river, sea. That part of the Orinoco between the

20

Apure and the Guaviare is often denoted by the name of Baraguan. A famous strait, which we have described above, bears also this name, which is no doubt a corruption of the word Paragua. Great rivers in every zone are called by the dwellers on their banks the river, without any particular denominations. If other names be added, they change in every province. Thus the Rio Turiva, near the Encaramada, has five names in the different parts of its course. The Upper Orinoco, or Paragua, is called by the Maquiritares (near Esmeralda) Maraguaca, on account of the lofty mountains of this name near Duida. Gili volume 1 pages 22 and 364. Caulin page 75. In most of the names of the rivers of America we recognize the root water. Thus yacu in the Peruvian, and veni in the Maypure tongues, signify water and river. In the Lule dialect I find fo, water; foyavolto, a river; foysi, a lake; as in Persian, ab is water; abi frat, the river Euphrates; abdan, a lake. The root water is preserved in the derivatives.) On this last tributary stream Diego de Ordaz received from the natives the first idea of civilized nations who inhabited the table-lands of the Andes of New Granada; of a very powerful prince with one eye (Indio tuerto), and of animals less than stags, but fit for riding like Spanish horses. Ordaz had no idea that these animals were llamas (ovejas del Peru). Must we admit that llamas, which were used in the Andes to draw the plough and as beasts of burden, but not for riding, were already common on the north and east of Quito? I find that Orellana saw these animals at the river Amazon, above the confluence of the Rio Negro, consequently in a climate very different from that of the table-land of the Andes. The table of an army of Omaguas mounted on llamas served to embellish the account given by the fellow-travellers of Felipe de Urre of their adventurous expedition to the Upper Caqueta. We cannot be sufficiently attentive to these traditions, which seem to prove that the domestic animals of Quito and Peru had already begun to descend the Cordilleras, and spread themselves by degrees in the eastern regions of South America.

Herrera, the treasurer of the expedition of Ordaz, was sent in 1553, by the governor Geronimo de Ortal, to pursue the discovery of the Orinoco and the Meta. He lost nearly thirteen months between Punta Barina and the confluence of the Carony in constructing flat-bottomed boats, and making the preparations indispensable for a long voyage. We cannot read without astonishment the narrative of those daring enterprises, in which three or four hundred horses were embarked to be put ashore whenever cavalry could act on one of the banks. We find in the expedition of Herrera the same stations which we already knew; the fortress of Paria, the Indian village of Uriaparia (no doubt below Imataca, on a point where the inundations of the delta prevented the Spaniards from being able to procure firewood), Caroa, in the province of Carora; the rivers Caranaca (Caura?) and Caxavana (Cuchivero?); the village of Cabritu (Cabruta), and the Raudal near the mouth of the Meta (probably the Raudal of Cariven and the Piedra de la Paciencia). As the Rio Meta, on account of the proximity of its sources and of its tributary streams to the auriferous Cordilleras of new Grenada (Cundinamarca), enjoyed great celebrity, Herrera attempted to go up this river. He there found nations more civilized than those of the Orinoco, but that fed on the flesh of mute dogs. Herrera was killed in battle by an arrow poisoned with the juice of curare (yierva); and when dying named Alvaro de Ordaz his lieutenant, who led the remains of the expedition (1535) to the fortress of Paria, after having lost the few horses which had resisted a campaign of eighteen months.

Confused reports which were circulated of the wealth of the inhabitants of the Meta, and the other tributary streams that descend from the eastern side of the Cordilleras of New Grenada, engaged successively Geronimo de Ortal, Nicolas Federmann, and Jorge de Espira (George von Speier), in 1535 and 1536, to undertake expeditions by land towards the south and south-west. From the promontory of Paria, as far as Cabo de la Vela, little figures of molten gold had been found in the hands of the natives, as early as the years 1498 and 1500. The principal markets for these amulets, which the women used as ornaments, were the villages of Curiana (Coro) and Cauchieto (Near the Rio la Hacha). The metal employed by the founders of Cauchieto came from a mountainous country more to the south. It may be conceived that the expeditions of Ordaz and Herrera served to increase the desire of drawing nearer to those auriferous countries. George von Speier left Coro (1535), and penetrated by the mountains of Merida to the banks of the Apure and the Meta. He passed these two rivers near their sources, where they have but little breadth. The Indians told him that, farther on, white men wandered about the plains. Speier, who imagined that he was not far from the banks of the Amazon, had no doubt that these wandering Spaniards were men unfortunately shipwrecked in the expedition of Ordaz. He crossed the savannahs of San Juan de los Llanos, which were said to abound in gold; and made a long stay at an Indian village called Pueblo de Nuestra Senora, and afterwards La Fragua, south-east of the Paramo de la Suma Paz. I have been on the western back of this group of mountains, at Fusagasuga, and there heard that the plains by which they are skirted toward the east still enjoy some celebrity for wealth among the natives. Speier found in the populous village of La Fragua a Casa del Sol (temple of the sun), and a convent of virgins similar to those of Peru and New Granada. Were these the consequence of a migration of religious rites towards the east? or must we admit that the plains of San Juan were their first cradle? Tradition, indeed, records that Bochica, the legislator of New Granada and high-priest of Iraca, had gone up from the plains of the east to the table-land of Bogota. But Bochica being at once the offspring and the symbol of the sun, his history may contain allegories that are merely astrological. Speier, pursuing his way toward the south, and crossing the two branches of the Guaviare, which are the Ariare and the Guayavero (Guayare or Canicamare), arrived on the banks of the great Rio Papamene or Caqueta. The resistance he met with during a whole year in the province de los Choques, put an end, in 1537, to this memorable expedition. Nicolas Federmann and Geronimo de Ortal (1536), who went from Macarapana and the mouth of the Rio Neveri, followed (1535) the traces of Jorge de Espira. The former sought for gold in the Rio Grande de la Magdalena; the latter endeavoured to discover a temple of the sun (Casa del Sol) on the banks of the Meta. Ignorant of the idiom of the natives, they seemed to see everywhere, at the foot of the Cordilleras, the reflexion of the greatness of the temples of Iraca (Sogamozo), which was then the centre of the civilization of Cundinamarca.

I have now examined, in a geographical point of view, the expeditions on the Orinoco, and in a western and southern direction on the eastern back of the Andes, before the tradition of El Dorado was spread among the conquistadores. This tradition, as we have noticed above, had its origin in the kingdom of Quito, where Luis Daza (1535) met with an Indian of New Grenada who had been sent by his prince (no doubt the zippa of Bogota, or the zaque of Tunja), to demand assistance from Atahualpa, inca of Peru. This ambassador boasted, as usual, the wealth of his country; but what particularly fixed the attention of the Spaniards who were assembled with Daza in the town of Tacunga (Llactacunga), was the history of a lord who, his body covered with powdered gold, went into a lake amid the mountains. This lake may have been the Laguna de Totta, a little to the east of Sogamozo (Iraca) and of Tunja (Hunca, the town of Huncahua), where two chiefs, ecclesiastical and secular, of the empire of Cundinamarca, or Cundirumarca, resided; but no historical remembrance being attached to this mountain lake, I rather suppose that it was the sacred lake of Guatavita, on the east of the mines of rock-salt of Zipaquira, into which the gilded lord was made to enter. I saw on its banks the remains of a staircase hewn in the rock, and serving for the ceremonies of ablution. The Indians said that powder of gold and golden vessels were thrown into this lake, as a sacrifice to the adoratorio de Guatavita. Vestiges are still found of a breach which was made by the Spaniards for the purpose of draining the lake. The temple of the sun at Sogamozo being pretty near the northern coasts of Terra Firma, the notions of the gilded man were soon applied to a high-priest of the sect of Bochica, or Indacanzas, who every morning, before he performed his sacrifice, caused powder of gold to be stuck upon his hands and face, after they had been smeared with grease. Other accounts, preserved in a letter of Oviedo addressed to the celebrated cardinal Bembo, say that Gonzalo Pizarro, when he discovered the province of cinnamon-trees, "sought at the same time a great prince, noised in those countries, who was always covered with powdered gold, so that from head to foot he resembled an image of gold fashioned by the hand of a skilful workman (a una figura d'oro lavorato di mano d'un buonissimo orefice). The powdered gold is fixed to the body by means of an odoriferous resin; but, as this kind of garment would be uneasy to him while he slept, the prince washes himself every evening, and is gilded anew in the morning, which proves that the empire of El Dorado is infinitely rich in mines." It seems probable that there was something in the ceremonies of the worship introduced by Bochica which gave rise to a tradition so generally spread. The strangest customs are found in the

New World. In Mexico the sacrificers painted their bodies and wore a kind of cape, with hanging sleeves of tanned human skin.

On the banks of the Caura, and in other wild parts of Guiana, where painting the body is used instead of tattooing, the nations anoint themselves with turtle-fat, and stick spangles of mica with a metallic lustre, white as silver and red as copper, on their skin, so that at a distance they seem to wear laced clothes. The fable of the gilded man is, perhaps, founded on a similar custom; and, as there were two sovereign princes in New Granada, the lama of Iraca and the secular chief or zaque of Tunja, we cannot be surprised that the same ceremony was attributed sometimes to the prince and sometimes to the high-priest. It is more extraordinary that, as early as the year 1535, the country of El Dorado was sought for on the east of the Andes. Robertson is mistaken in admitting that Orellana received the first notions of it (1540) on the banks of the Amazon. The history of Fray Piedro Simon, founded on the memoirs of Queseda, the conqueror of Cundirumarca, proves directly the contrary; and Gonzalo Diaz de Pineda, as early as 1536, sought for the gilded man beyond the plains of the province of Quixos. The ambassador of Bogota, whom Daza met with in the kingdom of Quito, had spoken of a country situate toward the east. Was this because the table-land of New Granada is not on the north, but on the north-east of Quito? We may venture to say that the tradition of a naked man covered with powdered gold must have belonged originally to a hot region, and not to the cold table-lands of Cundirumarca, where I often saw the thermometer sink below four or five degrees; however, on account of the extraordinary configuration of the country, the climate differs greatly at Guatavita, Tunja, Iraca, and on the banks of the Sogamozo. Sometimes, also, religious ceremonies are preserved which took rise in another zone; and the Muyscas, according to ancient traditions, made Bochica, their first legislator and the founder of their worship, arrive from the plains situate to the east of the Cordilleras. I shall not decide whether these traditions expressed an historical fact, or merely indicated, as we have already observed in another place, that the first Lama, who was the offspring and symbol of the sun, must necessarily have come from the countries of the East. Be it as it may, it is not less certain that the celebrity which the expeditions of Ordaz, Herrera, and Speier had already given to the Orinoco, the Meta, and the province of Papamene, situate between the sources of the Guaviare and Caqueta, contributed to fix the fable of El Dorado near to the eastern back of the Cordilleras.

The junction of three bodies of troops on the table-land of New Granada spread through all that part of America occupied by the Spaniards the news of an immensely rich and populous country which remained to be conquered. Sebastian de Belalcazar marched from Quito by way of Popayan (1536) to Bogota; Nicholas Federmann, coming from Venezuela, arrived from the east by the plains of Meta. These two captains found, already settled on the table-land of Cundirumarca, the famous Adelantado Gonzalo Ximenez de Queseda, one of whose descendants I saw near Zipaquira, with bare feet, attending cattle. The fortuitous meeting of the three conquistadores, one of the most extraordinary and dramatic events of the history of the conquest, took place in 1538. Belalcazar's narratives inflamed the imagination of warriors eager for adventurous enterprises; and the notions communicated to Luis Daza by the Indian of Tacunga were compared with the confused ideas which Ordaz had collected on the Meta respecting the treasures of a great king with one eye (Indio tuerto), and a people clothed, who rode upon llamas. An old soldier, Pedro de Limpias, who had accompanied Federmann to the table-land of Bogota, carried the first news of El Dorado to Coro, where the remembrance of the expedition of Speier (1535 to 1537) to the Rio Papamene was still fresh. It was from this same town of Coro that Felipe von Huten (Urre, Utre) undertook his celebrated voyage to the province of the Omaguas, while Pizarro, Orellana, and Hernan Perez de Quesada, brother of the Adelantado, sought for the gold country at the Rio Napo, along the river of the Amazons, and on the eastern chain of the Andes of New Grenada. The natives, in order to get rid of their troublesome guests, continually described Dorado as easy to be reached, and situate at no considerable distance. It was like a phantom that seemed to flee before the Spaniards, and to call on them unceasingly. It is in the nature of man, wandering on the earth, to figure to himself happiness beyond the region which he knows. El Dorado, similar to Atlas and the islands of the

Hesperides, disappeared by degrees from the domain of geography, and entered that of mythological fictions.

I shall not here relate the numerous enterprises which were undertaken for the conquest of this imaginary country. Unquestionably we are indebted to them in great part for our knowledge of the interior of America; they have been useful to geography, as errors and daring hypotheses are often to the search of truth: but in the discussion on which we are employed, it is incumbent on me to rest only upon those facts which have had the most direct influence on the construction of ancient and modern maps. Hernan Perez de Quesada, after the departure of his brother the Adelantado for Europe, sought anew (1539) but this time in the mountainous land north-east of Bogota, the temple of the sun (Casa del Sol), of which Geronimo de Ortal had heard spoken in 1536 on the banks of the Meta. The worship of the sun introduced by Bochica, and the celebrity of the sanctuary of Iraca, or Sogamozo, gave rise to those confused reports of temples and idols of massy gold; but on the mountains as in the plains, the traveller believed himself to be always at a distance from them, because the reality never corresponded with the chimerical dreams of the imagination. Francisco de Orellana, after having vainly sought El Dorado with Pizarro in the Provincia de los Canelos, and on the auriferous banks of the Napo, went down (1540) the great river of the Amazon. He found there, between the mouths of the Javari and the Rio de la Trinidad (Yupura?) a province rich in gold, called Machiparo (Muchifaro), in the vicinity of that of the Aomaguas, or Omaguas. These notions contributed to carry El Dorado toward the south-east, for the names Omaguas (Om-aguas, Aguas), Dit-Aguas, and Papamene, designated the same country—that which Jorge de Espira had discovered in his expedition to the Caqueta. The Omaguas, the Manaos or Manoas, and the Guaypes (Uaupes or Guayupes) live in the plains on the north of the Amazon. They are three powerful nations, the latter of which, stretching toward the west along the banks of the Guape or Uaupe, had been already mentioned in the voyages of Quesada and Huten. These two conquistadores, alike celebrated in the history of America, reached by different roads the llanos of San Juan, then called Valle de Nuestra Senora. Hernan Perez de Quesada (1541) passed the Cordilleras of Cundirumarca, probably between the Paramos of Chingasa and Suma Paz; while Felipe de Huten, accompanied by Pedro de Limpias (the same who had carried to Venezuela the first news of Dorado from the land of Bogota), directed his course from north to south, by the road which Speier had taken to the eastern side of the mountains. Huten left Coro, the principal seat of the German factory or company of Welser, when Henry Remboldt was its director. After having traversed (1541) the plains of Casanare, the Meta, and the Caguan, he arrived at the banks of the Upper Guaviare (Guayuare), a river which was long believed to be the source of the Orinoco, and the mouth of which I saw in passing by San Fernando de Atabapo to the Rio Negro. Not far from the right bank of the Guaviare, Huten entered Macatoa, the city of the Guapes. The people there were clothed, the fields appeared well cultivated; everything denoted a degree of civilization unknown in the hot region of America which extends to the east of the Cordilleras. Speier, in his expedition to the Rio Caqueta and the province of Papamene, had probably crossed the Guaviare far above Macatoa, before the junction of the two branches of this river, the Ariari and the Guayavero. Huten was told that on advancing more to the south-east he would enter the territory of the great nation of the Omaguas, the priest-king of which was called Quareca, and which possessed numerous herds of llamas. These traces of cultivation—these ancient resemblances to the table-land of Quito—appear to me very remarkable. It has already been said above that Orellana saw llamas at the dwelling of an Indian chief on the banks of the Amazon, and that Ordaz had heard mention made of them in the plains of Meta.

I pause where ends the domain of geography and shall not follow Huten in the description either of that town of immense extent, which he saw from afar; or of the battle of the Omaguas, where thirty-nine Spaniards (the names of fourteen are recorded in the annals of the time) fought against fifteen thousand Indians. These false reports contributed greatly to embellish the fable of El Dorado. The name of the town of the Omaguas is not found in the narrative of Huten; but the Manoas, from whom Father Fritz received, in the seventeenth century, plates of beaten gold, in his mission of Yurim-Aguas, are neighbours of the Omaguas. The name of Manoa subsequently passed from the country of the Amazons to an imaginary town, placed in El Dorado de la Parima. The celebrity attached to those countries between the Caqueta (Papamene) and the Guaupe (one of the tributary streams of the Rio Negro) excited Pedro de Ursua, in 1560, to that fatal expedition, which ended by the revolt of the tyrant Aguirre. Ursua, in going down the Caqueta to enter the river of the Amazons, heard of the province of Caricuri. This denomination clearly indicates the country of gold; for I find that this metal is called caricuri in the Tamanac, and carucuru in the Caribbee. Is it a foreign word that denotes gold among the nations of the Orinoco, as the words sugar and cotton are in our European languages? This would prove that these nations learned to know the precious metals among the foreign products which came to them from the Cordilleras,* or from the plains at the eastern back of the Andes. (* In Peruvian or Quichua (lengua del Inca) gold is called cori, whence are derived chichicori, gold in powder, and corikoya, gold-ore.)

We arrive now at the period when the fable of El Dorado was fixed in the eastern part of Guiana, first at the pretended lake Cassipa (on the banks of the Paragua, a tributary stream of the Carony), and afterwards between the sources of the Rio Essequibo and the Rio Branco. This circumstance has had the greatest influence on the state of geography in those countries. Antonio de Berrio, son-in-law* (* Properly casado con una sobrina. Fray Pedro Simon pages 597 and 608. Harris Coll. volume 2 page 212. Laet page 652. Caulin page 175. Raleigh calls Quesada Cemenes de Casada. He also confounds the periods of the voyages of Ordaz (Ordace), Orellana (Oreliano), and Ursua. See Empire of Guiana pages 13 to 20.) and sole heir of the great Adelantado Gonzalo Ximenez de Quesada, passed the Cordilleras to the east of Tunja,* (* No doubt between the Paramos of Chita and of Zoraca, taking the road of Chire and Pore. Berrio told Raleigh that he came from the Casanare to the Pato, from the Pato to the Meta, and from the Meta to the Baraguan (Orinoco). We must not confound this Rio Pato (a name connected no doubt with that of the ancient mission of Patuto) with the Rio Paute.) embarked on the Rio Casanare, and went down by this river, the Meta, and the Orinoco, to the island of Trinidad. We scarcely know this voyage except by the narrative of Raleigh; it appears to have preceded a few years the first foundation of Vieja Guayana, which was in the year 1591. A few years later (1595) Berrio caused his maese de campo, Domingo de Vera, to prepare in Europe an expedition of two thousand men to go up the Orinoco, and conquer El Dorado, which then began to be called the country of the Manoa, and even the Laguna de la gran Manoa. Rich landholders sold their farms, to take part in a crusade, to which twelve Observantin monks, and ten secular ecclesiastics were annexed. The tales related by one Martinez* (Juan Martin de Albujar?), who said he had been abandoned in the expedition of Diego de Ordaz, and led from town to town till he reached the capital of El Dorado, had inflamed the imagination of Berrio. (* I believe I can demonstrate that the fable of Juan Martinez, spread abroad by the narrative of Raleigh, was founded on the adventures of Juan Martin de Albujar, well known to the Spanish historians of the Conquest; and who, in the expedition of Pedro de Silva (1570), fell into the hands of the Caribs of the Lower Orinoco. This Albujar married an Indian woman and became a savage himself, as happens sometimes in our own days on the western limits of Canada and of the United States. After having long wandered with the Caribs, the desire of rejoining the Whites led him by the Rio Essequibo to the island of Trinidad. He made several excursions to Santa Fe de Bogota, and at length settled at Carora. (Simon page 591). I know not whether he died at Porto Rico; but it cannot be doubted that it was he who learned from the Carib traders the name of the Manoas [of Jurubesh]. As he lived on the banks of the Upper Carony and reappeared by the Rio Essequibo, he may have contributed also to place the lake Manoa at the isthmus of Rupunuwini. Raleigh makes his Juan Martinez embark below Morequito, a village at the east of that confluence of the Carony with the Orinoco. Thence he makes him dragged by the Caribs from town to town, till he finds at Manoa a relation of the inca Atabalipa (Atahualpa), whom he had known before at Caxamarca, and who had fled before the Spaniards. It appears that Raleigh had forgotten that the voyage of Ordaz (1531) was two years anterior to the death of Atahualpa and the entire destruction of the empire of Peru! He must have confounded the expedition of Ordaz with that of Silva (1570), in which Juan Martin de Albuzar partook. The latter, who related his tales at Santa Fe, at Venezuela, and perhaps at Porto Rico, must have combined what he had heard from the Caribs with what he had learned from the Spaniards respecting the town of the Omaguas seen by Huten; of the gilded man who sacrificed in a lake, and of the flight of the family of Atahualpa into the forests of Vilcabamba, and the eastern Cordillera of the Andes. Garcilasso volume 2 page 194.) It is difficult to distinguish what this conquistador had himself observed in going down the Orinoco from what he said he had collected in a pretended journal of Martinez, deposited at Porto Rico. It appears that in general at that period the same ideas prevailed respecting America as those which we have long entertained in regard to Africa; it was imagined that more civilization would be found towards the centre of the continent than on the coasts. Already Juan Gonzalez, whom Diego de Ordaz had sent in 1531 to explore the banks of the Orinoco, announced that "the farther you went up this river the more you saw the population increase." Berrio mentions the often-inundated province of Amapaja, between the confluence of the Meta and the Cuchivero, where he found many little idols of molten gold, similar to those which were fabricated at Cauchieto, east of Coro. He believed this gold to be a product of the granitic soil that covers the mountainous country between the Carichana, Uruana, and Cuchivero. In fact the natives have recently found a mass of native gold in the Quebrada del Tigre near the mission of Encaramada. Berrio mentions on the east of the province of Amapaja the Rio Carony (Caroly), which was said to issue from a great lake, because one of the tributary streams of the Carony, the Rio Paragua (river of the great water), had been taken for an inland sea, from ignorance of the Indian languages. Several of the Spanish historians believed that this lake, the source of the Carony, was the Grand Manoa of Berrio; but the notions he communicated to Raleigh show that the Laguna de Manoa (del Dorado, or de Parime) was supposed to be to the south of the Rio Paragua, transformed into Laguna Cassipa. "Both these basins had auriferous sands; but on the banks of the Cassipa was situate Macureguarai (Margureguaira), the capital of the cacique of Aromaja, and the first city of the imaginary empire of Guyana."

As these often-inundated lands have been at all times inhabited by nations of Carib race, who carried on a very active inland trade with the most distant regions, we must not be surprised that more gold was found here in the hands of the Indians than elsewhere. The natives of the coast did not employ this metal in the form of ornaments or amulets only; but also as a medium of exchange. It is not extraordinary, therefore, that gold has disappeared on the coast of Paria, and among the nations of the Orinoco, their inland communications have been impeded by the Europeans. The natives who have remained independent are in our days, no doubt, more wretched, more indolent, and in a ruder state, than they were before the conquest. The king of Morequito, whose son Raleigh took to England, had visited Cumana in 1594, to exchange a great quantity of images of massy gold for iron tools, and European merchandise. The unexpected appearance of an Indian chief augmented the celebrity of the riches of the Orinoco. It was supposed that El Dorado must be near the country from which the king of Morequito came; and as this country was often inundated, and rivers vaguely called great seas, or great basins of water, El Dorado must be on the banks of a lake. It was forgotten that the gold brought by the Caribs and other trading people was as little the produce of the soil as the diamonds of Brazil and India are the produce of the regions of Europe, where they are most abundant. The expedition of Berrio which had increased in number during the stay of the vessels at Cumana, La Margareta, and the island of Trinidad, proceeded by Morequito (near Vieja Guayana) towards the Rio Paragua, a tributary stream of the Carony; but sickness, the ferocity of the natives, and the want of subsistence, opposed invincible obstacles to the progress of the Spaniards. They all perished; except about thirty, who returned in a deplorable state to the post of Santo Thome.

These disasters did not calm the ardour displayed during the first half of the 17th century in the search of El Dorado. The Governor of the island of Trinidad, Antonio de Berrio, became the prisoner of Sir Walter Raleigh in the celebrated incursion of that navigator, in 1595, on the coast of Venezuela and at the mouths of the Orinoco. Raleigh collected from Berrio, and from other prisoners made by Captain Preston* at the taking of Caracas, all the information which had been obtained at that period on the countries situate to the south of Vieya Guayana. (* These prisoners belonged to the expedition of Berrio and of Hernandez de Serpa. The English landed at Macuto (then Guayca Macuto), whence a white man, Villalpando, led them by a mountain-path between Cumbre and the Silla (perhaps passing over the ridge of Galipano) to the town of Caracas. Simon page 594; Raleigh page 19. Those only who are acquainted with the situation can be sensible how difficult and daring this enterprise was.) He lent faith to the fables invented by Juan Martin de Albujar, and entertained no doubt either of the existence of the two lakes Cassipa and Rupunuwini, or of that of the great empire of the Inca, which, after the death of Atahualpa, the fugitive princes were supposed to have founded near the sources of the Essequibo. We are not in possession of a map that was constructed by Raleigh, and which he recommended to lord Charles Howard to keep secret. The geographer Hondius has filled up this void; and has even added to his map a table of longitudes and latitudes, among which figure the laguna del Dorado, and the Ville Imperiale de Manoas. Raleigh, when at anchor near the Punta del Gallo* in the island of Trinidad (* The northern part of La Punta de Icacos, which is the south-east cape of the island of Trinidad. Christopher Columbus cast anchor there on August 3, 1498. A great confusion exists in the denomination of the different capes of the island of Trinidad; and as recently, since the expedition of Fidalgo and Churruca, the Spaniards reckon the longitudes in South America west of La Punta de la Galera (latitude 10 degrees 50 minutes, longitude 63 degrees 20 minutes), it is important to fix the attention of geographers on this point. Columbus called the south-east cape of the island Punta Galera, on account of the form of a rock. From Punta de la Galera he sailed to the west and landed at a low cape, which he calls Punta del Arenal; this is our Punta de Icacos. In this passage, near a place (Punta de la Playa) where he stopped to take in water (perhaps at the mouth of the Rio Erin), he saw to the south, for the first time, the continent of America, which he called Isla Santa. It was, therefore, the eastern coast of the province of Cumana, to the east of the Cano Macareo, near Punta Redonda, and not the mountainous coast of Paria (Isla de Gracia, of Columbus), which was first discovered.), made his lieutenants explore the mouths of the Orinoco, principally those of Capuri, Grand Amana (Manamo Grande), and Macureo (Macareo). As his ships drew a great deal of water, he found it difficult to enter the bocas chicas, and was obliged to construct flat-bottomed barks. He remarked the fires of the Tivitivas (Tibitibies), of the race of the Guaraon Indians, on the tops of the mauritia palm-trees; and appears to have first brought the fruit to Europe (fructum squamosum, similem palmae pini). I am surprised, that he scarcely mentions the settlement, which had been made by Berrio under the name of Santo Thome (la Vieja Guayana.) This settlement however dates from 1591; and though, according to Fray Pedro Simon, "religion and policy prohibited all mercantile connection between Christians [Spaniards] and Heretics [the Dutch and English]," there was then carried on at the end of the sixteenth century, as in our days, an active contraband trade by the mouths of the Orinoco. Raleigh passed the river Europa (Guarapo), and "the plains of Saymas (Chaymas), which extend, keeping the same level, as far as Cumana and Caracas;" he stopped at Morequito (perhaps a little to the north of the site of the villa de Upata, in the missions of the Carony), where an old cacique confirmed to him all the reveries of Berrio on the irruption of foreign nations (Orejones and Epuremei) into Guiana. The Raudales or cataracts of the Caroli (Carony), a river which was at that period considered as the shortest way for reaching the towns of Macureguarai and Manoa, situate on the banks of lake Cassipa and of lake Rupunuwini or Dorado, put an end to this expedition.

Raleigh went scarcely the distance of sixty leagues along the Orinoco; but he names the upper tributary streams, according to the vague notions he had collected; the Cari, the Pao, the Apure (Capuri?) the Guarico (Voari?) the Meta,* and even, "in the province of Baraguan, the great cataract of Athule (Atures), which prevents all further navigation." (* Raleigh distinguishes the Meta from the Beta, which flows into the Baraguan (the Orinoco) conjointly with the Daune, near Athule; as he distinguishes the Casanare, a tributary stream of the Meta, and the Casnero, which comes from the south, and appears to be the Rio Cuchivero. All above the confluence of the Apure was then very confusedly known; and streams that flow into the tributary streams of the Orinoco were considered as flowing into this river itself. The Apure (Capuri) and Meta appeared long to be the same river on account of their proximity, and the numerous branches by which the Arauca and the Apure join each other. Is the name of Beta perchance connected with that of the nation of Betoyes, of the plains of the Casanare and the Meta? Hondius and the geographers who have followed him,

27

with the exception of De L'Isle (1700), and of Sanson (1656), place the province of Amapaja erroneously to the east of the Orinoco. We see clearly by the narrative of Raleigh (pages 26 and 72), that Amapaja is the inundated country between the Meta and the Guarico. Where are the rivers Dauney and Ubarro? The Guaviare appears to me to be the Goavar of Raleigh.) Notwithstanding Raleigh's exaggeration, so little worthy of a statesman, his narrative contains important materials for the history of geography. The Orinoco above the confluence of the Apure was at that period as little known to Europeans, as in our time the course of the Niger below Sego. The names of several very remote tributary streams were known, but not their situation; and when the same name, differently pronounced, or not properly apprehended by the ear, furnished different sounds, their number was multiplied. Other errors had perhaps their source in the little interest which Antonio de Berrio, the Spanish governor, felt in communicating true and precise notions to Raleigh, who indeed complains of his prisoner, "as being utterly unlearned, and not knowing the east from the west." I shall not here discuss the point how far the belief of Raleigh, in all he relates of inland seas similar to the Caspian sea; on "the imperial and golden city of Manoa," and on the magnificent palaces built by the emperor Inga of Guyana, in imitation of those of his ancestors at Peru, was real or pretended. The learned historian of Brazil, Mr. Southey, and the biographer of Raleigh, Sir G. Cayley, have recently thrown much light on this subject. It seems to me difficult to doubt of the extreme credulity of the chief of the expedition, and of his lieutenants. We see Raleigh adapted everything to the hypotheses he had previously formed. He was certainly deceived himself; but when he sought to influence the imagination of queen Elizabeth, and execute the projects of his own ambitious policy, he neglected none of the artifices of flattery. He described to the Queen "the transports of those barbarous nations at the sight of her picture;" he would have "the name of the august virgin, who knows how to conquer empires, reach as far as the country of the warlike women of the Orinoco and the Amazon;" he asserts that "at the period when the Spaniards overthrew the throne of Cuzco, an ancient prophecy was found, which predicted that the dynasty of the Incas would one day owe its restoration to Great Britain;" he advises that "on pretext of defending the territory against external enemies, garrisons of three or four thousand English should be placed in the towns of the Inca, obliging this prince to pay a contribution annually to Queen Elizabeth of three hundred thousand pounds sterling;" finally he adds, like a man who foresees the future, that "all the vast countries of South America will one day belong to the English nation."* (* "I showed them her Majesty's picture, which the Casigui so admired and honoured, as it had been easy to have brought them idolatrous thereof. And I further remember that Berreo confessed to me and others (which I protest before the majesty of God to be true), that there was found among prophecies at Peru (at such a time as the empire was reduced to the Spanish obedience) in their chiefest temple, among divers others which foreshowed the losse of the said empyre, that from Inglatierra those Ingas should be again in time to come restored. The Inga would yield to her Majesty by composition many hundred thousand pounds yearely as to defend him against all enemies abroad and defray the expenses of a garrison of 3000 or 4000 soldiers. It seemeth to me that this Empyre of Guiana is reserved for the English nation." (Raleigh pages 7, 17, 51 and 100.)

The four voyages of Raleigh to the Lower Orinoco succeeded each other from 1595 to 1617. After all these useless attempts the ardour of research after El Dorado has greatly diminished. No expeditions have since been formed by a numerous band of colonists; but some solitary enterprises have been encouraged by the governors of the provinces. The notions spread by the journeys of Father Acunha in 1688, and Father Fritz in 1637, to the auriferous land of the Manoas of Jurubesh, and to the Laguna de Ore, contributed to renew the ideas of El Dorado in the Portuguese and Spanish colonies north and south of the equator. At Cuenza, in the kingdom of Quito, I met with some men, who were employed by the bishop Marfil to seek at the east of the Cordilleras, in the plains of Macas, the ruins of the town of Logrono, which was believed to be situate in a country rich in gold. We learn by the journal of Hortsmann, which I have often quoted, that it was supposed, in 1740, El Dorado might be reached from Dutch Guiana by going up the Rio Essequibo. Don Manuel Centurion, the governor of Santo Thome del Angostura, displayed an extreme ardour for reaching the imaginary lake of Manoa. Arimuicaipi, an Indian of the nation of the

Ipurucotos, went down the Rio Carony, and by his false narrations inflamed the imagination of the Spanish colonists. He showed them in the southern sky the Clouds of Magellan, the whitish light of which he said was the reflection of the argentiferous rocks situate in the middle of the Laguna Parima. This was describing in a very poetical manner the splendour of the micaceous and talcy slates of his country! Another Indian chief, known among the Caribs of Essequibo by the name El Capitan Jurado, vainly attempted to undeceive the governor Centurion. Fruitless attempts were made by the Caura and the Rio Paragua; and several hundred persons perished miserably in these rash enterprises, from which, however, geography has derived some advantages. Nicolas Rodriguez and Antonio Santos (1775 to 1780) were employed by the Spanish governor. Santos, proceeding by the Carony, the Paragua, the Paraguamusi, the Anocapra, and the mountains of Pacaraymo and Quimiropaca, reached the Uraricuera and the Rio Branco. I found some valuable information in the journals of these perilous expeditions.

The maritime charts which the Florentine traveller, Amerigo Vespucci,* constructed in the early years of the sixteenth century, as Piloto mayor de la Casa de Contratacion of Seville, and in which he placed, perhaps artfully, the words Tierra de Amerigo, have not reached our times. (* He died in 1512, as Mr. Munoz has proved by the documents of the archives of Simancas. Hist. del Nuevo Mundo volume 1 page 17. Tiraboschi, Storia della Litteratura.) The most ancient monument we possess of the geography of the New Continent,* is the map of the world by John Ruysch, annexed to a Roman edition of Ptolemy in 1508. (* See the learned researches of M. Walckenaer, in the Bibliographie Universelle volume 6 page 209 article Buckinck. On the maps added to Ptolemy in 1506 we find no trace of the discoveries of Columbus.) We there find Yucatan and Honduras (the most southern part of Mexico)* figured as an island, by the name of Culicar. (* No doubt the lands between Uucatan, Cape Gracias a Dios, and Veragua, discovered by Columbus (1502 and 1503), by Solis, and by Pincon (1506).) There is no isthmus of Panama, but a passage, which permits of a direct navigation from Europe to India. The great southern island (South America) bears the name of Terra de Pareas, bounded by two rivers, the Rio Lareno and the Rio Formoso. These Pareas are, no doubt, the inhabitants of Paria, a name which Christopher Columbus had already heard in 1498, and which was long applied to a great part of America. Bishop Geraldini says clearly, in a letter addressed to Pope Leo X in 1516: Insula illa, quae Europa et Asia est major, quam indocti Continentem Asiae appellant, et alii Americam vel Pariam nuncupant [that island, larger than Europe and Asia joined together, which the unlearned call the continent of Asia, and others America or Paria].* (* Alexandri Geraldini Itinerarium page 250.) I find in the map of the world of 1508 no trace whatever of the Orinoco. This river appears, for the first time, by the name of Rio Dolce, on the celebrated map constructed in 1529 by Diego Ribeyro, cosmographer of the emperor Charles V, which was published, with a learned commentary, by M. Sprengel, in 1795. Neither Columbus (1498) nor Alonzo de Ojeda, accompanied by Amerigo Vespucci (1499), had seen the real mouth of the Orinoco; they confounded it with the northern opening of the Gulf of Paria, to which they attributed (by an exaggeration so common to the navigators of that time, an immense volume of fresh water. It was Vicente Yanez Pincon, who, after having discovered the mouth of the Rio Maranon,* first saw, in 1500, that of the Orinoco. (* The name of Maranon was known fifty-nine years before the expedition of Lopez de Aguirre; the denomination of the river is therefore erroneously attributed to the nickname of maranos (hogs), which this adventurer gave his companions in going down the river Amazon. Was not this vulgar jest rather an allusion to the Indian name of the river?) He called this river Rio Dolce—a name which, since Ribeyro, was long preserved on our maps, and which has sometimes been given erroneously to the Maroni and to the Essequibo.

The great Lake Parima did not appear on our maps* till after the first voyage of Raleigh. (* I find no trace of it on a very rare map, dedicated to Richard Hakluyt, and constructed on the meridian of Toledo. Novus Orbis, Paris 1587. In this map, published before the voyage of Quiros, a group of Islands is marked (Infortunatae Insulae) where the Friendly Islands actually are. Ortelius (1570) already knew them. Were they islands seen by Magellan?) It was Jodocus Hondius who, as early as the year 1599, fixed the ideas of geographers and figured the interior of Spanish Guiana as a country well known. He transformed the isthmus between the Rio Branco and the Rio Rupunuwini (one of the tributary streams of the Essequibo) into the lake Rupunuwini, Parima, or Dorado, two hundred leagues long, and forty broad, and bounded by the latitudes of 1 degree 45 minutes south, and 2 degrees north. This inland sea, larger than the Caspian, is sometimes traced in the midst of a mountainous country, without communication with any river;* (* See, for instance, Hondius, Nieuwe Caerte van het goudrycke landt Guiana, 1599; and Sanson's Map of America, in 1656 and 1669.) and sometimes the Rio Oyapok (Waiapago, Japoc, Viapoco) and the Rio de Cayana are made to issue from it.* (* Brasilia et Caribaua, auct. Hondio et Huelsen 1599.) The first of these rivers, confounded in the eighth article of the treaty of Utrecht with the Rio de Vicente Pincon (Rio Calsoene of D'Anville), has been, even down to the late congress of Vienna, the subject of interminable discussions between the French and Portuguese diplomatists.* (* I have treated this question in a Memoire sur la fixation des limites de La Guyane Francaise, written at the desire of the Portuguese government during the negotiations of Paris in 1817. (See Schoell, Archives polit. or Pieces inedites volume 1 pages 48 to 58.) Ribeyro, in his celebrated map of the world of 1529, places the Rio de Vicente Pincon south of the Amazon, near the Gulf of Maranhao. This navigator landed at this spot, after having been at Cape Saint Augustin, and before he reached the mouth of the Amazon. Herrera dec. I page 107. The narrative of Gomara, Hist. Nat. 1553 page 48, is very confused in a geographical point of view.) The second is an imaginary prolongation either of the Tonnegrande or of the Oyac (Wia?). The inland sea (Laguna Parime) was at first placed in such a manner that its western extremity coincided with the meridian of the confluence of the Apure and the Orinoco. By degrees it was advanced toward the east,* the western extremity being found to the south of the mouth of the Orinoco. (* Compare the maps of 1599 with those of Sanson (1656) and of Blaeuw (1633).) This change produced others in the respective situations of the lakes Parima and Cassipa, as well as in the direction of the course of the Orinoco. This great river is represented as running from its delta as far as beyond the Meta, from south to north, like the river Magdalena. The tributary streams, therefore, which were made to issue from the lake Cassipa, the Carony, the Arui, and the Caura, then took the direction of the latitude, while in nature they follow that of a meridian. Beside the lakes Parima and Cassipa, a third was traced upon the maps, from which the Aprouague (Apurwaca) was made to issue. It was then a general practice among geographers to attach all rivers to great lakes. By this means Ortelius joined the Nile to the Zaire or Rio Congo, and the Vistula to the Wolga and the Dnieper. North of Mexico, in the pretended kingdoms of Quivira and Cibola, rendered celebrated by the falsehoods of the monk Marcos de Niza, a great inland sea was imagined, from which the Rio Colorado of California was made to issue.* (* This is the Mexican Dorado, where it was pretended that vessels had been found on the coasts [of New Albion?] loaded with the merchandise of Catayo and China (Gomara, Hist. Gen. page 117), and where Fray Marcos (like Huten in the country of the Omaguas) had seen from afar the gilded roofs of a great town, one of the Siete Ciudades. The inhabitants have great dogs, en los quales quando se mudan cargan su menage. (Herrera dec. 6 pages 157 and 206.) Later discoveries, however, leave no doubt that there existed a centre of civilization in those countries.) A branch of the Rio Magdalena flowed to the Laguna de Maracaybo; and the lake of Xarayes, near which a southern Dorado was placed, communicated with the Amazon, the Miari* (Meary) (* As this river flows into the gulf of Maranhao (so named because some French colonists, Rifault, De Vaux, and Ravadiere, believed they were opposite the mouth of the Maranon or Amazon), the ancient maps call the Meary Maranon, or Maranham. See the maps of Hondius, and Paulo de Forlani. Perhaps the idea that Pincon, to whom the discovery of the real Maranon is due, had landed in these parts, since become celebrated by the shipwreck of Ayres da Cunha, has also contributed to this confusion. The Meary appears to me identical with the Rio de Vicente Pincon of Diego Ribeyro, which is more than one hundred and forty leagues from that of the modern geographers. At present the name of Maranon has remained at the same time to the river of the Amazons, and to a province much farther eastward, the capital of which is Maranhao, or St. Louis de Maranon.) and the Rio de San Francisco. These hydrographic reveries have for the most part disappeared; but the lakes Cassipa and Dorado have been long simultaneously preserved on our maps.

In following the history of geography we see the Cassipa, figured as a rectangular parallelogram, enlarge by degrees at the expense of El Dorado. While the latter is sometimes suppressed, no one ventures to touch the former,* which is the Rio Paragua (a tributary stream of the Caroni) enlarged by temporary inundations. (* Sanson, Course of the Amazon, 1680; De L'Isle, Amerique Merid. 1700. D'Anville, first edition of his America, 1748.) When D'Anville learned from the expedition of Solano that the sources of the Orinoco, far from lying to the west, on the back of the Andes of Pasto, came from the east, from the mountains of Parima, he restored in the second edition of his fine map of America (1760) the Laguna Parime, and very arbitrarily made it to communicate with three rivers, the Orinoco, the Rio Branco, and the Essequibo, by the Mazuruni and the Cujuni; assigning to it the latitude from 3 to 4 degrees north, which had till then been given to lake Cassipa.

I have now stated, as I announced above, the variable forms which geographical errors have assumed at different periods. I have explained what in the configuration of the soil, the course of the rivers, the names of the tributary streams, and the multiplicity of the portages, may have given rise to the hypothesis of an inland sea in the centre of Guiana. However dry discussions of this nature may appear, they ought not to be regarded as sterile and fruitless. They show travellers what remains to be discovered; and make known the degree of certainty which long-repeated assertions may claim. It is with maps, as with those tables of astronomical positions which are contained in our ephemerides, designed for the use of navigators: the most heterogeneous materials have been employed in their construction during a long space of time; and, without the aid of the history of geography, we could scarcely hope to discover at some future day on what authority every partial statement rests.

Before I resume the thread of my narrative, it remains for me to add a few general reflections on the auriferous lands situate between the Amazon and the Orinoco. We have just shown that the fable of El Dorado, like the most celebrated fables of the nations of the ancient world, has been applied progressively to different spots. We have seen it advance from the south-west to the north-east, from the oriental declivity of the Andes towards the plains of Rio Branco and the Essequibo, an identical direction with that in which the Caribs for ages conducted their warlike and mercantile expeditions. It may be conceived that the gold of the Cordilleras might be conveyed from hand to hand, through an infinite number of tribes, as far as the shore of Guiana; since, long before the fur-trade had attracted English, Russian, and American vessels to the north-west coast of America, iron tools had been carried from New Mexico and Canada beyond the Rocky Mountains. From an error in longitude, the traces of which we find in all the maps of the 16th century, the auriferous mountains of Peru and New Granada were supposed to be much nearer the mouths of the Orinoco and the Amazon than they are in fact. Geographers have the habit of augmenting and extending beyond measure countries that are recently discovered. In the map of Peru, published at Verona by Paulo di Forlani, the town of Quito is placed at the distance of 400 leagues from the coast of the South Sea, on the meridian of Cumana; and the Cordillera of the Andes there fills almost the whole surface of Spanish, French, and Dutch Guiana. This erroneous opinion of the breadth of the Andes has no doubt contributed to give so much importance to the granitic plains that extend on their eastern side. Unceasingly confounding the tributary streams of the Amazon with those of the Orinoco, or (as the lieutenants of Raleigh called it, to flatter their chief) the Rio Raleana, to the latter were attributed all the traditions which had been collected respecting the Dorado of Quixos, the Omaguas, and the Manoas.* (* The flight of Manco-Inca, brother of Atahualpa, to the east of the Cordilleras, no doubt gave rise to the tradition of the new empire of the Incas in Dorado. It was forgotten that Caxamarca and Cuzco, two towns where the princes of that unfortunate family were at the time of their emigration, are situate to the south of the Amazon, in the latitudes seven degrees eight minutes, and thirteen degrees twenty-one minutes south, and consequently four hundred leagues south-west of the pretended town of Manoa on the lake Parima (three degrees and a half north latitude). It is probable that, from the extreme difficulty of penetration into the plains east of the Andes, covered with forests, the fugitive princes never went beyond the banks of the Beni. The following is what I learnt with certainty respecting the emigration of the family of the Inca, some sad vestiges of which I

31

saw on passing by Caxamarca. Manco-Inca, acknowledged as the legitimate successor of Atahualpa, made war without success against the Spaniards. He retired at length into the mountains and thick forests of Vilcabamba, which are accessible either by Huamanga and Antahuaylla, or by the valley of Yucay, north of Cuzco. Of the two Sons of Manco-Inca, the eldest, Sayri-Tupac, surrendered himself to the Spaniards, upon the invitation of the viceroy of Peru, Hurtado de Mendoza. He was received with great pomp at Lima, was baptized there, and died peaceably in the fine valley of Yucay. The youngest son of Manco-Inca, Tupac-Amaru, was carried off by stratagem from the forests of Vilcabamba, and beheaded on pretext of a conspiracy formed against the Spanish usurpers. At the same period, thirty-five distant relations of the Inca Atahualpa were seized, and conveyed to Lima, in order to remain under the inspection of the Audiencia. (Garcilasso volume 2 pages 194, 480 and 501.) It is interesting to inquire whether any other princes of the family of Manco-Capac have remained in the forests of Vilcabamba, and if there still exist any descendants of the Incas of Peru between the Apurimac and the Beni. This supposition gave rise in 1741 to the famous rebellion of the Chuncoes, and to that of the Amages and Campoes led on by their chief, Juan Santos, called the false Atahualpa. The late political events of Spain have liberated from prison the remains of the family of Jose Gabriel Condorcanqui, an artful and intrepid man, who, under the name of the Inca Tupac-Amaru, attempted in 1781 that restoration of the ancient dynasty which Raleigh had projected in the time of Queen Elizabeth.) The geographer Hondius supposed that the Andes of Loxa, celebrated for their forests of cinchona, were only twenty leagues distant from the lake Parima, or the banks of the Rio Branco. This proximity procured credit to the tidings of the flight of the Inca into the forests of Guiana, and the removal of the treasures of Cuzco to the easternmost parts of that country. No doubt in going up towards the east, either by the Meta or by the Amazon, the civilization of the natives, between the Puruz, the Jupura, and the Iquiari, was observed to increase. They possessed amulets, little idols of molten gold, and chairs, elegantly carved; but these traces of dawning civilization are far distant from those cities and houses of stone described by Raleigh and those who followed him. We have made drawings of some ruins of great edifices east of the Cordilleras, when going down from Loxa towards the Amazon, in the province of Jaen de Bracamoros; and thus far the Incas had carried their arms, their religion, and their arts. The inhabitants of the Orinoco were also, before the conquest, when abandoned to themselves, somewhat more civilized than the independent hordes of our days. They had populous villages along the river, and a regular trade with more southern nations; but nothing indicates that they ever constructed an edifice of stone. We saw no vestige of any during the course of our journey.

Though the celebrity of the riches of Spanish Guiana is chiefly assignable to the geographical situation of the country and the errors of the old maps, we are not justified in denying the existence of any auriferous land in the tract of country of eighty-two thousand square leagues, which stretches between the Orinoco and the Amazon, on the east of the Andes of Quito and New Granada. What I saw of this country between the second and eighth degrees of latitude, and the sixty-sixth and seventy-first degrees of longitude, is entirely composed of granite, and of a gneiss passing into micaceous and talcous slate. These rocks appear naked in the lofty mountains of Parima, as well as in the plains of the Atabapo and the Cassiquiare. Granite predominates there over the other rocks; and though, in both continents, the granite of ancient formation is pretty generally destitute of gold-ore, we cannot thence conclude that the granite of Parima contains no vein, no stratum of auriferous quartz. On the east of the Cassiquiare towards the sources of the Orinoco, we observed that the number of these strata and these veins increased. The granite of these countries, by its structure, its mixture of hornblende, and other geological features alike important, appears to me to belong to a more recent formation, perhaps posterior to the gneiss, and analogous to the stanniferous granites, the hyalomictes, and the pegmatites. Now the least ancient granites are also the least destitute of metals; and several auriferous rivers and torrents in the Andes, in the Salzburg, Fichtelgebirge, and the table-land of the two Castiles, lead us to believe that these granites sometimes contain native gold, and portions of auriferous pyrites and galena disseminated throughout the whole rock, as is the case with tin and magnetic and micaceous iron. The group of the mountains of Parima, several summits of which attain the height of one thousand three hundred toises, was almost entirely unknown before our visit to the Orinoco. This group, however, is a hundred leagues long and eighty broad; and though wherever M. Bonpland and I traversed this vast group of mountains, its structure seemed to us extremely uniform, it would be wrong to affirm that it may not contain very metalliferous transition rocks and mica-slates superimposed on the granite.

I have already observed that the silvery lustre and frequency of mica have contributed to give Guiana great celebrity for metallic wealth. The peak of Calitamini, glowing every evening at sunset with a reddish fire, still attracts the attention of the inhabitants of Maypures. According to the fabulous stories of the natives, the islets of mica-slate, situate in lake Amucu, augment by their reflection the lustre of the nebulae of the southern sky. "Every mountain," says Raleigh, "every stone in the forests of the Orinoco, shines like the precious metals; if it be not gold, it is madre del oro (mother of gold)." Raleigh asserts that he brought back gangues of auriferous white quartz ("harde white sparr"); and to prove the richness of this ore he gives an account of the assays that were made by the officers of the mint at London.* (* Messrs. Westewood, Dimocke, and Bulmar.) I have no reason to believe that the chemists of that time sought to lead Queen Elizabeth into error, and I will not insult the memory of Raleigh by supposing, like his contemporaries,* that the auriferous quartz which he brought home had not been collected in America. (* See the defence of Raleigh in the preface to the Discovery of Guiana, 1596 pages 2 to 4.) We cannot judge of things from which we are separated by so long an interval of time. The gneiss of the littoral chain* contains traces of the precious metals (* In the southern branch of this chain which passes by Yusma, Villa de Cura and Ocumare, particularly near Buria, Los Teques and Los Marietas.); and some grains of gold have been found in the mountains of Parima, near the mission of Encaramada. How can we infer the absolute sterility of the primitive rocks of Guiana from testimony merely negative, from the circumstance that during a journey of three months we saw no auriferous vein appearing above the soil?

In order to bring together whatever may enlighten the government of this country on a subject so long disputed, I will enter upon a few more geological considerations. The mountains of Brazil, notwithstanding the numerous traces of embedded ore which they display between Saint Paul and Villa Rica, have furnished only stream-works of gold. More than six-sevenths of the seventy-eight thousand marks (52,000 pounds) of this metal, with which at the beginning of the 19th century America annually supplied the commerce of Europe, have come, not from the lofty Cordilleras of the Andes, but from the alluvial lands on the east and west of the Cordilleras. These lands are raised but little above the level of the sea, like those of Sonora in Mexico, and of Choco and Barbacoas in New Granada; or they stretch along in table-lands, as in the interior of Brazil.* (* The height of Villa Rica is six hundred and thirty toises; but the great table-land of the Capitania de Minas Geraes is only three hundred toises in height. See the profile which Colonel d'Eschwege has published at Weimar, with an indication of the rocks, in imitation of my profile of the Mexican table-land.) Is it not probable that some other depositions of auriferous earth extend toward the northern hemisphere, as far as the banks of the Upper Orinoco and the Rio Negro, two rivers which form but one basin with that of the Amazon? I observed, when speaking of El Dorado de Canelas, the Omaguas and the Iquiare, that almost all the rivers which flow from the west wash down gold in abundance, and very far from the Cordilleras. From Loxa to Popayan these Cordilleras are composed alternately of trachytes and primitive rocks. The plains of Ramora, of Logrono, and of Macas (Sevilla del Oro), the great Rio Napo with its tributary streams* (the Ansupi and the Coca, in the province of Quixos (* The little rivers Cosanga, Quixos, and Papallacta or Maspa, which form the Coca, rise on the eastern slope of the Nevado de Antisana. The Rio Ansupi brings down the largest grains of gold: it flows into the Napo, south of the Archidona, above the mouth of the Misagualli. Between the Misagualli and the Rio Coca, in the province of Avila, five other northern tributary streams of the Napo (the Siguna, Munino, Suno, Guataracu, and Pucono) are known as being singularly auriferous. These local details are taken from several manuscript reports of the Governor of Quixos, from which I traced the map of the countries east of the Antisana.)), the Caqueta de Mocoa as far as the mouth of the Fragua, in fine, all the country comprised between Jaen de Bracamoros and the Guaviare,* (* From Rio Santiago, a tributary stream of the Upper Maranon, to the Llanos of Caguan and of San Juan,) preserve their ancient celebrity for metallic wealth. More to the east, between the sources of the Guainia (Rio Negro), the Uaupes, the Iquiare, and the Yurubesh, we find a soil incontestably auriferous. There Acunha and Father Fritz placed their Laguna del Oro; and various accounts which I obtained at San Carlos from Portuguese Americans explain perfectly what La Condamine has related of the plates of beaten gold found in the hands of the natives. If we pass from the Iquiare to the left bank of the Rio Negro, we enter a country entirely unknown, between the Rio Branco, the sources of the Essequibo, and the mountains of Portuguese Guiana. Acunha speaks of the gold washed down by the northern tributary streams of the Lower Maranon, such as the Rio Trombetas (Oriximina), the Curupatuba, and the Ginipape (Rio de Paru). It appears to me a circumstance worthy of attention that all these rivers descend from the same table-land, the northern slope of which contains the lake Amucu, the Dorado of Raleigh and the Dutch, and the isthmus between the Rupunuri (Rupunuwini) and the Rio Mahu. There is no reason for denying the existence of auriferous alluvial lands far from the Cordilleras of the Andes on the north of the Amazon; as there are on the south in the mountains of Brazil. The Caribs of the Carony, the Cuyuni and the Essequibo, have practised on a small scale the washing of alluvial earth from the remotest times.* (* "On the north of the confluence of the Curupatuba and the Amazon," says Acunha, "is the mountain of Paraguaxo, which, when illumined by the sun, glows with the most beautiful colours; and thence from time to time issues a horrible noise (revienta con grandes struenos)." Is there a volcanic phenomenon in this eastern part of the New Continent? or is it the love of the marvellous, which has given rise to the tradition of the bellowings (bramidos) of Paraguaxo? The lustre emitted from the sides of the mountain recalls to mind what we have mentioned above of the miraculous rocks of Calitamini, and the island Ipomucena, in the imaginary Lake Dorado. In one of the Spanish letters intercepted at sea by Captain George Popham, in 1594, it is said, "Having inquired of the natives whence they obtained the spangles and powder of gold, which we found in their huts, and which they stick on their skin by means of some greasy substances, they told us that in a certain plain they tore up the grass, and gathered the earth in baskets, to subject it to the process of washing." Raleigh page 109. Can this passage be explained by supposing that the Indians sought thus laboriously, not for gold, but for spangles of mica, which the natives of Rio Caura still employ as ornaments, when they paint their bodies?) When we examine the structure of mountains and embrace in one point of view an extensive surface of the globe, distances disappear; and places the most remote insensibly draw near each other. The basin of the Upper Orinoco, the Rio Negro, and the Amazon is bounded by the mountains of Parime on the north, and by those of Minas Geraes, and Matogrosso on the south. The opposite slopes of the same valley often display an analogy in their geological relations.

I have described in this and the preceding volume the vast provinces of Venezuela and Spanish Guiana. While examining their natural limits, their climate, and their productions, I have discussed the influence produced by the configuration of the soil on agriculture, commerce, and the more or less rapid progress of society. I have successively passed over the three regions that succeed each other from north to south; from the Mediterranean of the West Indies to the forests of the Upper Orinoco and of the Amazon. The fertile land of the shore, the centre of agricultural riches, is succeeded by the Llanos, inhabited by pastoral tribes. These Llanos are in their turn bordered by the region of forests, the inhabitants of which enjoy, I will not say liberty, which is always the result of civilization, but a sort of savage independence. On the limit of these two latter zones the struggle now exists which will decide the emancipation and future prosperity of America. The changes which are preparing cannot efface the individual character of each region; but the manners and condition of the inhabitants will assume a more uniform colour. This consideration perhaps adds interest to a tour made in the beginning of the nineteenth century. We like to see, traced in the same picture, the civilized nations of the sea-shore, and the feeble remains of the natives of the Orinoco, who know no other worship than that of the powers of nature; and who, like the ancient Germans, deify the mysterious object which excites their simple admiration.* (* Deorum nominibus appellant secretum illud, quod sola reverentia vident. Tacitus Germania 9.)

CHAPTER 3.26.
THE LLANOS DEL PAO, OR EASTERN PART OF THE PLAINS OF
VENEZUELA. MISSIONS OF THE CARIBS. LAST VISIT TO THE COAST OF
NUEVA BARCELONA, CUMANA, AND ARAYA.

Night had set in when we crossed for the last time the bed of the Orinoco. We purposed to rest near the little fort San Rafael, and on the following morning at daybreak to set out on our journey through the plains of Venezuela. Nearly six weeks had elapsed since our arrival at Angostura; and we earnestly wished to reach the coast, with the view of finding, at Cumana, or at Nueva Barcelona, a vessel in which we might embark for the island of Cuba, thence to proceed to Mexico. After the sufferings to which we had been exposed during several months, whilst sailing in small boats on rivers infested by mosquitos, the idea of a sea voyage was not without its charms. We had no idea of ever again returning to South America. Sacrificing the Andes of Peru to the Archipelago of the Philippines (of which so little is known), we adhered to our old plan of remaining a year in New Spain, then proceeding in a galleon from Acapulco to Manila, and returning to Europe by way of Bassora and Aleppo. We imagined that, when we had once left the Spanish possessions in America, the fall of that ministry which had procured for us so many advantages, could not be prejudicial to the execution of our enterprise.

Our mules were in waiting for us on the left bank of the Orinoco. The collection of plants, and the different geological series which we had brought from the Esmeralda and Rio Negro, had greatly augmented our baggage; and, as it would have been dangerous to lose sight of our herbals, we expected to make a very slow journey across the Llanos. The heat was excessive, owing to the reverberation of the soil, which was almost everywhere destitute of vegetation; yet the centigrade thermometer during the day (in the shade) was only from thirty to thirty-four degrees, and during the night, from twenty-seven to twenty-eight degrees. Here, therefore, as almost everywhere within the tropics, it was less the absolute degree of heat than its duration that affected our sensations. We spent thirteen days in crossing the plains, resting a little in the Caribbee (Caraibes) missions and in the little town of Pao. The eastern part of the Llanos through which we passed, between Angostura and Nueva Barcelona, presents the same wild aspect as the western part, through which we had passed from the valleys of Aragua to San Fernando de Apure. In the season of drought, (which is here called summer,) though the sun is in the southern hemisphere, the breeze is felt with greater force in the Llanos of Cumana, than in those of Caracas; because those vast plains, like the cultivated fields of Lombardy, form an inland basin, open to the east, and closed on the north, south and west by high chains of primitive mountains. Unfortunately, we could not avail ourselves of this refreshing breeze, of which the Llaneros, or the inhabitants of the plains, speak with rapture. It was now the rainy season north of the equator; and though it did not rain in the plains, the change in the declination of the sun had for some time caused the action of the polar currents to cease. In the equatorial regions, where the traveller may direct his course by observing the direction of the clouds, and where the oscillations of the mercury in the barometer indicate the hour almost as well as a clock, everything is subject to a regular and uniform rule. The cessation of the breezes, the setting-in of the rainy season, and the frequency of electric explosions, are phenomena which are found to be connected together by immutable laws.

On entering the Llanos of Nueva Barcelona, we met with a Frenchman, at whose house we passed the first night, and who received us with the kindest hospitality. He was a native of Lyons, and he had left his country at a very early age. He appeared extremely indifferent to all that was passing beyond the Atlantic, or, as they say here, disdainfully enough, when speaking of Europe, on the other side of the great pool (al otro lado del charco). Our host was employed in joining large pieces of wood by means of a kind of glue called guayca. This substance, which is used by the carpenters of Angostura, resembles the best animal glue. It is found perfectly prepared between the bark and the alburnum of a creeper* of the family of the Combretaceae. (* Combretum guayca.) It probably resembles in its chemical properties birdlime, the vegetable principle obtained from the berries of the mistletoe, and the internal bark of the holly. An astonishing abundance of this glutinous matter issues from the twining branches of the vejuco de guayca when they are cut. Thus we find within the tropics a substance in a state of purity and deposited in peculiar organs, which in the temperate zone can be procured only by artificial means.

We did not arrive until the third day at the Caribbee missions of Cari. We observed that the ground was less cracked by the drought in this country than in the Llanos of Calabozo. Some showers had revived the vegetation. Small gramina and especially those herbaceous sensitive-plants so useful in fattening half-wild cattle, formed a thick turf. At great distances one from another, there arose a few fan-palms (Corypha tectorum), rhopalas* (chaparro (* The Proteaceae are not, like the Araucaria, an exclusively southern form. We found the Rhopala complicata and the R. obovata, in 2 degrees 30 minutes, and in 10 degrees of north latitude.)), and malpighias* with coriaceous and glossy leaves. (* A neighbouring genus, Byrsonima cocollobaefolia, B. laurifolia, near Matagorda, and B. ropalaefolia.) The humid spots are recognized at a distance by groups of mauritia, which are the sago-trees of those countries. Near the coast this palm-tree constitutes the whole wealth of the Guaraon Indians; and it is somewhat remarkable that we also found it one hundred and sixty leagues farther south, in the midst of the forests of the Upper Orinoco, in the savannahs that surround the granitic peak of Duida.* (* The moriche, like the Sagus Rumphii, is a palm-tree of the marshes, not a palm-tree of the coast, like the Chamaerops humilis, the common cocoa-tree, and the lodoicea.) It was loaded at this season with enormous clusters of red fruit, resembling fir-cones. Our monkeys were extremely fond of this fruit, which has the taste of an over-ripe apple. The monkeys were placed with our baggage on the backs of the mules, and they made great efforts to reach the clusters that hung over their heads. The plain was undulating from the effects of the mirage; and when, after travelling for an hour, we reached the trunks of the palm-trees, which appeared like masts in the horizon, we observed with astonishment how many things are connected with the existence of a single plant. The winds, losing their velocity when in contact with the foliage and the branches, accumulate sand around the trunk. The smell of the fruit and the brightness of the verdure attract from afar the birds of passage, which love to perch on the slender, arrow-like branches of the palm-tree. A soft murmuring is heard around; and overpowered by the heat, and accustomed to the melancholy silence of the plains, the traveller imagines he enjoys some degree of coolness on hearing the slightest sound of the foliage. If we examine the soil on the side opposite to the wind, we find it remains humid long after the rainy season. Insects and worms, everywhere else so rare in the Llanos, here assemble and multiply. This one solitary and often stunted tree, which would not claim the notice of the traveller amid the forests of the Orinoco, spreads life around it in the desert.

On the 13th of July we arrived at the village of Cari, the first of the Caribbee missions that are under the Observantin monks of the college of Piritu. We lodged as usual at the convent, that is, with the clergyman. Our host could scarcely comprehend how natives of the north of Europe could arrive at his dwelling from the frontiers of Brazil by the Rio Negro, and not by way of the coast of Cumana. He behaved to us in the most affable manner, at the same time manifesting that somewhat importunate curiosity which the appearance of a stranger, not a Spaniard, always excites in South America. He expressed his belief that the minerals we had collected must contain gold; and that the plants, dried with so much care, must be medicinal. Here, as in many parts of Europe, the sciences are thought worthy to occupy the mind only so far as they confer some immediate and practical benefit on society.

We found more than five hundred Caribs in the village of Cari; and saw many others in the surrounding missions. It is curious to observe this nomad people, recently attached to the soil, and differing from all the other Indians in their physical and intellectual powers. They are a very tall race of men, their height being from five feet six inches, to five feet ten inches. According to a practice common in America, the women are more sparingly clothed than the men. The former wear only the guajuco, or perizoma, in the form of a band. The men have the lower part of the body wrapped in a piece of blue cloth, so dark as to be almost black. This drapery is so ample that, on the lowering of the temperature towards evening, the Caribs throw it over their shoulders. Their bodies tinged with onoto,* (* Rocou, obtained from the Bixa orellana. This paint is called in the Carib tongue, bichet.) their tall figures, of a reddish copper-colour, and their picturesque drapery, when seen from a distance, relieved against the sky as a background, resemble antique statues of bronze. The men cut their hair in a very peculiar manner, very much in the style of the monks. A part of the forehead is shaved, which makes it appear extremely high, and a circular tuft of hair is left near the crown of the head. This resemblance between the Caribs and the monks is not the result of mission life. It is not caused, as had been erroneously supposed, by the desire of the natives to imitate their masters, the Franciscan monks. The tribes that have preserved their wild independence, between the sources of the Carony and the Rio Branco, are distinguished by the same cerquillo de frailes,* (* Circular tonsure of the friars.) which the early Spanish historians at the time of the discovery of America attributed to the nations of the Carib race. All the men of this race whom we saw either during our voyage on the Lower Orinoco, or in the missions of Piritu, differ from the other Indians not only in the tallness of their stature, but also in the regularity of their features. Their noses are smaller, and less flattened; the cheek-bones are not so high; and their physiognomy has less of the Mongol character. Their eyes, which are darker than those of the other hordes of Guiana, denote intelligence, and it may even be said, the habit of reflection. The Caribs have a gravity of manner, and a certain look of sadness which is observable among most of the primitive inhabitants of the New World. The expression of severity in their features is heightened by the practice of dyeing their eyebrows with the juice of caruto: they also lengthen their eyebrows, thereby giving them the appearance of being joined together; and they often mark their faces all over with black spots to give themselves a more fierce appearance. The Carib women are less robust and good-looking than the men, On them devolves almost the whole burden of domestic work, as well as much of the out-door labour. They asked us eagerly for pins, which they stuck under their lower lip, making the head of the pin penetrate deeply into the skin. The young girls are painted red, and are almost naked. Among the different nations of the old and the new worlds, the idea of nudity is altogether relative. A woman in some parts of Asia is not permitted to show the tips of her fingers; while an Indian of the Carib race is far from considering herself unclothed if she wear round her waist a guajuco two inches broad. Even this band is regarded as less essential than the pigment which covers the skin. To go out of the hut without being painted, would be to transgress all the rules of Carib decency.

The Indians of the missions of Piritu especially attracted our attention, because they belong to a nation which, by its daring, its warlike enterprises, and its mercantile spirit has exercised great influence over the vast country extending from the equator towards the northern coast. Everywhere on the Orinoco we beheld traces of the hostile incursions of the Caribs: incursions which heretofore extended from the sources of the Carony and the Erevato as far as the banks of the Ventuari, the Atacavi, and the Rio Negro. The Carib language is consequently the most general in this part of the world; it has even passed (like the language of the Lenni-Lenapes, or Algonkins, and the Natchez or Muskoghees, on the west of the Allegheny mountains) to tribes which have not a common origin.

When we survey that multitude of nations spread over North and South America, eastward of the Cordilleras of the Andes, we fix our attention particularly on those who, having long held dominion over their neighbours, have acted an important part on the stage of the world. It is the business of the historian to group facts, to distinguish masses, to ascend to the common sources of many migrations and popular movements. Great empires, the regular organization of a sacerdotal hierarchy, and the culture which that organization favours in the first ages of society, have existed only on the high mountains of the western world. In Mexico we see a vast monarchy enclosing small republics; at Cundinamarca and Peru we find pure theocracies. Fortified towns, highways and large edifices of stone, an extraordinary development of the feudal system, the separation of castes, convents of men and women, religious congregations regulated by discipline more or less severe, complicated divisions of time connected with the calendars, the zodiacs, and the astrology of the enlightened nations of Asia—all these phenomena in America belong to one region only, the long and narrow Alpine band extending from the thirtieth degree of north latitude to the twenty-fifth degree of south. The migration of nations in the ancient world was from east to west; the Basques or Iberians, the Celts, the Germans and the Pelasgi, appeared in succession. In the New World similar migrations flowed from north to south. Among the nations that inhabit the two hemispheres, the direction of this movement followed that of the mountains; but in the torrid zone the temperate table-lands of the Cordilleras had greater influence on the destiny of mankind, than the mountains of Asia and central Europe. As, properly speaking, only civilized nations have a history, the history of the Americans is necessarily no more than that of a small portion of the inhabitants of the mountains. Profound obscurity envelops the vast country which stretches from the eastern slope of the Cordilleras towards the Atlantic; and for this very reason, whatever in that country relates to the preponderance of one nation over others, to distant migrations, to the physiognomical features which denote a foreign race, excite our deepest interest.

Amidst the plains of North America, some powerful nation, which has disappeared, constructed circular, square, and octagonal fortifications; walls six thousand toises in length; tumuli from seven to eight hundred feet in diameter, and one hundred and forty feet in height, sometimes round, sometimes with several stories and containing thousands of skeletons. These skeletons are the remains of men less slender and more squat than the present inhabitants of those countries. Other bones wrapped in fabrics resembling those of the Sandwich and Feejee Islands are found in the natural grottoes of Kentucky. What is become of those nations of Louisiana anterior to the Lenni-Lenapes, the Shawanese, and perhaps even to the Sioux (Nadowesses, Nahcotas) of the Missouri, who are strongly mongolised; and who, it is believed, according to their own traditions, came from the coast of Asia? In the plains of South America we find only a very few hillocks of that kind called cerros hechos a mano;* (* Hills made by the hand, or artificial hills.) and nowhere any works of fortification analogous to those of the Ohio. However, on a vast space of ground, at the Lower Orinoco, as well as on the banks of the Cassiquiare and between the sources of the Essequibo and the Rio Branco, there are rocks of granite covered with symbolic figures. These sculptures denote that the extinct generations belonged to nations different from those which now inhabit the same regions. There seems to be no connection between the history of Mexico and that of Cundinamarca and of Peru; but in the plains of the east a warlike and long-dominant nation betrays in its features and its physical constitution traces of a foreign origin. The Caribs preserve traditions that seem to indicate ancient communications between North and South America. Such a phenomenon deserves particular attention. If it be true that savages are for the most part degenerate races, remnants escaped from a common wreck, as their languages, their cosmogonic fables, and numerous other indications seem to prove, it becomes doubly important to examine the course by which these remnants have been driven from one hemisphere to the other.

That fine race of people, the Caribs, now occupy only a small part of the country which they inhabited at the time of the discovery of America. The cruelties exercised by Europeans have entirely exterminated them from the West Indian Islands and the coasts of Darien; while under the government of the missions they have formed populous villages in the provinces of New Barcelona and Spanish Guiana. The Caribs who inhabit the Llanos of Piritu and the banks of the Carony and the Cuyuni may be estimated at more than thirty-five thousand. If we add to this number the independent Caribs who live westward of the mountains of Cayenne and Pacaraymo, between the sources of the Essequibo and the Rio Branco, we shall no doubt obtain a total of forty thousand individuals of pure race, unmixed with any other tribes of natives. Prior to my travels, the Caribs were mentioned in many geographical works as an extinct race. Writers unacquainted with the interior of the Spanish colonies of the continent supposed that the small islands of Dominica, Guadaloupe, and St. Vincent had been the principal abodes of that nation of which the only vestiges now remaining throughout the whole of the eastern West India Islands are skeletons petrified, or rather enveloped in a limestone containing madrepores.* (* These skeletons were discovered in 1805 by M. Cortez. They are encased in a formation of madrepore breccia, which the negroes call God's masonry, and which, like the travertin of Italy, envelops fragments of vases and other objects created by human skill. M. Dauxion Lavaysse and Dr. Koenig first made known in Europe this phenomenon which has greatly interested geologists.)

The name of Caribs, which I find for the first time in a letter of Peter Martyr d'Anghiera is derived from Calina and Caripuna, the l and p being transferred into r and b. It is very remarkable that this name, which Columbus heard pronounced by the people of Hayti, was known to exist at the same time among the Caribs of the islands and those of the continent. From the word Carina, or Calina, has been formed Galibi (Caribi). This is the distinctive denomination of a tribe in French Guiana,* who are of much more diminutive stature than the inhabitants of Cari, but speaking one of the numerous dialects of the Carib tongue. (* The Galibis (Calibitis), the Palicours, and the Acoquouas, also cut their hair in the style of the monks; and apply bandages to the legs of their children for the purpose of swelling the muscles. They have the same predilection for green stones (saussurite) which we observed among the Carib nations of the Orinoco. There exist, besides, in French Guiana, twenty Indian tribes which are distinguished from the Galibis though their language proves that they have a common origin.) The inhabitants of the islands are called Calinago in the language of the men; and in that of the women, Callipinan. The difference in the language of the two sexes is more striking among the people of the Carib race than among other American nations (the Omaguas, the Guaranis, and the Chiquitos) where it applies only to a limited number of ideas; for instance, the words mother and child. It may be conceived that women, from their separate way of life, frame particular terms which men do not adopt. Cicero observes* that old forms of language are best preserved by women because by their position in society they are less exposed to those vicissitudes of life, changes of place and occupation which tend to corrupt the primitive purity of language among men. (* Cicero, de Orat. lib. 3 cap. 12 paragraph 45 ed. Verburg. Facilius enim mulieres incorruptam antiquitatem conservant, quod multorum sermonis expertes ea tenent semper, quae prima didicerunt.) But in the Carib nations the contrast between the dialect of the two sexes is so great that to explain it satisfactorily we must refer to another cause; and this may perhaps be found in the barbarous custom, practised by those nations, of killing their male prisoners, and carrying the wives of the vanquished into captivity. When the Caribs made an irruption into the archipelago of the West India Islands, they arrived there as a band of warriors, not as colonists accompanied by their families. The language of the female sex was formed by degrees, as the conquerors contracted alliances with the foreign women; it was composed of new elements, words distinct from the Carib words,* which in the interior of the gynaeceums were transmitted from generation to generation, but on which the structure, the combinations, the grammatical forms of the language of the men exercised an influence. (* The following are examples of the difference between the language of the men (m), and the women (w); isle, oubao (m), acaera (w); man, ouekelli (m), eyeri (w); but, irhen (m), atica (w).) There was then manifested in a small community the peculiarity which we now find in the whole group of the nations of the New Continent. The American languages, from Hudson's Bay to the Straits of Magellan, are in general characterized by a total disparity of words combined with a great analogy in their structure. They are like different substances invested with analogous forms. If we recollect that this phenomenon extends over one-half of our planet, almost from pole to pole; if we consider the shades in the grammatical forms (the genders applied to the three persons of the verb, the reduplications, the frequentatives, the duals); it appears highly astonishing to find a uniform tendency in the development of intelligence and language among so considerable a portion of the human race.

We have just seen that the dialect of the Carib women in the West India Islands contains the vestiges of a language that was extinct. Some writers have imagined that this extinct language might be that of the Ygneris, or primitive inhabitants of the Caribbee Islands; others have traced in it some resemblance to the ancient idiom of Cuba, or to those of the Arowaks, and the Apalachites in Florida: but these hypotheses are all founded on a very imperfect knowledge of the idioms which it has been attempted to compare one with another.

The Spanish writers of the sixteenth century inform us that the Carib nations then extended over eighteen or nineteen degrees of latitude, from the Virgin Islands east of Porto Rico, to the mouths of the Amazon. Another prolongation toward the west, along the coast-chain of Santa Marta and Venezuela, appears less certain. Gomara, however, and the most ancient historians, give the name of Caribana, not, as it has since been applied, to the country between the sources of the Orinoco and the mountains of French Guiana,* (* This name is found in the map of Hondius, of 1599, which accompanies the Latin edition of the narrative of Raleigh's voyage. In the Dutch edition Nieuwe Caerte van het goudrycke landt Guiana, the Llanos of Caracas, between the mountains of Merida and the Rio Pao, bear the name of Caribana. We may remark here, what we observe so often in the history of geography, that the same denomination has spread by degrees from west to east.) but to the marshy plains between the mouths of the Rio Atrato and the Rio Sinu. I have visited those coasts in going from the Havannah to Porto Bello; and I there learned that the cape which bounds the gulf of Darien or Uraba on the east, still bears the name of Punta Caribana. An opinion heretofore prevailed pretty generally that the Caribs of the West India Islands derived their origin, and even their name, from these warlike people of Darien. "From the eastern shore springs Cape Uraba, which the natives call Caribana, whence the Caribs of the island are said to have received their present name."* (* Inde Vrabam ab orientali prehendit ora, quam appellant indigenae Caribana, unde Caribes insulares originem habere nomenque retinere dicuntur.) Thus Anghiera expresses himself in his Oceanica. He had been told by a nephew of Amerigo Vespucci that thence, as far as the snowy mountains of St. Marta, all the natives were e genere Caribium, vel Canibalium. I do not deny that Caribs may have had a settlement near the gulf of Darien, and that they may have been driven thither by the easterly currents; but it also may have happened that the Spanish navigators, little attentive to languages, gave the names Carib and Cannibal to every race of people of tall stature and ferocious character. Still it is by no means probable that the Caribs of the islands and of Parima took to themselves the name of the region which they had originally inhabited. On the east of the Andes and wherever civilization has not yet penetrated, it is the people who have given names to the places where they have settled.* (* These names of places can be perpetuated only where the nations succeed immediately to each other, and where the tradition is interrupted. Thus in the province of Quito many of the summits of the Andes bear names which belong neither to the Quichua (the language of Inca) nor to the ancient language of the Paruays, governed by the Conchocando of Lican.) The words Caribs and Cannibals appear significant; they are epithets referring to valour, strength and even superior intelligence.* (* Vespucci says: Charaibi magnae sapientiae viri.) It is worthy of remark that, at the arrival of the Portuguese, the Brazilians gave to their magicians the name of caraibes. We know that the Caribs of Parima were the most wandering people of America; possibly some wily individuals of that nation played the same part as the Chaldeans of the ancient continent. The names of nations readily become affixed to particular professions; and when, in the time of the Caesars, the superstitions of the East were introduced into Italy, the Chaldeans no more came from the banks of the Euphrates than our Gypsies (Egyptians or Bohemians) came from the banks of the Nile or the Elbe.

When a continent and its adjacent islands are peopled by one and the same race, we may choose between two hypotheses; supposing the emigration to have taken place either from the islands to the continent, or from the continent to the islands. The Iberians (Basques) who were settled at the same time in Spain and in the islands of the Mediterranean, afford an instance of this problem; as do also the Malays who appear to be indigenous in the peninsula of Malacca, and in the district of Menangkabao in the island of Sumatra.* (* Crawfurd, Indian Archipelago volume 2 page 371. I make use of the word indigenous (autocthoni) not to indicate a fact of creation, which does not belong to history, but simply to denote that we are ignorant of the autochthoni having been preceded by any other people.) The archipelago of the large and small West India Islands forms a narrow and broken neck of land, parallel with the isthmus of Panama, and supposed by some geographers to join the peninsula of Florida to the north-east extremity of South America. It is the eastern shore of an inland sea which may be considered as a basin with several outlets. This peculiar configuration of the land has served to support the different systems of migration, by which it has been attempted to explain the settlement of the nations of the Carib race in the islands and on the neighbouring continent. The Caribs of the continent admit that the small West India Islands were anciently inhabited by the Arowaks,* a warlike nation, the great mass of which still inhabit the insalubrious shores of Surinam and Berbice. (* Arouaques. The missionary Quandt (Nachricht von Surinam, 1807 page 47) calls them Arawackes.) They assert that the Arowaks, with the exception of the women, were all exterminated by Caribs, who came from the mouths of the Orinoco. In support of this tradition they refer to the traces of analogy existing between the language of the Arowaks and that of the Carib women; but it must be recollected that the Arowaks, though the enemies of the Caribs, belonged to the same branch of people; and that the same analogy exists between the Arowak and Carib languages as between the Greek and the Persian, the German and the Sanscrit. According to another tradition, the Caribs of the islands came from the south, not as conquerors, but because they were expelled from Guiana by the Arowaks, who originally ruled over all the neighbouring nations. Finally, a third tradition, much more general and more probable, represents the Caribs as having come from Florida, in North America. Mr. Bristock, a traveller who has collected every particular relating to these migrations from north to south, asserts that a tribe of Confachites (Confachiqui* (* The province of Confachiqui, which in 1541 became subject to a woman, is celebrated by the expedition of Hernando de Soto to Florida. Among the nations of the Huron tongue, and the Attakapas, the supreme authority was also often exercised by women.)) had long waged war against the Apalachites; that the latter, having yielded to that tribe the fertile district of Amana, called their new confederates Caribes (that is, valiant strangers); but that, owing to a dispute respecting their religious rites, the Confachite-Caribs were driven from Florida. They went first to the Yucayas or Lucayes Islands (to Cigateo and the neighbouring islands); thence to Ayay (Hayhay, now Santa Cruz), and to the lesser Caribbee Islands; and lastly to the continent of South America.* (* Rochefort, Hist. des Antilles volume 1 pages 326 to 353; Garcia page 322; Robertson book 3 note 69. The conjecture of Father Gili that the Caribs of the continent may have come from the islands at the time of the first conquest of the Spaniards (Saggio volume 3 page 204), is at variance with all the statements of the early historians.) It is supposed that this event took place toward the year 1100 of our era. In the course of this long migration the Caribs had not touched at the larger islands; the inhabitants of which however also believed that they came originally from Florida. The islanders of Cuba, Hayti, and Boriken (Porto Rico) were, according to the uniform testimony of the first conquistadores, entirely different from the Caribs; and at the period of the discovery of America, the latter had already abandoned the group of the lesser Lucayes Islands; an archipelago in which there prevailed that variety of languages always found in lands peopled by shipwrecked men and fugitives.* (* La gente de las islas Yucayas era (1492) mas blanca y de major policia que la de Cuba y Haiti. Havia mucha diversidad de lenguas. [The people of the Lucayes were (1492) of fairer complexion and of more civilized manners than those of Cuba and Hayti. They had a great diversity of languages.] Gomara, Hist. de Ind. fol. 22.)

The dominion so long exercised by the Caribs over a great part of the continent, joined to the remembrance of their ancient greatness, has inspired them with a sentiment of dignity and national superiority which is manifest in their manners and their discourse. "We alone are a nation," say they proverbially; "the rest of mankind (oquili) are made to serve us." This contempt of the Caribs for their enemies is so strong that I saw a child of ten years of age foam with rage on being called a Cabre or Cavere; though he had never in his life seen an individual of that unfortunate race of people who gave their name to the town of Cabruta (Cabritu); and who, after long resistance, were almost entirely exterminated by the Caribs. Thus we find among half savage hordes, as in the most civilized part of Europe, those inveterate animosities which have caused the names of hostile nations to pass into their respective languages as insulting appellations.

The missionary of the village of Cari led us into several Indian huts, where extreme neatness and order prevailed. We observed with pain the torments which the Carib mothers inflict on their infants for the purpose not only of enlarging the calf of the leg, but also of raising the flesh in alternate stripes from the ankle to the top of the thigh. Narrow ligatures, consisting of bands of leather, or of woven cotton, are fixed two or three inches apart from each other, and being tightened more and more, the muscles between the bands become swollen. The monks of the missions, though ignorant of the works or even of the name of Rousseau, attempt to oppose this ancient system of physical education: but in vain. Man when just issued from the woods and supposed to be so simple in his manners, is far from being tractable in his ideas of beauty and propriety. I observed, however, with surprise, that the manner in which these poor children are bound, and which seems to obstruct the circulation of the blood, does not operate injuriously on their muscular movements. There is no race of men more robust and swifter in running than the Caribs.

If the women labour to form the legs and thighs of their children so as to produce what painters call undulating outlines, they abstain (at least in the Llanos), from flattening the head by compressing it between cushions and planks from the most tender age. This practice, so common heretofore in the islands and among several tribes of the Caribs of Parima and French Guiana, is not observed in the missions which we visited. The men there have foreheads rounder than those of the Chaymas, the Otomacs, the Macos, the Maravitans and most of the inhabitants of the Orinoco. A systematizer would say that the form is such as their intellectual faculties require. We were so much the more struck by this fact as some of the skulls of Caribs engraved in Europe, for works on anatomy, are distinguished from all other human skulls by the extremely depressed forehead and acute facial angle. In some osteological collections skulls supposed to be those of Caribs of the island of St. Vincent are in fact skulls shaped by having been pressed between planks. They have belonged to Zambos (black Caribs) who are descended from Negroes and true Caribs.* (* These unfortunate remnants of a nation heretofore powerful were banished in 1795 to the Island of Rattam in the Bay of Honduras because they were accused by the English Government of having connexions with the French. In 1760 an able minister, M. Lescallier, proposed to the Court of Versailles to invite the Red and Black Caribs from St. Vincent to Guiana and to employ them as free men in the cultivation of the land. I doubt whether their number at that period amounted to six thousand, as the island of St. Vincent contained in 1787 not more than fourteen thousand inhabitants of all colours.) The barbarous habit of flattening the forehead is practised by several nations,* of people not of the same race; and it has been observed recently in North America; but nothing is more vague than the conclusion that some degree of conformity in customs and manners proves identity of origin. (* For instance the Tapoyranas of Guiana (Barrere page 239), the Solkeeks of Upper Louisiana (Walckenaer, Cosmos page 583). Los Indios de Cumana, says Gomara (Hist. de Ind.), aprietan a los ninos la cabeca muy blando, pero mucho, entre dos almohadillas de algodon para ensancharlos la cara, que lo tienen por hermosura. Las donzellas traen senogiles muy apretados par debaxo y encima de las rodillas, para que los muslos y pantorillas engorden mucho. [The Indians of Cumana press down the heads of young infants tightly between cushions stuffed with cotton for the purpose of giving width to their faces, which they regard as a beauty. The young girls wear very tight bandages round their knees in order to give thickness to the thighs and calves of the legs.]) On observing the spirit of order and submission which prevails in the Carib missions, the traveller can scarcely persuade himself that he is among cannibals. This American word, of somewhat doubtful signification, is probably derived from the language of Hayti, or that of Porto Rico; and it has passed into the languages of Europe, since the end of the fifteenth century, as synonymous with that of anthropophagi. "These newly discovered man-eaters, so greedy of human flesh, are called Caribes or Cannibals,"* says Anghiera, in the third decade of his Oceanica, dedicated to Pope Leo X. (* Edaces humanarum carnium novi helluones anthropophagi, Caribes alias Canibales appellati.) There can be little doubt that the Caribs of the islands, when a conquering people, exercised cruelties upon the Ygneris, or ancient inhabitants of the West Indies, who were weak and not very warlike; but we must also admit that these cruelties were exaggerated by the early travellers, who heard only the narratives of the old enemies of the Caribs. It is not always the vanquished solely, who are calumniated by their contemporaries; the insolence of the conquerors is punished by the catalogue of their crimes being augmented.

All the missionaries of the Carony, the Lower Orinoco and the Llanos del Cari whom we had an opportunity of consulting assured us that the Caribs are perhaps the least anthropophagous nations of the New Continent. They extend this remark even to the independent hordes who wander on the east of the Esmeralda, between the sources of the Rio Branco and the Essequibo. It may be conceived that the fury and despair with which the unhappy Caribs defended themselves against the Spaniards, when in 1504 a royal decree declared them slaves, may have contributed to acquire for them a reputation for ferocity. The first idea of attacking this nation and depriving it of liberty and of its natural rights originated with Christopher Columbus, who was not in all instances so humane as he is represented to have been. Subsequently the licenciado Rodrigo de Figueroa was appointed by the court, in 1520, to determine the tribes of South America, who were to be regarded as of Carib race, or as cannibals; and those who were Guatiaos,* that is, Indians of peace, and friends of the Castilians. (* I had some trouble in discovering the origin of this denomination which has become so important from the fatal decrees of Figueroa. The Spanish historians often employ the word guatiao to designate a branch of nations. To become a guatiao of any one seems to have signified, in the language of Hayti, to conclude a treaty of friendship. In the West India Islands, as well as in the archipelago of the South Sea, names were exchanged in token of alliance. Juan de Esquivel (1502) se hice guatiao del cacique Cotubanama; el qual desde adelante se llamo Juan de Esquivel, porque era liga de perpetua amistad entre los Indios trocarse los nombres: y trocados quedaban guatiaos, que era tanto coma confederados y hermanos en armas. Ponce de Leon se hace guatiao con el poderoso cacique Agueinaha." Herrera dec. 1 pages 129, 159 and 181. [Juan de Esquivel (1502) became the guatiao of the cacique Cotubanama; and thenceforth the latter called himself Juan de Esquivel, for among the Indians the exchange of names was a bond of perpetual friendship. Those who exchanged names became guaitaos, which meant the same as confederates or brethren-in-arms. Ponce de Leon became guatiao with the powerful cacique Agueinaha.] One of the Lucayes Islands, inhabited by a mild and pacific people, was heretofore called Guatao; but we will not insist on the etymology of this word, because the languages of the Lucayes Islands differed from those of Hayti.) The ethnographic document called El Auto de Figueroa is one of the most curious records of the barbarism of the first conquistadores. Without any attention to the analogy of languages, every nation that could be accused of having devoured a prisoner after a battle was arbitrarily declared of Carib race. The inhabitants of Uriapari (on the peninsula of Paria) were named Caribs; the Urinacos (settled on the banks of the Lower Orinoco, or Urinucu), Guatiaos. All the tribes designated by Figueroa as Caribs were condemned to slavery; and might at will be sold, or exterminated by war. In these sanguinary struggles, the Carib women, after the death of their husbands, defended themselves with such desperation that Anghiera says they were taken for tribes of Amazons. But amidst the cruelties exercised on the Caribs, it is consolatory to find, that there existed some courageous men who raised the voice of humanity and justice. Some of the monks embraced an opinion different from that which they had at first adopted. In an age when there could be no hope of founding public liberty on civil institutions, an attempt was at least made to defend individual liberty. "That is a most holy law (ley sanctissima)," says Gomara, in 1551, "by which our emperor has prohibited the reducing of the Indians to slavery. It is just that men, who are all born free, should not become the slaves of one another."

During our abode in the Carib missions, we observed with surprise the facility with which young Indians of eighteen years of age, when appointed to the post of alguazil, would harangue the municipality for whole hours in succession. Their tone of voice, their gravity of deportment, the gestures which accompanied their speech, all denoted an intelligent people capable of a high degree of civilization. A Franciscan monk, who knew enough of the Carib language to preach in it occasionally, pointed out to us that the long and harmonious periods which occur in the discourses of the Indians are never confused or obscure. Particular inflexions of the verb indicate beforehand the nature of the object, whether it be animate or inanimate, singular or plural. Little annexed forms (suffixes) mark the gradations of sentiment; and here, as in every language formed by a free development, clearness is the result of that regulating instinct which characterises human intelligence in the various stages of barbarism and cultivation. On holidays, after the celebration of mass, all the inhabitants of the village assemble in front of the church. The young girls place at the feet of the missionary faggots of wood, bunches of plantains, and other provision of which he stands in need for his household. At the same time the governador, the alguazil, and other municipal officers, all of whom are Indians, exhort the natives to labour, proclaim the occupations of the ensuing week, reprimand the idle, and flog the untractable. Strokes of the cane are received with the same insensibility as that with which they are given. It were better if the priest did not impose these corporal punishments at the instant of quitting the altar, and if he were not, in his sacerdotal habits, the spectator of this chastisement of men and women; but this abuse is inherent in the principle on which the strange government of the missions is founded. The most arbitrary civil power is combined with the authority exercised by the

43

priest over the little community; and, although the Caribs are not cannibals, and we would wish to see them treated with mildness and indulgence, it may be conceived that energetic measures are sometimes necessary to maintain tranquillity in this rising society.

The difficulty of fixing the Caribs to the soil is the greater, as they have been for ages in the habit of trading on the rivers. We have already described this active people, at once commercial and warlike, occupied in the traffic of slaves, and carrying merchandize from the coasts of Dutch Guiana to the basin of the Amazon. The travelling Caribs were the Bokharians of equinoctial America. The necessity of counting the objects of their little trade, and transmitting intelligence, led them to extend and improve the use of the quipos, or, as they are called in the missions, the cordoncillos con necos (cords with knots). These quipos or knotted cords are found in Canada, in Mexico (where Boturini procured some from the Tlascaltecs), in Peru, in the plains of Guiana, in central Asia, in China, and in India. As rosaries, they have become objects of devotion in the hands of the Christians of the East; as suampans, they have been employed in the operations of manual arithmetic by the Chinese, the Tartars, and the Russians. The independent Caribs who inhabit the little-known country situated between the sources of the Orinoco and those of the rivers Essequibo, Carony, and Parima, are divided into tribes; and, like the nations of the Missouri, of Chili, and of ancient Germany, form a political confederation. This system is most in accordance with the spirit of liberty prevailing amongst those warlike hordes who see no advantage in the ties of society but for common defence. The pride of the Caribs leads them to withdraw themselves from every other tribe; even from those to whom, by their language, they have some affinity.

They claim the same separation in the missions, which seldom prosper when any attempt is made to associate them with other mixed communities, that is, with villages where every hut is inhabited by a family belonging to another nation and speaking another language. The authority of the chiefs of the independent Caribs is hereditary in the male line only, the children of sisters being excluded from the succession. This law of succession which is founded on a system of mistrust, denoting no great purity of manners, prevails in India; among the Ashantees (in Africa); and among several tribes of the savages of North America.* (* Among the Hurons (Wyandots) and the Natchez the succession to the magistracy is continued by the women: it is not the son who succeeds, but the son of the sister, or of the nearest relation in the female line. This mode of succession is said to be the most certain because the supreme power remains attached to the blood of the last chief; it is a practice that insures legitimacy. Ancient traces of this strange mode of succession, so common in Africa and in the East Indies, exist in the dynasty of the kings of the West India Islands.) The young chiefs and other youths who are desirous of marrying, are subject to the most extraordinary fasts and penances, and are required to take medicines prepared by the marirris or piaches, called in the transalleghenian countries, war-physic. The Carribbee marirris are at once priests, jugglers and physicians; they transmit to their successors their doctrine, their artifices, and the remedies they employ. The latter are accompanied by imposition of hands, and certain gestures and mysterious practices, apparently connected with the most anciently known processes of animal magnetism. Though I had opportunities of seeing many persons who had closely observed the confederated Caribs, I could not learn whether the marirris belong to a particular caste. It is observed in North America that, among the Shawanese,* (* People that came from Florida, or from the south (shawaneu) to the north.) divided into several tribes, the priests, who preside at the sacrifices, must be (as among the Hebrews) of one particular tribe, that of the Mequachakes. Any facts that may hereafter be discovered in America respecting the remains of a sacerdotal caste appears to me calculated to excite great interest, on account of those priest-kings of Peru, who styled themselves the children of the Sun; and of those sun-kings among the Natchez, who recall to mind the Heliades of the first eastern colony of Rhodes.

On quitting the mission of Cari, we had some difficulties to settle with our Indian muleteers. They had discovered that we had brought skeletons with us from the cavern of Ataruipe; and they were fully persuaded that the beasts of burden which carried the bodies of their old relations would perish on the journey.* (* See volume 2.24.) Every precaution we had taken was useless; nothing escapes a Carib's penetration and keen sense of smell, and it required all the authority of the missionary to forward our passage. We had to cross the Rio

44

Cari in a boat, and the Rio de agua clara, by fording, or, it may almost be said, by swimming. The quicksands of the bed of this river render the passage very difficult at the season when the waters are high. The strength of the current seems surprising in so flat a country; but the rivers of the plains are precipitated, to quote a correct observation of Pliny the younger,[*] "less by the declivity of their course than by their abundance, and as it were by their own weight." (* Epist. lib. 8 ep. 8. Clitumnus non loci devexitate, sed ipsa sui copia et quasi pondere impellitur.) We had two bad stations, one at Matagorda and the other at Los Riecetos, before we reached the little town of Pao. We beheld everywhere the same objects; small huts constructed of reeds, and roofed with leather; men on horseback armed with lances, guarding the herds; herds of cattle half wild, remarkable for their uniform colour, and disputing the pasturage with horses and mules. No sheep or goats are found on these immense plains. Sheep do not thrive well in equinoctial America, except on table-lands above a thousand toises high, where their fleece is long and sometimes very fine. In the burning climate of the plains, where the wolves give place to jaguars, these small ruminating animals, destitute of means of defence, and slow in their movements, cannot be preserved in any considerable numbers.

We arrived on the 15th of July at the Fundacion, or Villa, del Pao, founded in 1744, and situated very favourably for a commercial station between Nueva Barcelona and Angostura. Its real name is El Concepcion del Pao. Alcedo, La Cruz, Olmedilla, and many other geographers, have mistaken the situation of this small town of the Llanos of Barcelona, confounding it either with San Juan Bauptisto del Pao of the Llanos of Caracas, or with El Valle del Pao de Zarate. Though the weather was cloudy I succeeded in obtaining some heights of alpha Centauri, serving to determine the latitude of the place; which is 8 degrees 37 minutes 57 seconds. Some altitudes of the sun gave me 67 degrees 8 minutes 12 seconds for the longitude, supposing Angostura to be 66 degrees 15 minutes 21 seconds. The astronomical determinations of Calabozo and Concepcion del Pao are very important to the geography of this country, where, in the midst of savannahs, fixed points are altogether wanting. Some fruit-trees grow in the vicinity of Pao: they are rarely seen in the Llanos. We even found some cocoa-trees, which appeared very vigorous, notwithstanding the great distance of the sea. I was the more struck with this fact because doubts have recently been started respecting the veracity of travellers, who assert that they have seen the cocoa-tree, which is a palm of the shore, at Timbuctoo, in the centre of Africa. We several times saw cocoa-trees amid the cultivated spots on the banks of the Rio Magdalena, more than a hundred leagues from the coast.

Five days, which to us appeared very tedious, brought us from Villa del Pao to the port of Nueva Barcelona. As we advanced the sky became more serene, the soil more dusty, and the atmosphere more hot. The heat from which we suffered is not entirely owing to the temperature of the air, but is produced by the fine sand mingled with it; this sand strikes against the face of the traveller, as it does against the ball of the thermometer. I never observed the mercury rise in America, amid a wind of sand, above 45.8 degrees centigrade. Captain Lyon, with whom I had the pleasure of conversing on his return from Mourzouk, appeared to me also inclined to think that the temperature of fifty-two degrees, so often felt in Fezzan, is produced in great part by the grains of quartz suspended in the atmosphere. Between Pao and the village of Santa Cruz de Cachipo, founded in 1749, and inhabited by five hundred Caribs, we passed the western elongation of the little table-land, known by the name of Mesa de Amana. This table-land forms a point of partition between the Orinoco, the Guarapiche, and the coast of New Andalusia. Its height is so inconsiderable that it would scarcely be an obstacle to the establishment of inland navigation in this part of the Llanos. The Rio Mano however, which flows into the Orinoco above the confluence of the Carony, and which D'Anville (I know not on what authority) has marked in the first edition of his great map as issuing from the lake of Valencia, and receiving the waters of the Guayra, could never have served as a natural canal between two basins of rivers. No bifurcation of this kind exists in the Llano. A great number of Carib Indians, who now inhabit the missions of Piritu, were formerly on the north and east of the table-land of Amana, between Maturin, the mouth of the Rio Arco, and the Guarapiche. The incursions of Don Joseph Careno, one of the most enterprising governors of the province of Cumana, occasioned a general migration of independent Caribs toward the banks of the Lower Orinoco in 1720.

The whole of this vast plain consists of secondary formations which to the southward rest immediately on the granitic mountains of the Orinoco. On the north-west they are separated by a narrow band of transition-rocks from the primitive mountains of the shore of Caracas. This abundance of secondary rocks, covering without interruption a space of more than seven thousand square leagues,* is a phenomenon the more remarkable in that region of the globe, because in the whole of the Sierra da la Parima, between the right bank of the Orinoco and the Rio Negro, there is, as in Scandinavia, a total absence of secondary formations. (* Reckoning only that part of the Llanos which is bounded by the Rio Apure on the south, and by the Sierra Nevada de Merida and the Parima de las Rosas on the west.) The red sandstone, containing some vestiges of fossil wood (of the family of monocotyledons) is seen everywhere in the plains of Calabozo: farther east it is overlaid by calcareous and gypseous rocks which conceal it from the research of the geologist. The marly gypsum, of which we collected specimens near the Carib mission of Cachipo, appeared to me to belong to the same formation as the gypsum of Ortiz. To class it according to the type of European formations I would range it among the gypsums, often muriatiferous, that cover the Alpine limestone or zechstein. Farther north, in the direction of the mission of San Josef de Curataquiche, M. Bonpland picked up in the plain some fine pieces of riband jasper, or Egyptian pebbles. We did not see them in their native place enchased in the rock, and cannot determine whether they belong to a very recent conglomerate or to that limestone which we saw at the Morro of Nueva Barcelona, and which is not transition limestone though it contains beds of schistose jasper (kieselschiefer).

We rested on the night of the 16th of July in the Indian village of Santa Cruz de Cachipo. This mission, founded in 1749 by several Carib families who inhabited the inundated and unhealthy banks of the Lagunetas de Auache, is opposite the confluence of the Zir Puruay with the Orinoco. We lodged at the house of the missionary, Fray Jose de las Piedras; and, on examining the registers of the parish, we saw how rapidly the prosperity of the community has been advanced by his zeal and intelligence. Since we had reached the middle of the plains, the heat had increased to such a degree that we should have preferred travelling no more during the day; but we were without arms and the Llanos were then infested by large numbers of robbers who attacked and murdered the whites who fell into their hands. Nothing can be worse than the administration of justice in these colonies. We everywhere found the prisons filled with malefactors on whom sentence is not passed till after the lapse of seven or eight years. Nearly a third of the prisoners succeed in making their escape; and the unpeopled plains, filled with herds, furnish them with booty. They commit their depredations on horseback in the manner of the Bedouins. The insalubrity of the prisons would be attended with fatal results but that these receptacles are cleared from time to time by the flight of the prisoners. It also frequently happens that sentences of death, tardily pronounced by the Audiencia of Caracas, cannot be executed for want of a hangman. In these cases the barbarous custom is observed of pardoning one criminal on condition of his hanging the others. Our guides related to us that, a short time before our arrival on the coast of Cumana, a Zambo, known for the great ferocity of his manners, determined to screen himself from punishment by turning executioner. The preparations for the execution however, shook his resolution; he felt a horror of himself, and preferring death to the disgrace of thus saving his life, he called again for his irons which had been struck off. He did not long remain in prison, and he underwent his sentence through the baseness of one of his accomplices. This awakening of a sentiment of honour in the soul of a murderer is a psychologic phenomenon worthy of reflection. The man who had so often shed the blood of travellers in the plains recoiled at the idea of becoming the passive instrument of justice in inflicting upon others a punishment which he felt that he himself deserved.

If, even in the peaceful times when M. Bonpland and myself had the good fortune to travel through North and South America, the Llanos were the refuge of malefactors who had committed crimes in the missions of the Orinoco, or who had escaped from the prisons on the coast, how much worse must that state of things have been rendered by discord during the continuance of that sanguinary struggle which has terminated in conferring freedom and independence on those vast regions! Our European wastes and heaths are but a feeble image of the savannahs of the New Continent which for the space of eight or ten thousand square leagues are smooth as the surface of the sea. The immensity of their extent insures impunity to robbers, who conceal themselves more effectually in the savannahs than in our mountains and forests; and it is easy to conceive that even a European police would not be very effective in regions where there are travellers and no roads, herds and no herdsmen, and farms so solitary that notwithstanding the powerful action of the mirage, a journey of several days may be made without seeing one appear within the horizon.

Whilst traversing the Llanos of Caracas, New Barcelona, and Cumana, which succeed each other from west to east, from the snowy mountains of Merida to the Delta of the Orinoco, we feel anxious to know whether these vast tracts of land are destined by nature to serve eternally for pasture or whether they will at some future time be subject to the plough and the spade. This question is the more important as the Llanos, situated at the two extremities of South America, are obstacles to the political union of the provinces they separate. They prevent the agriculture of the coast of Venezuela from extending towards Guiana and they impede that of Potosi from advancing in the direction of the mouth of the Rio de la Plata. The intermediate Llanos preserve, together with pastoral life, somewhat of a rude and wild character which separates and keeps them remote from the civilization of countries anciently cultivated. Thus it has happened that in the war of independence they have been the scene of struggle between the hostile parties; and that the inhabitants of Calabozo have almost seen the fate of the confederate provinces of Venezuela and Cundinamarca decided before their walls. In assigning limits to the new states and to their subdivisions, it is to be hoped there may not be cause hereafter to repent having lost sight of the importance of the Llanos, and the influence they may have on the disunion of communities which important common interests should bring together. These plains would serve as natural boundaries like the seas or the virgin forests of the tropics, were it not that armies can cross them with greater facility, as their innumerable troops of horses and mules and herds of oxen furnish every means of conveyance and subsistence.

What we have seen of the power of man struggling against the force of nature in Gaul, in Germany and recently (but still beyond the tropics) in the United States, scarcely affords any just measure of what we may expect from the progress of civilization in the torrid zone. Forests disappear but very slowly by fire and the axe when the trunks of trees are from eight to ten feet in diameter; when in falling they rest one upon another, and the wood, moistened by almost continual rains, is excessively hard. The planters who inhabit the Llanos or Pampas do not generally admit the possibility of subjecting the soil to cultivation; it is a problem not yet solved. Most of the savannahs of Venezuela have not the same advantage as those of North America. The latter are traversed longitudinally by three great rivers, the Missouri, the Arkansas, and the Red River of Nachitoches; the savannahs of Araura, Calabozo, and Pao are crossed in a transverse direction only by the tributary streams of the Orinoco, the most westerly of which (the Cari, the Pao, the Acaru, and the Manapire) have very little water in the season of drought. These streams scarcely flow at all toward the north; so that in the centre of the Llanos there remain vast tracts of land called bancos and mesas* frightfully parched. (* The Spanish words banco and mesa signify literally bench and table. In the Llanos of South America little elevations rising slightly above the general elevation of the plain are called bancos and mesas from their supposed resemblance to benches and tables.) The eastern parts, fertilized by the Portuguesa, the Masparro, and the Orivante, and by the tributary streams of those three rivers, are most susceptible of cultivation. The soil is sand mixed with clay, covering a bed of quartz pebbles. The vegetable mould, the principal source of the nutrition of plants, is everywhere extremely thin. It is scarcely augmented by the fall of the leaves, which, in the forests of the torrid zone, is less periodically regular than in temperate climates. During thousands of years the Llanos have been destitute of trees and brushwood; a few scattered palms in the savannah add little to that hydruret of carbon, that extractive matter, which, according to the experiments of Saussure, Davy, and Braconnot, gives fertility to the soil. The social plants which almost exclusively predominate in the steppes, are monocotyledons; and it is known how much grasses impoverish the soil into which their fibrous roots penetrate. This action of the killingias, paspalums and cenchri, which form the turf, is everywhere the same; but where the rock is ready to pierce the earth this varies according as it rests on red sandstone, or on compact limestone and gypsum; it varies according as periodical inundations accumulate mud on the lower grounds or as the shock of the waters carries away from the small elevations the little soil that has covered them. Many solitary cultivated spots already exist in the midst of the pastures where running water and tufts of the mauritia palm have been found. These farms, sown with maize, and planted with cassava, will multiply considerably if trees and shrubs be augmented.

The aridity and excessive heat of the mesas do not depend solely on the nature of their surface and the local reverberation of the soil; their climate is modified by the adjacent regions; by the whole of the Llano of which they form a part. In the deserts of Africa or Arabia, in the Llanos of South America, in the vast heaths extending from the extremity of Jutland to the mouth of the Scheldt, the stability of the limits of the desert, the savannahs, and the downs, depends chiefly on their immense extent and the nakedness these plains have acquired from some revolution destructive of the ancient vegetation of our planet. By their extent, their continuity, and their mass they oppose the inroads of cultivation and preserve, like inland gulfs, the stability of their boundaries. I will not enter upon the great question, whether in the Sahara, that Mediterranean of moving sands, the germs of organic life are increased in our days. In proportion as our geographical knowledge has extended we have discovered in the eastern part of the desert islets of verdure; oases covered with date-trees crowd together in more numerous archipelagos, and open their ports to the caravans; but we are ignorant whether the form of the oases have not remained constantly the same since the time of Herodotus. Our annals are too incomplete to enable us to follow Nature in her slow and gradual progress. From these spaces entirely bare whence some violent catastrophe has swept away the vegetable covering and the mould; from those deserts of Syria and Africa which, by their petrified wood, attest the changes they have undergone; let us turn to the grass-covered Llanos and to the consideration of phenomena that come nearer the circle of our daily observations. Respecting the possibility of a more general cultivation of the steppes of America, the colonists settled there, concur in the opinions I have deduced from the climatic action of these steppes considered as surfaces, or continuous masses. They have observed that downs enclosed within cultivated and wooded land sooner yield to the labours of the husbandman than soils alike circumscribed, but forming part of a vast surface of the same nature. This observation is extremely just whether in reference to soil covered with heath, as in the north of Europe; with cistuses, mastic-trees, or palmettos, as in Spain; or with cactuses, argemones, or brathys, as in equinoctial America. The more space the association occupies the more resistance do the social plants oppose to the labourer. With this general cause others are combined in the Llanos of Venezuela; namely the action of the small grasses which impoverish the soil; the total absence of trees and brushwood; the sandy winds, the heat of which is increased by contact with a surface absorbing the rays of the sun during twelve hours, and unshaded except by the stalks of the aristides, chanchuses, and paspalums. The progress observable on the vegetation of large trees and the cultivation of dicotyledonous plants in the vicinity of towns, (for instance around Calabozo and Pao) prove what may be gained upon the Llano by attacking it in small portions, enclosing it by degrees, and dividing it by coppices and canals of irrigation. Possibly the influence of the winds which render the soil sterile might be diminished by sowing on a large scale, for example, over fifteen or twenty acres, the seeds of the psidium, the croton, the cassia, or the tamarind, which prefer dry, open spots. I am far from believing that the savannahs will ever disappear entirely; or that the Llanos, so useful for pasturage and the trade in cattle, will ever be cultivated like the valleys of Aragua or other parts near the coast of Caracas and Cumana: but I am persuaded that in the lapse of ages a considerable portion of these plains, under a government favourable to industry, will lose the wild aspect which has characterized them since the first conquest by Europeans.

After three days' journey we began to perceive the chain of the mountains of Cumana, which separates the Llanos, or, as they are often called here, the great sea of verdure,* from the coast of the Caribbean Sea. (* Los Llanos son como un mar de yerbas— The Llanos are like a vast sea of grass—is an observation often repeated in these regions.) If the Bergantin be more than eight hundred toises high, it may be seen supposing only an ordinary refraction of one fourteenth of the arch, at the distance of twenty-seven nautical leagues; but the state of the atmosphere long concealed from us the majestic view of this curtain of mountains. It appeared at first like a fog-bank which hid the stars near the pole at their rising and setting; gradually this body of vapour seemed to augment and condense, to assume a bluish tint, and become bounded by sinuous and fixed outlines. The same effects which the mariner observes on approaching a new land present themselves to the traveller on the borders of the Llano. The horizon began to enlarge in some part and the vault of heaven seemed no longer to rest at an equal distance on the grass-covered soil. A llanero, or inhabitant of the Llanos, is happy when, as expressed in the simple phraseology of the country, he can see everywhere well around him. What appears to European eyes a covered country, slightly undulated by a few scattered hills, is to him a rugged region bristled with mountains. After having passed several months in the thick forests of the Orinoco, in places where one is accustomed, when at any distance from the river, to see the stars only in the zenith, as through the mouth of a well, a journey in the Llanos is peculiarly agreeable and attractive. The traveller experiences new sensations; and, like the Llanero, he enjoys the happiness of seeing well around him. But this enjoyment, as we ourselves experienced, is not of long duration. There is doubtless something solemn and imposing in the aspect of a boundless horizon, whether viewed from the summits of the Andes or the highest Alps, amid the expanse of the ocean or in the vast plains of Venezuela and Tucuman. Infinity of space, as poets in every language say, is reflected within ourselves; it is associated with ideas of a superior order; it elevates the mind which delights in the calm of solitary meditation. It is true, also, that every view of unbounded space bears a peculiar character. The prospect surveyed from a solitary peak varies according as the clouds reposing on the plain extend in layers, are conglomerated in groups, or present to the astonished eye, through broad openings, the habitations of man, the labour of agriculture, or the verdant tint of the aerial ocean. An immense sheet of water, animated by a thousand various beings even to its utmost depths, changing perpetually in colour and aspect, moveable at its surface like the element that agitates it, all charm the imagination during long voyages by sea; but the dusty and creviced Llano, throughout a great part of the year, has a depressing influence on the mind by its unchanging monotony. When, after eight or ten days' journey, the traveller becomes accustomed to the mirage and the brilliant verdure of a few tufts of mauritia* (* The fan-palm, or sago-tree of Guiana.) scattered from league to league, he feels the want of more varied impressions. He loves again to behold the great tropical trees, the wild rush of torrents or hills and valleys cultivated by the hand of the labourer. If the deserts of Africa and of the Llanos or savannahs of the New Continent filled a still greater space than they actually occupy, nature would be deprived of many of the beautiful products peculiar to the torrid zone.* (* In calculating from maps on a very large scale I found the Llanos of Cumana, Barcelona, and Caracas, from the delta of the Orinoco to the northern bank of the Apure, 7200 square leagues, the Llanos between the Apure and Putumayo, 21,000 leagues; the Pampas on the north-west of Buenos Ayres, 40,000 square leagues; the Pampas south of the parallel of Buenos Ayres, 37,000 square leagues. The total area of the Llanos of South America, covered with gramina, is consequently 105,200 square leagues, twenty leagues to an equatorial degree.) The heaths of the north, the steppes of the Volga and the Don, are scarcely poorer in species of plants and animals than are the twenty-eight thousand square leagues of savannahs extending in a semicircle from north-east to south-west, from the mouths of the Orinoco to the banks of the Caqueta and the Putumayo, beneath the finest sky in the world, and in the land of plantains and bread-fruit trees. The influence of the equinoctial climate, everywhere else so vivifying, is not felt in places where the great associations of gramina almost exclude every other plant. Judging from the aspect of the soil we might have believed ourselves to be in the temperate zone and even still farther northward but that a few scattered palms, and at nightfall the fine constellations of the southern sky (the Centaur, Canopus, and the innumerable nebulae with which the Ship is resplendent), reminded us that we were only eight degrees distant from the equator.

A phenomenon which fixed the attention of De Luc and which in these latter years has furnished a subject of speculation to geologists, occupied us much during our journey across the Llanos. I allude not to those blocks of primitive rock which occur, as in the Jura, on the slope of limestone mountains, but to those enormous blocks of granite and syenite which, in limits very distinctly marked by nature, are found scattered on the north of Holland, Germany and the countries of the Baltic. It seems to be now proved that, distributed as in radii, they came at the time of the ancient revolutions of our globe from the Scandinavian peninsula southward; and that they did not primitively belong to the granitic chains of the Harz and Erzgeberg, which they approach without, however, reaching their foot.* (* Leopold von Buch, Voyage en Norwege volume 1 page 30.) I was surprised at not seeing one of these blocks in the Llanos of Venezuela, though these immense plains are bounded on the south by the Sierra Parima, a group of mountains entirely granitic and exhibiting in its denticulated and often columnar peaks traces of the most violent destruction. Northward the granitic chain of the Silla de Caracas and Porto Cabello are separated from the Llanos by a screen of mountains that are schistose between Villa de Cura and Parapara, and calcareous between the Bergantin and Caripe. I was no less struck by this absence of blocks on the banks of the Amazon. La Condamine affirms that from the Pongo de Manseriche to the Strait of Pauxis not the smallest stone is to be found. Now the basin of the Rio Negro and of the Amazon is also a Llano, a plain like those of Venezuela and Buenos Ayres. The difference consists only in the state of vegetation. The two Llanos situated at the northern and southern extremities of South America are covered with gramina; they are treeless savannahs; but the intermediate Llano, that of the Amazon, exposed to almost continual equatorial rains, is a thick forest. I do not remember having heard that the Pampas of Buenos Ayres or the savannahs of the Missouri* and New Mexico contain granitic blocks. (* Are there any isolated blocks in North America northward of the great lakes?) The absence of this phenomenon appears general in the New World as it probably also is in Sahara, in Africa; for we must not confound the rocky masses that pierce the soil in the midst of the desert, and of which travellers often make mention, with mere scattered fragments. These facts seem to prove that the blocks of Scandinavian granite which cover the sandy countries on the south of the Baltic, and those of Westphalia and Holland, must be traced to some local revolution. The ancient conglomerate (red sandstone) which covers a great part of the Llanos of Venezuela and of the basin of the Amazon contains no doubt fragments of the same primitive rocks which constitute the neighbouring mountains; but the convulsions of which these mountains exhibit evident marks, do not appear to have been attended with circumstances favourable to the removal of great blocks. This geognostic phenomenon was to me the more unexpected since there exists nowhere in the world so smooth a plain entirely granitic. Before my departure from Europe I had observed with surprise that there were no primitive blocks in Lombardy and in the great plain of Bavaria which appears to be the bottom of an ancient lake, and which is situated two hundred and fifty toises above the level of the ocean. It is bounded on the north by the granites of the Upper Palatinate; and on the south by Alpine limestone, transition-thonschiefer, and the mica-slates of the Tyrol.

We arrived, on the 23rd of July, at the town of Nueva Barcelona, less fatigued by the heat of the Llanos, to which we had been long accustomed, than annoyed by the winds of sand which occasion painful chaps in the skin. Seven months previously, in going from Cumana to Caracas, we had rested a few hours at the Morro de Barcelona, a fortified rock, which, near the village of Pozuelos, is joined to the continent only by a neck of land. We were received with the kindest hospitality in the house of Don Pedro Lavie, a wealthy merchant of French extraction. This gentleman, who was accused of having given refuge to the unfortunate Espana when a fugitive on these coasts in 1796, was arrested by order of the Audiencia, and conveyed as a prisoner to Caracas. The friendship of the governor of Cumana and the remembrance of the services he had rendered to the rising commerce of those countries contributed to procure his liberty. We had endeavoured to alleviate his captivity by visiting him in prison; and we had now the satisfaction of finding him in the midst of his family. Illness under which he was suffering had been aggravated by confinement; and he sank into the grave without seeing the dawn of those days of independence, which his friend Don Joseph Espana had predicted on the scaffold prior to his execution. "I die," said that man, who was formed for the accomplishment of grand projects, "I die an ignominious death; but my fellow citizens will soon piously collect my ashes, and my name will reappear with glory." These remarkable words were uttered in the public square of Caracas, on the 8th of May, 1799.

In 1790 Nueva Barcelona contained scarcely ten thousand inhabitants, and in 1800, its population was more than sixteen thousand. The town was founded in 1637 by a Catalonian conquistador, named Juan Urpin. A fruitless attempt was then made, to give the whole province the name of New Catalonia. As our maps often mark two towns, Barcelona and Cumanagoto, instead of one, and as the two names are considered as synonymous, it may be well to explain the cause of this error. Anciently, at the mouth of the Rio Neveri, there was an Indian town, built in 1588 by Lucas Faxardo, and named San Cristoval de los Cumanagotos. This town was peopled solely by natives who came from the saltworks of Apaicuare. In 1637 Urpin founded, two leagues farther inland, the Spanish town of Nueva Barcelona, which he peopled with some of the inhabitants of Cumanagoto, together with some Catalonians. For thirty-four years, disputes were incessantly arising between the two neighbouring communities till in 1671, the governor Angulo succeeded in persuading them to establish themselves on a third spot, where the town of Barcelona now stands. According to my observations it is situated in latitude 10 degrees 6 minutes 52 seconds.* (* These observations were made on the Plaza Major. They are merely the result of six circum-meridian heights of Canopus, taken all in one night. In Las Memorias de Espinosa the latitude is stated to be 10 degrees 9 minutes 6 seconds. The result of M. Ferrer's observations made it 10 degrees 8 minutes 24 seconds.) The ancient town of Cumanagoto is celebrated in the country for a miraculous image of the Virgin,* which the Indians say was found in the hollow trunk of an old tutumo, or calabash-tree (Crescentia cujete). (* La milagrosa imagen de Maria Santissima del Socorro, also called La Virgen del Tutumo.) This image was carried in procession to Nueva Barcelona; but whenever the clergy were dissatisfied with the inhabitants of the new city, the Virgin fled at night, and returned to the trunk of the tree at the mouth of the river. This miracle did not cease till a fine convent (the college of the Propaganda) was built, to receive the Franciscans. In a similar case, the Bishop of Caracas caused the image of Our Lady de los Valencianos to be placed in the archives of the bishopric, where she remained thirty years under seal.

The climate of Barcelona is not so hot as that of Cumana but it is extremely damp and somewhat unhealthy in the rainy season. M. Bonpland had borne very well the irksome journey across the Llanos; and had recovered his strength and activity. With respect to myself, I suffered more at Barcelona than I did at Angostura, immediately after our passage on the rivers. One of those extraordinary tropical rains during which, at sunset, drops of enormous size fall at great distances from one another, caused me to experience sensations which seemed to threaten an attack of typhus, a disease then prevalent on that coast. We remained nearly a month at Barcelona where we found our friend Fray Juan Gonzales, of whom I have often spoken, and who had traversed the Upper Orinoco before us. He expressed regret that we had not been able to prolong our visit to that unknown country; and he examined our plants and animals with that interest which must be felt by even the most uninformed man for the productions of a region he has long since visited. Fray Juan had resolved to go to Europe and to accompany us as far as the island of Cuba. We were together for the space of seven months, and his society was most agreeable: he was cheerful, intelligent and obliging. How little did we anticipate the sad fate that awaited him. He took charge of a part of our collections; and a friend of his own confided to his care a child who was to be conveyed to Spain for its education. Alas! the collection, the child and the young ecclesiastic were all buried in the waves.

South-east of Nueva Barcelona, at the distance of two leagues, there rises a lofty chain of mountains, abutting on the Cerro del Bergantin, which is visible at Cumana. This spot is known by the name of the hot waters, (aguas calientes). When I felt my health sufficiently restored, we made an excursion thither on a cool and misty morning. The waters, which are loaded with sulphuretted hydrogen, issue from a quartzose sandstone, lying on compact limestone, the same as that we had examined at the Morro. We again found in this limestone intercalated beds of black hornstein, passing into kieselschiefer. It is not, however, a transition rock; by its position, its division into small strata, its whiteness and its dull and conchoidal fractures (with very flattened cavities), it rather approximates to the limestone of Jura. The real kieselschiefer and Lydian-stone have not been observed hitherto except in the transition-slates and limestones. Is the sandstone whence the springs of the Bergantin issue of the same formation as the sandstone of the Imposible and the Tumiriquiri? The temperature of the thermal waters is only 43.2 degrees centigrade (the atmosphere being 27). They flow first to the distance of forty toises over the rocky surface of the ground; then they rush down into a natural cavern; and finally they pierce through the limestone to issue out at the foot of the mountain on the left bank of the little river Narigual. The springs, while in contact with the oxygen of the atmosphere, deposit a good deal of sulphur. I did not collect, as I had done at Mariara, the bubbles of air that rise in jets from these thermal waters. They no doubt contain a large quantity of nitrogen because the sulphuretted hydrogen decomposes the mixture of oxygen and nitrogen dissolved in the spring. The sulphurous waters of San Juan which issue from calcareous rock, like those of the Bergantin, have also a low temperature (31.3 degrees); while in the same region the temperature of the sulphurous waters of Mariara and Las Trincheras (near Porto Cabello), which gush immediately from gneiss-granite, is 58.9 degrees the former, and 90.4 degrees the latter. It would seem as if the heat which these springs acquire in the interior of the globe diminishes in proportion as they pass from primitive to secondary superposed rocks.

Our excursion to the Aguas Calientes of Bergantin ended with a vexatious accident. Our host had lent us one of his finest saddle-horses. We were warned at the same time not to ford the little river of Narigual. We passed over a sort of bridge, or rather some trunks of trees laid closely together, and we made our horses swim, holding their bridles. The horse I had ridden suddenly disappeared after struggling for some time under water: all our endeavours to discover the cause of this accident were fruitless. Our guides conjectured that the animal's legs had been seized by the caymans which are very numerous in those parts. My perplexity was extreme: delicacy and the affluent circumstances of my host forbade me to think of repairing his loss; and M. Lavie, more considerate of our situation than sensible of his own misfortune, endeavoured to tranquillize us by exaggerating the facility with which fine horses were procurable from the neighbouring savannahs.

The crocodiles of the Rio Neveri are large and numerous, especially near the mouth of the river; but in general they are less fierce than the crocodiles of the Orinoco. These animals manifest in America the same contrasts of ferocity as in Egypt and Nubia: this fact is obvious when we compare with attention the narratives of Burckhardt and Belzoni. The state of cultivation in different countries and the amount of population in the proximity of rivers modify the habits of these large saurians: they are timid when on dry ground and they flee from man, even in the water, when they are not in want of food and when they perceive any danger in attacking. The Indians of Nueva Barcelona convey wood to market in a singular manner. Large logs of zygophyllum and caesalpinia* are thrown into the river and carried down by the stream, while the owners of the wood swim here and there to float the pieces that are stopped by the windings of the banks. (* The Lecythis ollaria, in the vicinity of Nueva Barcelona, furnishes excellent timber. We saw trunks of this tree seventy feet high. Around the town, beyond that arid zone of cactus which separates Nueva Barcelona from the steppe, grow the Clerodendrum tenuifolium, the Ionidium itubu, which resembles the Viola, and the Allionia violacea.) This could not be done in the greater part of those American rivers in which crocodiles are found. The town of Barcelona has not, like Cumana, an Indian suburb; and the only natives who are seen there are inhabitants of the neighbouring missions or of huts scattered in the plain. Neither the one nor the other are of Carib race, but a mixture of the Cumanagotos, Palenkas and Piritus; short, stunted, indolent and addicted to drinking. Fermented cassava is here the favourite beverage; the wine of the palm-tree, which is used on the Orinoco, being almost unknown on the coast. It is curious to observe that men in different zones, to satisfy the passion of inebriety, employ not only all the families of monocotyledonous and dicotyledonous plants, but even the poisonous Agaric (Amanita muscaria) of which, with disgusting economy, the Coriacs have learnt to drink the same juice several times during five successive days.* (* Mr. Langsdor (Wetterauisches Journal part 1 page 254) first made known this very extraordinary physiological phenomenon, which I prefer describing in Latin: Coriaecorum gens, in ora Asiae septentrioni opposita, potum sibi excogitavit ex succo inebriante agarici muscarii. Qui succus (aeque ut asparagorum), vel per humanum corpus transfusus, temulentiam nihilominus facit. Quare gens misera et inops, quo rarius mentis sit suae, propriam urinam bibit identidem: continuoque mingens rursusque hauriens eundem succum (dicas, ne ulla in parte mundi desit ebrietas), pauculis agaricis producere in diem quintum temulentiam potest.)

The packet boats (correos) from Corunna bound for the Havannah and Mexico had been due three months; and it was believed they had been taken by the English cruisers stationed on this coast. Anxious to reach Cumana, in order to avail ourselves of the first opportunity that might offer for our passage to Vera Cruz, we hired an open boat called a lancha, a sort of craft employed habitually in the latitudes east of Cape Codera where the sea is scarcely ever rough. Our lancha, which was laden with cacao, carried on a contraband trade with the island of Trinidad. For this reason the owner imagined we had nothing to fear from the enemy's vessels, which then blockaded all the Spanish ports. We embarked our collection of plants, our instruments and our monkeys; and, the weather being delightful, we hoped to make a very short passage from the mouth of the Rio Neveri to Cumana: but we had scarcely reached the narrow channel between the continent and the rocky isles of Borracha and the Chimanas, when to our great surprise we came in sight of an armed boat, which, whilst hailing us from a great distance, fired some musket-shot at us. The boat belonged to a privateer of Halifax; and I recognized among the sailors a Prussian, a native of Memel. I had found no opportunity, since my arrival in America, of expressing myself in my native language, and I could have wished to have spoken it on a less unpleasant occasion. Our protestations were without effect: we were carried on board the privateer, and the captain, affecting not to recognize the passports delivered by the governor of Trinidad for the illicit trade, declared us to be a lawful prize. Being a little in the habit of speaking English, I entered into conversation with the captain, begging not to be taken to Nova Scotia, but to be put on shore on the neighbouring coast. While I endeavoured, in the cabin, to defend my own rights and those of the owner of the lancha, I heard a noise on deck. Something was whispered to the captain, who left us in consternation. Happily for us, an English sloop of war, the Hawk, was cruising in those parts, and had signalled the captain to bring to; but the signal not being promptly answered, a gun was fired from the sloop and a midshipman sent on board our vessel. He was a polite young man, and gave me hopes that the lancha, which was laden with cacao, would be given up, and that on the following day we might pursue our voyage. In the meantime he invited me to accompany him on board the sloop, assuring me that his commander, Captain Garnier, would furnish me with better accommodation for the night than I should find in the vessel from Halifax.

I accepted these obliging offers and was received with the utmost kindness by Captain Garnier, who had made the voyage to the north-west coast of America with Vancouver, and who appeared to be highly interested in all I related to him respecting the great cataracts of Atures and Maypures, the bifurcation of the Orinoco and its communication with the Amazon. He introduced to me several of his officers who had been with Lord Macartney in China. I had not, during the space of a year, enjoyed the society of so many well-informed persons. They had learned from the English newspapers the object of my enterprise. I was treated with great confidence and the commander gave me up his own state-room. They gave me at parting the astronomical Ephemerides for those years which I had not been able to procure in France or Spain. I am indebted to Captain Garnier for the observations I was enabled to make on the satellites beyond the equator and I feel it a duty to record here the gratitude I feel for his kindness. Coming from the forests of Cassiquiare, and having been confined during whole months to the narrow circle of missionary life, we felt a high gratification at meeting for the first time with men who had sailed round the world, and whose ideas were enlarged by so extensive and varied a course. I quitted the English vessel with impressions which are not yet effaced from my remembrance, and which rendered me more than ever satisfied with the career on which I had entered.

We continued our passage on the following day; and were surprised at the depth of the channels between the Caracas Islands, where the sloop worked her way through them almost touching the rocks. How much do these calcareous islets, of which the form and direction call to mind the great catastrophe that separated from them the mainland, differ in aspect from the volcanic archipelago on the north of Lanzerote where the hills of basalt seem to have been heaved up from the bottom of the sea! Numbers of pelicans and of flamingos, which fished in the nooks or harassed the pelicans in order to seize their prey, indicated our approach to the coast of Cumana. It is curious to observe at sunrise how the sea-birds suddenly appear and animate the scene, reminding us, in the most solitary regions, of the activity of our cities at the dawn of day. At nine in the morning we reached the gulf of Cariaco which serves as a roadstead to the town of Cumana. The hill, crowned by the castle of San Antonio, stood out, prominent from its whiteness, on the dark curtain of the inland mountains. We gazed with interest on the shore, where we first gathered plants in America, and where, some months later, M. Bonpland had been in such danger. Among the cactuses, that rise in columns twenty feet high, appear the Indian huts of the Guaykeries. Every part of the landscape was familiar to us; the forest of cactus, the scattered huts and that enormous ceiba, beneath which we loved to bathe at the approach of night. Our friends at Cumana came out to meet us: men of all castes, whom our frequent herborizations had brought into contact with us, expressed the greater joy at sight of us, as a report that we had perished on the banks of the Orinoco had been current for several months. These reports had their origin either in the severe illness of M. Bonpland, or in the fact of our boat having been nearly lost in a gale above the mission of Uruana.

We hastened to visit the governor, Don Vicente Emparan, whose recommendations and constant solicitude had been so useful to us during the long journey we had just terminated. He procured for us, in the centre of the town, a house which, though perhaps too lofty in a country exposed to violent earthquakes, was extremely useful for our instruments. We enjoyed from its terraces a majestic view of the sea, of the isthmus of Araya, and the archipelago of the islands of Caracas, Picuita and Borracha. The port of Cumana was every day more and more closely blockaded, and the vain expectation of the arrival of Spanish packets detained us two months and a half longer. We were often nearly tempted to go to the Danish islands which enjoyed a happy neutrality; but we feared that, if we left the Spanish colonies, we might find some obstacles to our return. With the ample freedom which in a moment of favour had been granted to us, we did not consider it prudent to hazard anything that might give umbrage to the local authorities. We employed our time in completing the Flora of Cumana, geologically examining the eastern part of the peninsula of Araya, and observing many eclipses of satellites, which confirmed the longitude of the place already obtained by other means. We also made experiments on the extraordinary refractions, on evaporation and on atmospheric electricity.

The living animals which we had brought from the Orinoco were objects of great curiosity to the inhabitants of Cumana. The capuchin of the Esmeralda (Simia chiropotes), which so much resembles man in the expression of its physiognomy; and the sleeping monkey (Simia trivirgata), which is the type of a new group; had never yet been seen on that coast. We destined them for the menagerie of the Jardin des Plantes at Paris. The arrival of a French squadron which had failed in an attack upon Curacao furnished us, unexpectedly, with an excellent opportunity for sending them to Guadaloupe; and General Jeannet, together with the commissary Bresseau, agent of the executive power at the Antilles, promised to convey them. The monkeys and birds died at Guadaloupe but fortunately the skin of the Simia chiropotes, the only one in Europe, was sent a few years ago to the Jardin des Plantes, where the couxio (Simia satanas) and the stentor or alouate of the steppes of Caracas (Simia ursina) had been already received. The arrival of so great a number of French military officers and the manifestation of political and religious opinions not altogether conformable with the interests of the governments of Europe excited singular agitation in the population of Cumana. The governor treated the French authorities with the forms of civility consistent with the friendly relations subsisting at that period between France and Spain. In the streets the coloured people crowded round the agent of the French Directory, whose dress was rich and theatrical. White men, too, with indiscreet curiosity, whenever they could make themselves understood, made enquiries concerning the degree of influence granted by the republic to the colonists in the government of Guadaloupe. The king's officers doubled their zeal in furnishing provision for the little squadron. Strangers, who boasted that they were free, appeared to these people troublesome guests; and in a country of which the growing prosperity depended on clandestine communication with the islands, and on a freedom of trade forced from the ministry, the European Spaniards extolled the wisdom of the old code of laws (leyes de Indias) which permitted the entrance of foreign vessels into their ports only in extreme cases of want or distress. These contrasts between the restless desires of the colonists and the distrustful apathy of the government, throw some light on the great political events which, after long preparation, have separated Spain from her colonies.

We again passed a few agreeable days, from the third to the fifth of November, at the peninsula of Araya, situated beyond the gulf of Cariaco, opposite to Cumana.* (* I have already described the pearls of Araya; its sulphurous deposits and submarine springs of liquid and colourless petroleum. See volume 1.5.) We were informed that the Indians carried to the town from time to time considerable quantities of native alum, found in the neighbouring mountains. The specimens shown to us sufficiently indicated that it was neither alunite, similar to the rock of Tolfa and Piombino, nor those capillary and silky salts of alkaline sulphate of alumina and magnesia that line the clefts and cavities of rocks, but real masses of native alum, with a conchoidal or imperfectly lamellar fracture. We were led to hope that we should find the mine of alum (mina de alun) in the slaty cordillera of Maniquarez, and so new a geological phenomenon was calculated to rivet our attention. The priest Juan Gonzales, and the treasurer, Don Manuel Navarete, who had been useful to us from our first arrival on this coast, accompanied us in our little excursion. We disembarked near Cape Caney and again visited the ancient salt-pit (which is converted into a lake by the irruption of the sea), the fine ruins of the castle of Araya and the calcareous mountain of the Barigon, which, from its steepness on the western side is somewhat difficult of access. Muriatiferous clay mixed with bitumen and lenticular gypsum and sometimes passing to a darkish brown clay, devoid of salt, is a formation widely spread through this peninsula, in the island of Margareta and on the opposite continent, near the castle of San Antonio de Cumana. Probably the existence of this formation has contributed to produce those ruptures and rents in the ground which strike the eye of the geologist when he stands on one of the eminences of the peninsula of Araya. The cordillera of this peninsula, composed of mica-slate and clay-slate, is separated on the north from the chain of mountains of the island of Margareta (which are of a similar composition) by the channel of Cubagua; and on the south it is separated from the lofty calcareous chain of the continent, by the gulf of Cariaco. The whole intermediate space appears to have been heretofore filled with muriatiferous clay; and no doubt the continual erosions of the ocean have removed this formation and converted the plain, first into lakes, then into gulfs, and finally into navigable channels. The account of what has passed in the most modern times at the foot of the castle of Araya, the irruption of the sea into the ancient salt-pit, the formation of the laguna de Chacopata and a lake, four leagues in length, which cuts the island of Margareta nearly into two parts, afford evident proofs of these successive erosions. In the singular configuration of the coasts in the Morro of Chacopata; in the little islands of the Caribbees, the Lobos and Tunal; in the great island of Coche, and the capes of Carnero and Mangliers there still seem to be apparent the remains of an isthmus which, stretching from north to south, formerly joined the peninsula of Araya to the island of Margareta. In that island a neck of very low land, three thousand toises long, and less than two hundred toises broad, conceals on the northern sides the two hilly groups, known by the names of La Vega de San Juan and the Macanao. The Laguna Grande of Margareta has a very narrow opening to the south and small boats pass by portage over the neck of land or northern dyke. Though the waters on these shores seem at present to recede from the continent it is nevertheless very probable that in the lapse of ages, either by an earthquake or by a sudden rising of the ocean, the long island of Margareta will be divided into two rocky islands of a trapezoidal form.

The limestone of the Barigon, which is a part of the great formation of sandstone or calcareous breccia of Cumana, is filled with fossil shells in as perfect preservation as those of other tertiary limestones in France and Italy. We detached some blocks containing oysters eight inches in diameter, pectens, venuses, and lithophyte polypi. I recommend to naturalists better versed in the knowledge of fossils than I then was, to examine with care this mountainous coast (which is easy of access to European vessels) in their way to Cumana, Guayra or Curacao. It would be curious to discover whether any of these shells and these species of petrified zoophytes still inhabit the seas of the West Indies, as M. Bonpland conjectured, and as is the case in the island of Timor and perhaps in Guadaloupe.

We sailed on the 4th of November, at one o'clock in the morning, in search of the mine of native alum. I took with me the chronometer and my large Dollond telescope, intending to observe at the Laguna Chica (Small Lake), east of the village of Maniquarez, the immersion of the first satellite of Jupiter; this design, however, was not accomplished, contrary winds having prevented our arrival before daylight. The spectacle of the phosphorescence of the ocean and the sports of the porpoises which surrounded our canoe somewhat atoned for this disappointment. We again passed those spots where springs of petroleum gush from mica-slate at the bottom of the sea and the smell of which is perceptible from a considerable distance. When it is recollected that farther eastward, near Cariaco, the hot and submarine waters are sufficiently abundant to change the temperature of the gulf at its surface, we cannot doubt that the petroleum is the effect of distillation at an immense depth, issuing from those primitive rocks beneath which lies the focus of all volcanic commotion.

The Laguna Chica is a cove surrounded by perpendicular mountains, and connected with the gulf of Cariaco only by a narrow channel twenty-five fathoms deep. It seems, like the fine port of Acapulco, to owe its existence to the effect of an earthquake. A beach shows that the sea is here receding from the land, as on the opposite coast of Cumana. The peninsula of Araya, which narrows between Cape Mero and Cape las Minas to one thousand four hundred toises, is little more than four thousand toises in breadth near the Laguna Chica, reckoning from one sea to the other. We had to cross this distance in order to find the native alum and to reach the cape called the Punta de Chuparuparu. The road is difficult only because no path is traced; and between precipices of some depth we were obliged to step over ridges of bare rock, the strata of which are much inclined. The principal point is nearly two hundred and twenty toises high; but the mountains, as it often happens in a rocky isthmus, display very singular forms. The Paps (tetas) of Chacopata and Cariaco, midway between the Laguna Chica and the town of Cariaco, are peaks which appear isolated when viewed from the platform of the castle of Cumana. The vegetable earth in this country is only thirty toises above sea level. Sometimes there is no rain for the space of fifteen months; if, however, a few drops fall immediately after the flowering of the melons and gourds, they yield fruit weighing from sixty to seventy pounds, notwithstanding the apparent dryness of the air. I say apparent dryness, for my hygrometric observations prove that the atmosphere of Cumana and Araya contains nearly nine-tenths of the quantity of watery vapour necessary to its perfect saturation. It is this air, at once hot and humid, that nourishes those vegetable reservoirs, the cucurbitaceous plants, the agaves and melocactuses half-buried in the sand. When we visited the peninsula the preceding year there was a great scarcity of water; the goats for want of grass died by hundreds. During our stay at the Orinoco the order of the seasons seemed to be entirely changed. At Araya, Cochen, and even in the island of Margareta it had rained abundantly; and those showers were remembered by the inhabitants in the same way as a fall of aerolites would be noted in the recollection of the naturalists of Europe.

The Indian who was our guide scarcely knew in what direction we should find the alum; he was ignorant of its real position. This ignorance of localities characterises almost all the guides here, who are chosen from among the most indolent class of the people. We wandered for eight or nine hours among rocks totally bare of vegetation. The mica-slate passes sometimes to clay-slate of a darkish grey. I was again struck by the extreme regularity in the direction and inclination of the strata. They run north 50 degrees east, inclining from 60 to 70 degrees north-west. This is the general direction which I had observed in the gneiss-granite of Caracas and the Orinoco, in the hornblende-slates of Angostura, and even in the greater part of the secondary rocks we had just examined. The beds, over a vast extent of land, make the same angle with the meridian of the place; they present a parallelism, which may be considered as one of the great geologic laws capable of being verified by precise measures. Advancing toward Cape Chuparuparu, the veins of quartz that cross the mica-slate increase in size. We found some from one to two toises broad, full of small fasciculated crystals of rutile titanite. We sought in vain for cyanite, which we had discovered in some blocks near Maniquarez. Farther on the mica-state presents not veins, but little beds of graphite or carburetted iron. They are from two to three inches thick and have precisely the same direction and inclination as the rock. Graphite, in primitive soils, marks the first appearance of carbon on the globe—that of carbon uncombined with hydrogen. It is anterior to the period when the surface of the earth became covered with monocotyledonous plants. From the summit of those wild mountains there is a majestic view of the island of Margareta. Two groups of mountains already mentioned, those of Macanao and La Vega de San Juan, rise from the bosom of the waters. The capital of the island, La Asuncion, the port of Pampatar, and the villages of Pueblo de la Mar, Pueblo del Norte and San Juan belong to the second and most easterly of these groups. The western group, the Macanao, is almost entirely uninhabited. The isthmus that divides these large masses of mica-slate was scarcely visible; its form appeared changed by the effect of the mirage and we recognized the intermediate part, through which runs the Laguna Grande, only by two small hills of a sugarloaf form, in the meridian of the Punta de Piedras. Nearer we look down on the small desert archipelago of the four Morros del Tunal, the Caribbee and the Lobos Islands.

After much vain search we at length found, before we descended to the northern coast of the peninsula of Araya, in a ravine of very difficult access (Aroyo del Robalo), the mineral which had been shown to us at Cumana. The mica-slate changed suddenly into carburetted and shining clay-slate. It was an ampelite; and the waters (for there are small springs in those parts, and some have recently been discovered near the village of Maniquarez) were impregnated with yellow oxide of iron and had a styptic taste. We found the sides of the neighbouring rocks lined with capillary sulphate of alumina in effervescence; and real beds, two inches thick, full of native alum, extending as far as the eye could reach in the clay slate. The alum is greyish white, somewhat dull on the surface and of an almost glassy lustre internally. Its fracture is not fibrous but imperfectly conchoidal. It is slightly translucent when its fragments are thin; and has a sweetish and astringent taste without any bitter mixture. When on the spot, I proposed to myself the question whether this alum, so pure, and filling beds in the clay-slate without leaving the smallest void, be of a formation contemporary with the rock, or whether it be of a recent, and in some sort secondary, origin, like the muriate of soda, found sometimes in small veins, where strongly concentrated springs traverse beds of gypsum or clay. In these parts nothing seems to indicate a process of formation likely to be renewed in our days. The slaty rock exhibits no open cleft; and none is found parallel with the direction of the slates. It may also be inquired whether this aluminous slate be a transition-formation lying on the primitive mica-slate of Araya, or whether it owe its origin merely to a change of composition and texture in the beds of mica-slate. I lean to the latter proposition; for the transition is progressive, and the clay-slate (thonschiefer) and mica-slate appear to me to constitute here but one formation. The presence of cyanite, rutile-titanite, and garnets, and the absence of Lydian stone, and all fragmentary or arenaceous rocks, seem to characterise the formation we describe as primitive. It is asserted that even in Europe ampelite and green stone are found, though rarely, in slates anterior to transition-slate.

When, in 1785, after an earthquake, a great rocky mass was broken off in the Aroyo del Robalo, the Guaykeries of Los Serritos collected fragments of alum five or six inches in diameter, extremely pure and transparent. It was sold in my time at Cumana to the dyers and tanners, at the price of two reals* per pound, while alum from Spain cost twelve reals. (* The real is about 6 1/2 English pence.) This difference of price was more the result of prejudice and of the impediments to trade, than of the inferior quality of the alum of the country, which is fit for use without undergoing any purification. It is also found in the chain of mica-slate and clay-slate, on the north-west coast of the island of Trinidad, at Margareta and near Cape Chuparuparu, north of the Cerro del Distiladero.* (* Another place was mentioned to us, west of Bordones, the Puerto Escondido. But that coast appeared to me to be wholly calcareous; and I cannot conceive where could be the situation of ampelite and native alum on this point. Was it in the beds of slaty clay that alternate with the alpine limestone of

57

Cumanacoa? Fibrous alum is found in Europe only in formations posterior to those of transition, in lignites and other tertiary formations belonging to the lignites.) The Indians, who are naturally addicted to concealment, are not inclined to make known the spots whence they obtain native alum; but it must be abundant, for I have seen very considerable quantities of it in their possession at a time.

South America at present receives its alum from Europe, as Europe in its turn received it from the natives of Asia previous to the fifteenth century. Mineralogists, before my travels, knew no substances which, without addition, calcined or not calcined, could directly yield alum (sulphate of alumina and potash), except rocks of trachytic formation, and small veins traversing beds of lignite and bituminous wood. Both these substances, so different in their origin, contain all that constitutes alum, that is to say, alumina, sulphuric acid and potash. The ores of Tolfa, Milo and Nipoligo; those of Montione, in which silica does not accompany the alumina; the siliceous breccia of Mont Dore, which contains sulphur in its cavities; the alumiferous rocks of Parad and Beregh in Hungary, which belong also to trachytic and pumice conglomerates, may no doubt be traced to the penetration of sulphurous acid vapours. They are the products of a feeble and prolonged volcanic action, as may be easily ascertained in the solfataras of Puzzuoli and the Peak of Teneriffe. The alumite of Tolfa, which, since my return to Europe, I have examined on the spot, conjointly with Gay-Lussac, has, by its oryctognostic characters and its chemical composition, a considerable affinity to compact feldspar, which constitutes the basis of so many trachytes and transition-porphyries. It is a siliciferous subsulphate of alumina and potash, a compact feldspar, with the addition of sulphuric acid completely formed in it. The waters circulating in these alumiferous rocks of volcanic origin do not, however, deposit masses of native alum, to yield which the rocks must be roasted. I know not of any deposits analogous to those I brought from Cumana; for the capillary and fibrous masses found in veins traversing beds of lignites (as on the banks of the Egra, between Saatz and Commothau in Bohemia), or efflorescing in cavities (as at Freienwalde in Brandenburg, and at Segario in Sardinia), are impure salts, often destitute of potash, and mixed with the sulphates of ammonia and magnesia. A slow decomposition of the pyrites, which probably act as so many little galvanic piles, renders the waters alumiferous, that circulate across the bituminous lignites and carburetted clays. These waters, in contact with carbonate of lime, even give rise to the deposits of subsulphate of alumina (destitute of potash), found near Halle, and formerly believed erroneously to be pure alumina belonging, like the porcelain earth (kaolin) of Morl, to porphyry of red sandstone. Analogous chemical actions may take place in primitive and transition slates as well as in tertiary formations. All slates, and this fact is very important, contain nearly five per cent of potash, sulphuret of iron, peroxide of iron, carbon, etc. The contact of so many moistened heterogeneous substances must necessarily lead them to a change of state and composition. The efflorescent salts that abundantly cover the aluminous slates of Robalo, show how much these chemical effects are favoured by the high temperature of the climate; but, I repeat, in a rock where there are no crevices, no vacuities parallel to the direction and inclination of the strata, native alum, semitransparent and of conchoidal fracture, completely filling its place (its beds), must be regarded as of the same age with the rock in which it is contained. The term contemporary formation is here taken in the sense attached to it by geologists, in speaking of beds of quartz in clay-slate, granular limestone in mica-slate or feldspar in gneiss.

After having for a long time wandered over barren scenes amidst rocks entirely devoid of vegetation, our eyes dwelt with pleasure on tufts of malpighia and croton, which we found in descending toward the coast. These arborescent crotons were of two new species,* very remarkable for their form, and peculiar to the peninsula of Araya. (* Croton argyrophyllus and C. marginatus.) We arrived too late at the Laguna Chica to visit another rock situated farther east and celebrated by the name of the Laguna Grande, or the Laguna del Obispo.* (* Great Lake, or the Bishop's Lake.) We contented ourselves with admiring it from the height of the mountains that command the view; and, excepting the ports of Ferrol and Acapulco, there is perhaps none presenting a more extraordinary configuration. It is an inland gulf two miles and a half long from east to west, and one mile broad. The rocks of mica-slate that form the entrance of the port leave a free passage only two hundred and fifty toises broad. The water is everywhere from fifteen to twenty-five fathoms deep. Probably the government of Cumana will one day take advantage of the possession of this inland gulf and of that of Mochima,* eight leagues east of the bad road of Nueva Barcelona. (* This is a long narrow gulf, three miles from north to south, similar to the fiords of Norway.) The family of M. Navarete were waiting for us with impatience on the beach; and, though our boat carried a large sail, we did not arrive at Maniquarez before night.

We prolonged our stay at Cumana only a fortnight. Having lost all hope of the arrival of a packet from Corunna, we availed ourselves of an American vessel, laden at Nueva Barcelona with salt provision for the island of Cuba. We had now passed sixteen months on this coast and in the interior of Venezuela, and on the 16th of November we parted from our friends at Cumana to make the passage for the third time across the gulf of Cariaco to Nueva Barcelona. The night was cool and delicious. It was not without emotion that we beheld for the last time the disc of the moon illuminating the summit of the cocoa-trees that surround the banks of the Manzanares. The breeze was strong and in less than six hours we anchored near the Morro of Nueva Barcelona, where the vessel which was to take us to the Havannah was ready to sail.

CHAPTER 3.27.
POLITICAL STATE OF THE PROVINCES OF VENEZUELA. EXTENT OF TERRITORY. POPULATION. NATURAL PRODUCTIONS. EXTERNAL TRADE. COMMUNICATIONS BETWEEN THE DIFFERENT PROVINCES COMPRISING THE REPUBLIC OF COLUMBIA.

Before I quit the coasts of Terra Firma and draw the attention of the reader to the political importance of Cuba, the largest of the West India Islands, I will collect into one point of view all those facts which may lead to a just appreciation of the future relations of commercial Europe with the united Provinces of Venezuela. When, soon after my return to Germany, I published the Essai Politique sur la Nouvelle-Espagne, I at the same time made known some of the facts I had collected in relation to the territorial riches of South America. This comparative view of the population, agriculture and commerce of all the Spanish colonies was formed at a period when the progress of civilization was restrained by the imperfection of social institutions, the prohibitory system and other fatal errors in the science of government. Since the time when I developed the immense resources which the people of both North and South America might derive from their own position and their relations with commercial Europe and Asia, one of those great revolutions which from time to time agitate the human race has changed the state of society in the vast regions through which I travelled. The continental part of the New World is at present in some sort divided between three nations of European origin; one (and that the most powerful) is of Germanic race: the two others belong by their language, their literature, and their manners to Latin Europe. Those parts of the old world which advance farthest westward, the Spanish Peninsula and the British Islands, are those of which the colonies are most extensive; but four thousand leagues of coast, inhabited solely by the descendants of Spaniards and Portuguese, attest the superiority which in the fifteenth and sixteenth centuries the peninsular nations had acquired, by their maritime expeditions, over the navigators of other countries. It may be fairly asserted that their languages, which prevail from California to the Rio de la Plata and along the back of the Cordilleras, as well as in the forests of the Amazon, are monuments of national glory that will survive every political revolution.

The inhabitants of Spanish and Portuguese America form together a population twice as numerous as the inhabitants of English race. The French, Dutch, and Danish possessions of the new continent are of small extent; but, to complete the general view of the nations which may influence the destiny of the other hemisphere, we ought not to forget the colonists of Scandinavian origin who are endeavouring to form settlements from the peninsula of Alashka as far as California; and the free Africans of Hayti who have verified the prediction made by the Milanese traveller Benzoni in 1545. The situation of these Africans in an island more than three times the size of Sicily, in the middle of the West Indian Mediterranean, augments their political importance. Every friend of humanity prays for the development of the civilization which is advancing in so calm and unexpected a manner. As yet Russian America is less like an agricultural colony than the factories established by Europeans on the coast of Africa, to the great misfortune of the natives; they contain only military posts, stations of fishermen, and Siberian hunters. It is a curious phenomenon to find the rites of the Greek Church established in one part of America and to see two nations which inhabit the eastern and western extremities of Europe (the Russians and the Spaniards) thus bordering on each other on a continent on which they arrived by opposite routes; but the almost savage state of the unpeopled coasts of Ochotsk and Kamtschatka, the want of resources furnished by the ports of Asia, and the barbarous system hitherto adopted in the Scandinavian colonies of the New World, are circumstances which will hold them long in infancy. Hence it follows that if in the researches of political economy we are accustomed to survey masses only, we cannot but admit that the American continent is divided, properly speaking, between three great nations of English, Spanish, and Portuguese race. The first of these three nations, the Anglo-Americans, is, next to the English of Europe, that whose flag waves over the greatest extent of sea. Without any distant colonies, its commerce has acquired a growth attained in the old world by that nation alone which communicated to North America its language, its literature, its love of labour, its predilection for liberty, and a portion of its civil institutions.

The English and Portuguese colonists have peopled only the coasts which lie opposite to Europe; the Castilians, on the contrary, in the earliest period of the conquest, crossed the chain of the Andes and made settlements in the most western regions. There only, at Mexico, Cundinamarca, Quito and Peru, they found traces of ancient civilization, agricultural nations and flourishing empires. This circumstance, together with the increase of the native mountain population, the almost exclusive possession of great metallic wealth, and the commercial relations established from the beginning of the sixteenth century with the Indian archipelago, have given a peculiar character to the Spanish possessions in equinoctial America. In the East Indies, the people who fell into the hands of the English and Portuguese settlers were wandering tribes or hunters. Far from forming a portion of the agricultural and laborious population, as on the tableland of Anahuac, at Guatimala and in Upper Peru, they generally withdrew at the approach of the whites. The necessity of labour, the preference given to the cultivation of the sugar-cane, indigo, and cotton, the cupidity which often accompanies and degrades industry, gave birth to that infamous slave-trade, the consequences of which have been alike fatal to the old and the new world. Happily, in the continental part of Spanish America, the number of African slaves is so inconsiderable that, compared with the slave population of Brazil, or with that of the southern part of the United States, it is found to be in the proportion of one to fourteen. The whole of the Spanish colonies, without excluding the islands of Cuba and Porto Rico, have not, over a surface which exceeds at least by one-fifth that of Europe, as many negroes as the single state of Virginia. The Spanish Americans, in the union of New Spain and Guatimala, present an example, unique in the torrid zone, namely, a nation of eight millions of inhabitants governed conformably with European institutions and laws, cultivating sugar, cacao, wheat and grapes, and having scarcely a slave brought from Africa.

The population of the New Continent as yet surpasses but little that of France or Germany. It doubles in the United States in twenty-three or twenty-five years; and at Mexico, even under the government of the mother country, it doubles in forty or forty-five years. Without indulging too flattering hopes of the future, it may be admitted that in less than a century and a half the population of America will equal that of Europe. This noble rivalry in civilization and the arts of industry and commerce, far from impoverishing the old continent, as has often been supposed it might at the expense of the new one, will augment the wants of the consumer, the mass of productive labour, and the activity of exchange. Doubtless, in consequence of the great revolutions which human society undergoes, the public fortune, the common patrimony of civilization, is found differently divided among the nations of the old and the new world: but by degrees the equilibrium is restored; and it is a fatal, I had almost said an impious prejudice, to consider the growing prosperity of any other part of our planet as a calamity to Europe. The independence of the colonies will not contribute to isolate them from the old civilized nations, but will rather bring all more closely together. Commerce tends to unite countries which a jealous policy has long separated. It is the nature of civilization to go forward without any tendency to decline in the spot that gave it birth. Its progress from east to west, from Asia to Europe, proves nothing against this axiom. A clear light loses none of its brilliancy by being diffused over a wider space. Intellectual cultivation, that fertile source of national wealth, advances by degrees and extends without being displaced. Its movement is not a migration: and though it may seem to be such in the east, it is because barbarous hordes possessed themselves of Egypt, Asia Minor, and of once free Greece, the forsaken cradle of the civilization of our ancestors.

The barbarism of nations is the consequence of oppression exercised by internal despotism or foreign conquest; and it is always accompanied by progressive impoverishment, by a diminution of the public fortune. Free and powerful institutions, adapted to the interests of all, remove these dangers; and the growing civilization of the world, the competition of labour and of trade, are not the ruin of states whose welfare flows from a natural source. Productive and commercial Europe will profit by the new order of things in Spanish America, as it would profit from events that might put an end to barbarism in Greece, on the northern coast of Africa and in other countries subject to Ottoman tyranny. What most menaces the prosperity of the ancient continent is the prolongation of those intestine struggles which check production and diminish at the same time the number and wants of consumers. This struggle, begun in Spanish America six years after my departure, is drawing gradually to an end. We shall soon see both shores of the Atlantic peopled by independent nations, ruled by different forms of Government, but united by the remembrance of a common origin, uniformity of language, and the wants which civilization creates. It may be said that the immense progress of the art of navigation has contracted the boundaries of the seas. The Atlantic already assumes the form of a narrow channel which no more removes the New World from the commercial states of Europe, than the Mediterranean, in the infancy of navigation, removed the Greeks of Peloponnesus from those of Ionia, Sicily, and the Cyrenaic region.

I have thought it right to enter into these general considerations on the future connection of the two continents, before tracing the political sketch of the provinces of Venezuela. These provinces, governed till 1810 by a captain-general residing at Caracas, are now united to the old viceroyalty of New Grenada, or Santa Fe, under the name of the Republic of Columbia. I will not anticipate the description which I shall have hereafter to give of New Grenada; but, in order to render my observations on the statistics of Venezuela more useful to those who would judge of the political importance of the country and the advantages it may offer to the trade of Europe, even in its present unadvanced state of cultivation, I will describe the United Provinces of Venezuela in their relations with Cundinamarca, or New Grenada, and as forming part of the new state of Columbia. M. Bonpland and I passed nearly three years in the country which now forms the territory of the republic of Columbia; sixteen months in Venezuela and eighteen in New Grenada. We crossed the territory in its whole extent; on one hand from the mountains of Paria as far as Esmeralda on the Upper Orinoco, and San Carlo del Rio Negro, situated near the frontiers of Brazil; and on the other, from Rio Sinu and Carthagena as far as the snowy summits of Quito, the port of Guayaquil on the coast of the Pacific, and the banks of the Amazon in the province of Jaen de Bracamoros. So long a stay and an expedition of one thousand three hundred leagues in the interior of the country, of which more than six hundred and fifty were by water, have furnished me with a pretty accurate knowledge of local circumstances.

I am aware that travellers, who have recently visited America, regard its progress as far more rapid than my statistical researches seem to indicate. For the year 1913 they promise one hundred and twelve millions of inhabitants in Mexico, of which they believe that the population is doubled every twenty-two years; and during the same interval one hundred and forty millions in the United States. These numbers, I confess, do not appear to me to be alarming from the motives that may excite fear among the disciples of Malthus. It is possible that some time or other, two or three hundred millions of men may find subsistence in the vast extent of the new continent between the lake of Nicaragua and lake Ontario. I admit that the United States will contain above eighty millions of inhabitants a hundred years hence, allowing a progressive change in the period of doubling from twenty-five to thirty-five and forty years; but, notwithstanding the elements of prosperity to be found in equinoctial America, I doubt whether the increase of the population in Venezuela,

Spanish Guiana, New Grenada and Mexico can be in general so rapid as in the United States. The latter, which are situated entirely in the temperate zone, destitute of high chains of mountains, embrace an immense extent of country easy of cultivation. The hordes of Indian hunters flee both from the colonists, whom they abhor, and the methodist missionaries, who oppose their taste for indolence and a vagabond life. The more fertile land of Spanish America produces indeed on the same surface a greater amount of nutritive substances. On the table lands of the equinoctial regions wheat doubtless yields annually from twenty to twenty-four for one; but Cordilleras furrowed by almost inaccessible crevices, bare and arid steppes, forests that resist both the axe and fire, and an atmosphere filled with venomous insects, will long present powerful obstacles to agriculture and industry. The most active and enterprising colonists cannot, in the mountainous districts of Merida, Antioquia, and Los Pastos, in the llanos of Venezuela and Guaviare, in the forests of the Rio Magdalena, the Orinoco, and the province of Las Esmeraldas, west of Quito, extend their agricultural conquests as they have done in the woody plains westward of the Alleghenies, from the sources of the Ohio, the Tennessee and the Alabama, as far as the banks of the Missouri and the Arkansas. Calling to mind the account of my voyage on the Orinoco, it may be easy to appreciate the obstacles which nature opposes to the efforts of man in hot and humid climates. In Mexico, large extents of soil are destitute of springs; rain seldom falls, and the want of navigable rivers impedes communication. As the ancient native population is agricultural, and had been so long before the arrival of the Spaniards, the lands most easy of access and cultivation have already their proprietors. Fertile tracts of country, at the disposal of the first occupier, or ready to be sold in lots for the profit of the state, are much less common than Europeans imagine. Hence it follows that the progress of colonization cannot be everywhere as free and rapid in Spanish America as it has hitherto been in the western provinces of the United States. The population of that union is composed wholly of whites, and of negros, who, having been torn from their country, or born in the New World, have become the instruments of the industry of the whites. In Mexico, Guatimala, Quito, and Peru, on the contrary, there exist in our day more than five millions and a half of natives of copper-coloured race, whose isolated position, partly forced and partly voluntary, together with their attachment to ancient habits, and their mistrustful inflexibility of character, will long prevent their participation in the progress of the public prosperity, notwithstanding the efforts employed to disindianize them.

I dwell on the differences between the free states of temperate and equinoctial America, to show that the latter have to contend against obstacles connected with their physical and moral position; and to remind the reader that the countries embellished with the most varied and precious productions of nature, are not always susceptible of an easy, rapid, and uniformly extended cultivation. If we consider the limits which the population may attain as depending solely on the quantity of subsistence which the land is capable of producing, the most simple calculations would prove the preponderance of the communities established in the fine regions of the torrid zone; but political economy, or the positive science of government, is distrustful of ciphers and vain abstractions. We know that by the multiplication of one family only, a continent previously desert may reckon in the space of eight centuries more than eight millions of inhabitants; and yet these estimates, founded on the hypothesis of a continuous doubling in twenty-five or thirty years, are contradicted by the history of every country already advanced in civilization. The destinies which await the free states of Spanish America are too glorious to require to be embellished by illusions and chimerical calculations.

Among the thirty-four million inhabitants spread over the vast surface of continental America, in which estimate are comprised the savage natives, we distinguish, according to the three preponderant races, sixteen millions and a half in the possessions of the Spanish Americans, ten millions in those of the Anglo-Americans, and nearly four millions in those of the Portuguese Americans. The population of these three great divisions is, at the present time, in the proportion of 4, 2 1/2, 1; while the extent of surface over which the population is spread is, as the numbers 1.5, 0.7, 1. The area of the United States* is nearly one-fourth greater than that of Russia west of the Ural mountains; and Spanish America is in the same proportion more extensive than the whole of Europe. (* Notwithstanding the political changes which have taken place in the South American colonies, I shall throughout this work designate the country inhabited by the Spanish Americans by the denomination of Spanish America. I call the country of the Anglo-Americans the United States, without adding of North America, although other United States exist in South America. It is embarrassing to speak of nations who play a great part on the scene of the world without having collective names. The term American can no longer be applied solely to the citizens of the United States of North America; and it were to be wished that the nomenclature of the independent nations of the New Continent should be fixed in a manner at once convenient, harmonious, and precise.) The United States contain five-eighths of the proportion of the Spanish possessions, and yet their area is not one-half so large. Brazil comprehends tracts of country so desert toward the west that over an extent only a third less than that of Spanish America its population is in the proportion of one to four. The following table contains the results of an attempt which I made, conjointly with M. Mathieu, member of the Academy of Sciences, and of the Bureau des Longitudes, to estimate with precision the extent of the surface of the various states of America. We made use of maps on which the limits had been corrected according to the statements published in my Recueil d'Observations Astronomiques. Our scales were, generally speaking, so large that spaces from four to five leagues square were not omitted. We observed this degree of precision that we might not add the uncertainty of the measure of triangles, trapeziums, and the sinuosities of the coasts, to the uncertainty of geographical statements.

TABLE OF GREAT POLITICAL DIVISIONS.
COLUMN 1 : NAME.
COLUMN 2 : SURFACE IN SQUARE LEAGUES OF 20 TO AN EQUINOCTIAL DEGREE.
COLUMN 3 : POPULATION (1823).

Surface Pop.
1. Possessions of the Spanish Americans : 371,380 : 16,785,000.
   Mexico or New Spain : 75,830 : 6,800,000.
Guatemala : 16,740 : 1,600,000.
Cuba and Porto Rico : 4,430 : 800,000.
Columbia—Venezuela : 33,700 : 785,000.
Columbia—New Grenada and Quito : 58,250 : 2,000,000.
Peru : 41,420 : 1,400,000.
Chili : 14,240 : 1,100,000.
Buenos Ayres : 126,770 : 2,300,000.
   2. Possessions of the Portuguese
Americans (Brazil) : 256,990 : 4,000,000.
   3. Possessions of the
Anglo-Americans (United States) : 174,300 : 10,220,000.

From the statistical researches which have been made in several countries of Europe, important results have been obtained by a comparison of the relative population of maritime and inland provinces. In Spain these relations are to one another as nine to five; in the United Provinces of Venezuela, and, above all, in the ancient Capitania-General of Caracas, they are as thirty-five to one. How powerful soever may be the influence of commerce on the prosperity of states, and the intellectual development of nations, it would be wrong to attribute in America, as we do in Europe, to that cause alone the differences just mentioned. In Spain and Italy, if we except the fertile plains of Lombardy, the inland districts are arid and abounding in mountains or high table-lands: the meteorological circumstances on which the fertility of the soil depends are not the same in the lands bordering on the sea, as they are in the central provinces. Colonization in America has generally begun on the coast, and advanced slowly towards the interior; such is its progress in Brazil and in Venezuela. It is only where the coast is unhealthy, as in Mexico and New Grenada, or sandy and exempt from rain as in Peru, that the population is concentrated on the mountains, and the table-lands of the interior. These local circumstances are too often overlooked in considerations on the future fate of the Spanish colonies; they communicate a peculiar character to some of those countries, the physical and moral analogies of which are less striking than is commonly supposed. Considered with reference to the distribution of the population, the two provinces of New Grenada and Venezuela, which have been united in one political body, exhibit the most complete contrast. Their capitals (and the position of capitals always denotes where population is most concentrated) are at such unequal distances from the trading coasts of the Caribbean Sea, that the town of Caracas, to be placed on the same parallel with Santa-Fe de Bogota, must be transplanted southward to the junction of the Orinoco with the Guaviare, where the mission of San Fernando de Atabapo is situated.

The republic of Columbia is, with Mexico and Guatemala, the only state of Spanish America which occupies at once the coasts opposite to Europe and to Asia. From Cape Paria to the western extremity of Veragua is a distance of 400 sea leagues: and from Cape Burica to the mouth of Rio Tumbez the distance is 260. The shore possessed by the republic of Columbia consequently equals in length the line of coasts extending from Cadiz to Dantzic, or from Ceuta to Jaffa. This immense resource for national industry is combined with a degree of cultivation of which the importance has not hitherto been sufficiently acknowledged. The isthmus of Panama forms part of the territory of Columbia, and that neck of land, if traversed by good roads and stocked with camels, may one day serve as a portage for the commerce of the world, even though the plains of Cupica, the bay of Mandinga or the Rio Chagre should not afford the possibility of a canal for the passage of vessels proceeding from Europe to China,* or from the United States to the north-west coast of America. (* The old vice-royalty of Buenos Ayres extended also along a small portion of the South Sea coast.)

When considering the influence which the configuration of countries (that is, the elevation and the form of coasts) exercises in every district on the progress of civilization and the destiny of nations, I have pointed out the disadvantages of those vast masses of triangular continents, which, like Africa and the greater part of South America, are destitute of gulfs and inland seas. It cannot be doubted that the existence of the Mediterranean has been closely connected with the first dawn of human cultivation among the nations of the west, and that the articulated form of the land, the frequency of its contractions and the concatenation of peninsulas favoured the civilization of Greece, Italy, and perhaps of all Europe westward of the meridian of the Propontis. In the New World the uninterruptedness of the coasts and the monotony of their straight lines are most remarkable in Chili and Peru. The shore of Columbia is more varied, and its spacious gulfs, such as that of Paria, Cariaco, Maracaybo, and Darien, were, at the time of the first discovery better peopled than the rest and facilitated the interchange of productions. That shore possesses an incalculable advantage in being washed by the Caribbean Sea, a kind of inland sea with several outlets, and the only one pertaining to the New Continent. This basin, whose various shores form portions of the United States, of the republic of Columbia, of Mexico and several maritime powers of Europe, gives birth to a peculiar and exclusively American system of trade. The south-east of Asia with its neighbouring archipelago and, above all, the state of the Mediterranean in the time of the Phoenician and Greek colonies, prove that the nearness of opposite coasts, not having the same productions and not inhabited by nations of different races, exercises a happy influence on commercial industry and intellectual cultivation. The importance of the inland Caribbean Sea, bounded by Venezuela on the south, will be further augmented by the progressive increase of population on the banks of the Mississippi; for that river, the Rio del Norte and the Magdalena are the only great navigable streams which the Caribbean Sea receives. The depth of the American rivers, their immense branches, and the use of steam-boats, everywhere facilitated by the proximity of forests, will, to a certain extent, compensate for the obstacles which the uniform line of the coasts and the general configuration of the continent oppose to the progress of industry and civilization.

On comparing the extent of the territory with the absolute population, we obtain the result of the connection of those two elements of public prosperity, a connection that constitutes the relative population of every state in the New World. We shall find to every square sea league, in Mexico, 90; in the United States, 58; in the republic of Columbia, 30; and in Brazil, 15 inhabitants; while Asiatic Russia furnishes 11; the whole Russian Empire, 87; Sweden with Norway, 90; European Russia, 320; Spain, 763; and France, 1778. But these estimates of relative population, when applied to countries of immense extent, and of which a great part is entirely uninhabited, merely furnish mathematical abstractions of but little value. In countries uniformly cultivated—in France, for example—the number of inhabitants to the square league, calculated by separate departments, is in general only a third, more or less, than the relative population of the sum of all the departments. Even in Spain the deviations from the average number rise, with few exceptions, only from half to double. In America, on the contrary, it is only in the Atlantic states, from South Carolina to New Hampshire, that the population begins to spread with any uniformity. In that most civilized portion of the New World, from 130 to 900 inhabitants are reckoned to the square league, while the relative population on all the Atlantic states, considered together, is 240. The extremes (North Carolina and Massachusetts) are only in the relation of 1 to 7, nearly as in France, where the extremes, in the departments of the Hautes Alpes and the Cote-du-Nord are also in the relation of 1 to 6.7. The variations from the average number, which we generally find restricted to narrow limits in the civilized countries of Europe, exceed all measure in Brazil, in the Spanish colonies and even in the confederation of the United States, in its whole extent. We find in Mexico in some of the intendencias, for example, La Sonora and Durango, from 9 to 15 inhabitants to the square league, while in others, on the central table-land, there are more than 500. The relative population of the country situated between the eastern bank of the Mississippi and the Atlantic states is scarcely 47; while that of Connecticut, Rhode island, and Massachusetts is more than 800. Westward of the Mississippi as well as in the interior of Spanish Guiana there are not two inhabitants to the square league over much larger extents of territory than Switzerland or Belgium. The state of these countries is like that of the Russian Empire, where the relative population of some of the Asiatic governments (Irkutsk and Tobolsk) is to that of the best cultivated European districts as 1 to 300.

The enormous difference existing, in countries newly cultivated, between the extent of territory and the number of inhabitants, renders these partial estimates necessary. When we learn that New Spain and the United States, taking their entire extent at 75,000 and 174,000 square sea-leagues, give respectively 90 and 58 souls to each league, we no more obtain a correct idea of that distribution of the population on which the political power of nations depends, than we should of the climate of a country, that is to say, of the distribution of the heat in the different seasons, by the mere knowledge of the mean temperature of the whole year. If we take from the United States all their possessions west of the Mississippi, their relative population would be 121 instead of 58 to the square league; consequently much greater than that of New Spain. Taking from the latter country the Provincias internas (north and north-east of Nueva Galicia) we should find 190 instead of 90 souls to the square league.

The provinces of Caracas, Maracaybo, Cumana and Barcelona, that is, the maritime provinces of the north, are the most populous of the old Capitania-General of Caracas; but, in comparing this relative population with that of New Spain, where the two intendencias of Mexico and Puebla alone contain, on an extent scarcely equal to the superficies of the province of Caracas, a greater population than that of the whole republic of Columbia, we see that some Mexican intendencias which, with respect to the concentration of their culture, occupy but the seventh or eighth rank (Zacatecas and Guadalajara), contain more inhabitants to the square league than the province of Caracas. The average of the relative population of Cumana, Barcelona, Caracas and Maracaybo, is fifty-six; and, as 6200 square leagues, that is, one half of the extent of these four provinces are almost desert Llanos, we find, in reckoning the superficies and the scanty population of the plains, 102 inhabitants to the square league. An analogous modification gives the province of Caracas alone a relative population of 208, that is, only one-seventh less than that of the Atlantic States of North America.

As in political economy numerical statements become instructive only by a comparison with analogous facts I have carefully examined what, in the present state of the two continents, might be considered as a small relative population in Europe, and a very great relative population in America. I have, however, chosen examples only from among the provinces which have a continued surface of more than 600 square leagues in order to exclude the accidental accumulations of population which occur around great cities; for instance, on the coast of Brazil, in the valley of Mexico, on the table-lands of Santa Fe de Bogota and Cuzco; or finally, in the smaller West India Islands (Barbadoes, Martinique and

St. Thomas) of which the relative population is from 3000 to 4700 inhabitants to the square league, and consequently equal to the most fertile parts of Holland, France and Lombardy.

MINIMUM OF EUROPE:
INHABITANTS TO THE SQUARE LEAGUE.

The four least populous Governments of European Russia:

Archangel : 10.

Olonez : 42.

Wologda and Astracan : 52.

Finland : 106.

The least populous Province of Spain, that of Cuenca : 311.

The Duchy of Luneburg (on account of the heaths) : 550.

The least populous Department of Continental France : 758. (Hautes Alps)

Departments of France thinly peopled (the Creuse, : 1300. the Var and the Aude)

MAXIMUM OF AMERICA.

The central part of the Intendencias of : 1300.

Mexico and Puebla, above

In the United States, Massachusetts, but having only 522 square leagues of surface : 900.

Massachusetts, Rhode Island, and Connecticut, together : 840.

The whole Intendencia of Puebla : 540.

The whole Intendencia of Mexico : 460.

These two Mexican Intendencias together are nearly a third of the superficial extent of France, with a suitable population (in 1823 nearly 2,800,000 souls) to prevent the towns of Mexico and Puebla from having a sensible influence on the relative population.

Northern part of the Province of Caracas : 208. (without the Llanos)

This table shows that those parts of America which we now consider as the most populous attain the relative population of the kingdom of Navarre, of Galicia and the Asturias, which, next to the province of Guipuscoa, and the kingdom of Valencia, reckon the greatest number of inhabitants to the square league in all Spain; the maximum of America is, however, below the relative population of the whole of France (1778 to the square league), and would, in the latter country, be considered as a very thin population. If, taking a survey of the whole surface of America, we direct our attention to the Capitania-General of Venezuela, we find that the most populous of its subdivisions, the province of Caracas, considered as a whole, without excepting the Llanos, has, as yet, only the relative population of Tennessee; and that this province, without the Llanos, furnishes in its northern part, or more than 1800 square leagues, the relative population of South Carolina. Those 1800 square leagues, the centre of agriculture, are twice as numerously peopled as Finland, but still a third less than the province of Cuenca, which is the least populous of all Spain. We cannot dwell on this result without a painful feeling. Such is the state to which colonial politics and maladministration have, during three centuries, reduced a country which, for natural wealth, may vie with all that is most wonderful on earth. For a region equally desert, we must look either to the frozen regions of the north, or westward of the Allegheny mountains towards the forests of Tennessee, where the first clearings have only begun within the last eighty years!

The most cultivated part of the province of Caracas, the basin of the lake of Valencia, commonly called Los Valles de Aragua, contained in 1810 nearly 2000 inhabitants to the square league. Supposing a relative population three times less, and taking off from the whole surface of the Capitania-General nearly 24,000 square leagues as being occupied by the Llanos and the forests of Guiana, and, therefore, presenting great obstacles to agricultural labourers, we should still obtain a population of six millions for the remaining 9700 square leagues. Those who, like myself, have lived long within the tropics, will find no exaggeration in these calculations; for I suppose for the portion the most easily cultivated a relative population equal to that in the intendencias of Puebla and Mexico,* full of barren mountains, and extending towards the coast of the Pacific over regions almost desert. (* These two Intendencias contain together 5520 square leagues and a relative population of

508 inhabitants to the square sea-league.) If the territories of Cumana, Barcelona, Caracas, Maracaybo, Varinas and Guiana should be destined hereafter to enjoy good provincial and municipal institutions as confederate states, they will not require a century and a half to attain a population of six millions of inhabitants. Venezuela, the eastern part of the republic of Columbia, would not, even with nine millions, have a more considerable population than Old Spain; and can it be doubted that that part of Venezuela which is most fertile and easy of cultivation, that is, the 10,000 square leagues remaining after deducting the Llanos and the almost impenetrable forests between the Orinoco and the Cassiquiare, could support in the fine climate of the tropics as many inhabitants as 10,000 square leagues of Estramadura, the Castiles, and other provinces of the table-land of Spain? These predictions are by no means problematical, inasmuch as they are founded on physical analogies and on the productive power of the soil; but before we can indulge the hope that they will be actually accomplished, we must be secure of another element less susceptible of calculation—that national wisdom which subdues hostile passions, destroys the germs of civil discord and gives stability to free and energetic institutions.

When we take a view of the soil of Venezuela and New Grenada we perceive that no other country of Spanish America furnishes commerce with such various and rich productions of the vegetable kingdom. If we add the harvests of the province of Caracas to those of Guayaquil, we find that the republic of Columbia alone can furnish nearly all the cacao annually demanded by Europe. The union of Venezuela and New Grenada has also placed in the hands of one people the greater part of the quinquina exported from the New Continent. The temperate mountains of Merida, Santa Fe, Popayan, Quito and Loxa produce the finest qualities of this febrifugal bark hitherto known. I might swell the list of these valuable productions by the coffee and indigo of Caracas, so long esteemed in commerce; the sugar, cotton and flour of Bogota; the ipecacuanha of the banks of the Magdelena; the tobacco of Varinas; the Cortex Angosturae of Caroni; the balsam of the plains of Tolu; the skins and dried provisions of the Llanos; the pearls of Panama, Rio Hacha and Marguerita; and finally the gold of Popayan and the platinum which is nowhere found in abundance but at Choco and Barbacoa: but conformably with the plan I have adopted, I shall confine myself to the old Capitania-General of Caracas.

Owing to a peculiar disposition of the soil in Venezuela the three zones of agricultural, pastoral and hunting-life succeed each other from north to south along the coast in the direction of the equator. Advancing in that direction we may be said to traverse, in respect to space, the different stages through which the human race has passed in the lapse of ages, in its progress towards cultivation and in laying the foundations of civilized society. The region of the coast is the centre of agricultural industry; the region of the Llanos serves only for the pasturage of the animals which Europe has given to America and which live there in a half-wild state. Each of those regions includes from seven to eight thousand square leagues; further south, between the delta of the Orinoco, the Cassiquiare and the Rio Negro, lies a vast extent of land as large as France, inhabited by hunting nations, covered with thick forests and impassable swamps. The productions of the vegetable kingdom belong to the zones at each extremity; the intermediary savannahs, into which oxen, horses, and mules were introduced about the year 1548, afford food for some millions of those animals. At the time when I visited Venezuela the annual exportation from thence to the West India Islands amounted to 30,000 mules, 174,000 ox-hides and 140,000 arrobas (of twenty-five pounds) of tasajo,* or dried meat slightly salted. (* The back of the animal is cut in slices of moderate thickness. An ox or cow of the weight of 25 arrobas produces only 4 to 5 arrobas of tasajo or tasso. In 1792 the port of Barcelona alone exported 98,017 arrobas to the island of Cuba. The average price is 14 reals and varies from 10 to 18 (the real is worth about 6 1/2 pence English). M. Urquinasa estimates the total exportation of Venezuela in 1809 at 200,000 arrobas of tasajo.) It is not from the advancement of agriculture or the progressive encroachments on the pastoral lands that the hatos (herds and flocks) have diminished so considerably within twenty years; it is rather owing to the disorders of every kind that have prevailed, and the want of security for property. The impunity conceded to the skin-stealers and the accumulation of marauders in the savannahs preceded that destruction of cattle caused by the ravages of civil war and the supplies required for troops.

A very considerable number of goat-skins is exported to the island of Marguerita, Punta Araya and Corolas; sheep abound only in Carora and Tocuyo. The consumption of meat being immense in this country the diminution of animals has a greater influence here than in any other district on the well-being of the inhabitants. The town of Caracas, of which the population in my time was one-tenth of that of Paris, consumed more than one-half the quantity of beef annually used in the capital of France.

I might add to the productions of the vegetable and animal kingdoms of Venezuela the enumeration of the minerals, the working of which is worthy the attention of the government; but having from my youth been engaged in the practical labours of mines I know how vague and uncertain are the judgments formed of the metallic wealth of a country from the mere appearance of the rocks and of the veins in their beds. The utility of such labours can be determined only by well directed experiments by means of shafts or galleries. All that has been done in researches of this kind, under the dominion of the mother-country, has left the question wholly undecided and the most exaggerated ideas have been recently spread through Europe concerning the riches of the mines of Caracas. The common denomination of Columbia given to Venezuela and New Grenada has doubtless contributed to foster those illusions. It cannot be doubted that the gold-washings of New Grenada furnished, in the last years of public tranquillity, more than 18,000 marks of gold; that Choco and Barbacoa supply platinum in abundance; the valley of Santa Rosa in the province of Antioquia, the Andes of Quindiu and Gauzum near Cuenca, yield sulphuretted mercury; the table-land of Bogota (near Zipaquira and Canoas), fossil-salt and pit-coal; but even in New Grenada subterranean labours on the silver and gold veins have hitherto been very rare. I am far, however, from wishing to discourage the miners of those countries: I merely conceive that for the purpose of proving to the old world the political importance of Venezuela, the amazing territorial wealth of which is founded on agriculture and the produce of pastoral life, it is not necessary to describe as realities, or as the acquisitions of industry, what is, as yet, founded solely on hopes and probabilities more or less uncertain. The republic of Columbia also possesses on its coast, on the island of Marguerita, on the Rio Hacha and in the gulf of Panama pearl fisheries of ancient celebrity. In the present state of things, however, fishing for these pearls is an object of as little importance as the exportation of the metals of Venezuela. The existence of metallic veins on several points of the coast cannot be doubted. Mines of gold and silver were worked at the beginning of the conquest at Buria, near Barquesimeto, in the province of Los Mariches, at Baruta, on the south of Caracas, and at Real de Santa Barbara near the Villa de Cura. Grains of gold are found in the whole mountainous territory between Rio Yaracuy, the Villa de San Felipe and Nirgua, as well as between Guigue and Los Moros de San Juan. M. Bonpland and myself, during our long journey, saw nothing in the gneiss granite of Spanish Guiana to confirm the old faith in the metallic wealth of that district; yet it seems certain from several historical notices that there exist two groups of auriferous alluvial land; one between the sources of the Rio Negro, the Uaupes and the Iquiare; the other between the sources of the Essequibo, the Caroni and the Rupunuri. Hitherto only one working is found in Venezuela, that of Aroa: it furnished, in 1800, near 1500 quintals of copper of excellent quality. The green-stone rocks of the transition mountains of Tucutunemo (between Villa de Cura and Parapara) contain veins of malachite and copper pyrites. The indications of both ochreous and magnetic iron in the coast-chain, the native alum of Chuparipari, the salt of Araya, the kaolin of the Silla, the jade of the Upper Orinoco, the petroleum of Buen-Pastor and the sulphur of the eastern part of New Andalusia equally merit the attention of the government.

It is easy to ascertain the existence of some mineral substances which afford hopes of profitable working but it requires great circumspection to decide whether the mineral be sufficiently abundant and accessible to cover the expense.* (* In 1800 a day-labourer (peon) employed in working the ground gained in the province of Caracas 15 sous, exclusive of his food. A man who hewed building timber in the forests on the coast of Paria was paid at Cumana 45 to 50 sous a day, without his food. A carpenter gained daily from 3 to 6 francs in New Andalusia. Three cakes of cassava (the bread of the country), 21 inches in diameter, 1 1/2 lines thick, and 2 1/2 pounds weight, cost at Caracas one half-real, or 6 1/2 sous. A man eats daily not less than 2 sous' worth of cassava, that food being constantly mixed with bananas, dried meat (tasajo) and panelon, or unrefined sugar.) Even in the eastern part of South America gold and silver are found dispersed in a manner that surprises the European geologist; but that dispersion, together with the divided and entangled state of the veins and the appearance of some metals only in masses, render the working extremely expensive. The example of Mexico sufficiently proves that the interest attached to the labours of the mines is not prejudicial to agricultural pursuits, and that those two branches of industry may simultaneously promote each other. The failure of the attempts made under the intendant, Don Jose Avalo, must be attributed solely to the ignorance of the persons employed by the Spanish government who mistook mica and hornblende for metallic substances. If the government would order the Capitania-General of Caracas to be carefully examined during a series of years by men of science, well versed in geognosy and chemistry, the most satisfactory results might be expected.

The description above given of the productions of Venezuela and the development of its coast sufficiently shows the importance of the commerce of that rich country. Even under the thraldom of the colonial system, the value of the exported products of agriculture and of the gold-washings amount to eleven or twelve millions of piastres in the countries at present united under the denomination of the Republic of Columbia. The exports of the Capitania-General of Caracas alone, exclusive of the precious metals which are the objects of regular working, was (with the contraband) from five to six millions of piastres at the beginning of the nineteenth century. Cumana, Barcelona, La Guayra, Porto Cabello and Maracaybo are the most important parts of the coast; those that lie most eastward have the advantage of an easier communication with the Virgin Islands, Guadaloupe, Martinique and St. Vincent. Angostura, the real name of which is Santo Tome de Nueva Guiana, may be considered as the port of the rich province of Varinas. The majestic river on whose banks this town is built, affords by its communications with the Apure, the Meta and the Rio Negro the greatest advantages for trade with Europe.

The shores of Venezuela, from the beauty of their ports, the tranquillity of the sea by which they are washed and the fine timber that covers them, possess great advantages over the shores of the United States. In no part of the world do we find firmer anchorage or better positions for the establishment of ports. The sea of this coast is constantly calm, like that which extends from Lima to Guayaquil. The storms and hurricanes of the West Indies are never felt on the Costa Firme; and when, after the sun has passed the meridian, thick clouds charged with electricity accumulate on the mountains of the coasts, a pilot accustomed to these latitudes knows that this threatening aspect of the sky denotes only a squall. The virgin-forests near the sea, in the eastern part of New Andalusia, present valuable resources for the establishment of dockyards. The wood of the mountains of Paria may vie with that of the island of Cuba, Huasacualco, Guayaquil and San Blas. The Spanish Government at the close of the last century fixed its attention on this important object. Marine engineers were sent to mark the finest trunks of Brazil-wood, mahogany, cedrela and laurinea between Angostura and the mouth of the Orinoco, as well as on the banks of the Gulf of Paria, commonly called the Golfo triste. It was not intended to establish docks on that spot, but to hew the weighty timber into the forms necessary for ship-building, and to transport it to Caraque, near Cadiz. Though trees fit for masts are not found in this country, it was nevertheless hoped that the execution of this project would considerably diminish the importation of timber from Sweden and Norway. The experiment of forming this establishment was tried in a very unhealthy spot, the valley of Quebranta, near Guirie; I have already adverted to the causes of its destruction. The insalubrity of the place would, doubtless, have diminished in proportion as the forest (el monte virgen) should have been removed from the dwellings of the inhabitants. Mulattos, and not whites, ought to have been employed in hewing the wood, and it should have been remembered that the expense of the roads (arastraderos) for the transport of the timber, when once laid out, would not have been the same, and that, by the increase of the population, the price of day labour would progressively have diminished. It is for ship-builders alone, who determine the localities, to judge whether, in the present state of things, the freight of merchant-vessels be not far too high to admit of sending to Europe large quantities of roughly-hewn wood; but it cannot be doubted that Venezuela possesses on its maritime coast, as well as on the banks of the

Orinoco, immense resources for ship-building. The fine ships which have been launched from the dockyards of the Havannah, Guayaquil and San Blas have, no doubt, cost more than those constructed in Europe; but from the nature of tropical wood they possess the advantages of hardness and amazing durability.

The great struggle during which Venezuela has fought for independence has lasted more than twelve years. That period has been no less fruitful than civil commotions usually are in heroic and generous actions, guilty errors and violent passions. The sentiment of common danger has strengthened the ties between men of various races who, spread over the plains of Cumana or insulated on the table-land of Cundinamarca, have a physical and moral organization as different as the climates in which they live. The mother-country has several times regained possession of some districts; but as revolutions are always renewed with more violence when the evils that produce them can no longer be remedied these conquests have been transitory. To facilitate and give greater energy to the defence of this country the governments have been concentrated, and a vast state has been formed, extending from the mouth of the Orinoco to the other side of the Andes of Riobamba and the banks of the Amazon. The Capitania-General of Caracas has been united to the Vice-royalty of New Grenada, from which it was only separated entirely in 1777. This union, which will always be indispensable for external safety, this centralization of powers in a country six times larger than Spain, has been prompted by political views. The tranquil progress of the new government has justified the wisdom of those views, and the Congress will find still fewer obstacles in the execution of its beneficent projects for national industry and civilization, in proportion as it can grant increased liberty to the provinces, must render the people sensible to the advantages of institutions which they have purchased at the price of their blood. In every form of government, in republics as well as in limited monarchies, improvements, to be salutary, must be progressive. New Andalusia, Caracas, Cundinamarca, Popayan and Quito, are not confederate states like Pennsylvania, Virginia and Maryland. Without juntas, or provincial legislatures, all those countries are directly subject to the congress and government of Columbia. In conformity with the constitutional act, the intendants and governors of the departments and provinces are nominated by the president of the republic. It may be naturally supposed that such dependence has not always been deemed favourable to the liberty if the communes, which love to discuss their own local interests. The ancient kingdom of Quito, for instance, is connected by the habits and language of its mountainous inhabitants with Peru and New Grenada. If there were a provincial junta, if the congress alone determined the taxes necessary for the defence and general welfare of Columbia, the feeling of an individual political existence would render the inhabitants less interested in the choice of the spot which is the seat of the central government. The same argument applies to New Andalusia or Guiana which are governed by intendants named by the president. It may be said that these provinces have hitherto been in a position differing but little from those territories of the United States which have a population below 60,000 souls. Peculiar circumstances, which cannot be justly appreciated at such a distance, have doubtless rendered great centralization necessary in the civil administration; every change would be dangerous as long as the state has external enemies; but the forms useful for defence are not always those which, after the struggle, sufficiently favour individual liberty and the development of public prosperity.

The powerful union of North America has long been insulated and without contact with any states having analogous institutions. Although the progress America is making from east to west is considerably retarded near the right bank of the Mississippi, she will advance without interruption towards the internal provinces of Mexico, and will there find a European people of another race, other manners, and a different religious faith. Will the feeble population of those provinces, belonging to another dawning federation, resist; or will it be absorbed by the torrent from the east and transformed into an Anglo-American state, like the inhabitants of Lower Louisiana? The future will soon solve this problem. On the other hand, Mexico is separated from Columbia only by Guatimala, a country and extreme fertility which has recently assumed the denomination of the republic of Central America. The political divisions between Oaxaca and Chiapa, Costa Rica and Veragua, are not founded either on the natural limits or the manners and languages of the natives, but solely on the habit of dependence on the Spanish chiefs who resided at Mexico, Guatimala or Santa Fe de Bogota. It seems natural that Guatimala should one day join the isthmuses of Veragua and Panama to the isthmus of Costa Rica; and that Quito should connect New Grenada with Peru, as La Paz, Charcas and Potosi link Peru with Buenos-Ayres. The intermediate parts from Chiapa to the Cordilleras of Upper Peru form a passage from one political association to another, like those transitory forms which link together the various groups of the organic kingdom in nature. In neighbouring monarchies the provinces that adjoin each other present those striking demarcations which are the effect of great centralization of power in federal republics, states situated at the extremities of each system are some time before they acquire a stable equilibrium. It would be almost a matter of indifference to the provinces between Arkansas and the Rio del Norte whether they send their deputies to Mexico or to Washington. Were Spanish America one day to show a more uniform tendency towards the spirit of federalism, which the example of the United States has created on several points, there would result from the contact of so many systems or groups of states, confederations variously graduated. I here only touch on the relations that arise from this assemblage of colonies on an uninterrupted line of 1600 leagues in length. We have seen in North America, one of the old Atlantic states divided into two, and each having a different representation. The separation of Maine and Massachusetts in 1820 was effected in the most peaceable manner. Schisms of this kind will, it may be feared, render such changes turbulent. It may also be observed that the importance of the geographical divisions of Spanish America, founded at the same time on the relations of local position and the habits of several centuries, have prevented the mother-country from retarding the separation of the colonies by attempting to establish Spanish princes in the New World. In order to rule such vast possessions it would have been requisite to form six or seven centres of government; and that multiplicity of centres was hostile to the establishment of new dynasties at the period when they might still have been salutary to the mother country.

Bacon somewhere observes that it would be happy if nations would always follow the example of time, the greatest of all innovators, but who acts calmly and almost without being perceived. This happiness does not belong to colonies when they reach the critical juncture of emancipation; and least of all to Spanish America, engaged in the struggle at first not to obtain complete independence, but to escape from a foreign yoke. May these party agitations be succeeded by a lasting tranquillity! May the germ of civil discord, disseminated during three centuries to secure the dominion of the mother-country, gradually perish; and may productive and commercial Europe be convinced that to perpetuate the political agitations of the New World would be to impoverish herself by diminishing the consumption of her productions and losing a market which already yields more than seventy millions of piastres. Many years must no doubt elapse before seventeen millions of inhabitants, spread over a surface one-fifth greater than the whole of Europe, will have found a stable equilibrium in governing themselves. The most critical moment is that when nations, after long oppression, find themselves suddenly at liberty to promote their own prosperity. The Spanish Americans, it is unceasingly repeated, are not sufficiently advanced in intellectual cultivation to be fitted for free institutions. I remember that at a period not very remote, the same reasoning was applied to other nations who were said to have made too great an advance in civilization. Experience, no doubt, proves that nations, like individuals, find that intellect and learning do not always lead to happiness; but without denying the necessity of a certain mass of knowledge and popular instruction for the stability of republics or constitutional monarchies, we believe that stability depends much less on the degree of intellectual improvement than on the strength of the national character; on that balance of energy and tranquillity of ardour and patience which maintains and perpetuates new institutions; on the local circumstances in which a nation is placed; and on the political relations of a country with neighbouring states.

CHAPTER 3.28.
PASSAGE FROM THE COAST OF VENEZUELA TO THE HAVANNAH. GENERAL VIEW OF THE POPULATION OF THE WEST INDIA ISLANDS, COMPARED WITH THE POPULATION OF THE NEW CONTINENT, WITH

RESPECT TO DIVERSITY OF RACES, PERSONAL LIBERTY, LANGUAGE, AND WORSHIP.

We sailed from Nueva Barcelona on the 24th of November at nine o'clock in the evening; and we doubled the small rocky island of Borachita. The night was marked by coolness which characterizes the nights of the tropics, and the agreeable effect of which can only be conceived by comparing the nocturnal temperature, from 23 to 24 degrees centigrade, with the mean temperature of the day, which in those latitudes is generally, even on the coast, from 28 to 29 degrees. Next day, soon after the observation of noon, we reached the meridian of the island of Tortugas. It is destitute of vegetation; and like the little islands of Coche and Cabagua is remarkable for its small elevation above the level of the sea.

In the forenoon of the 26th we began to lose sight of the island of Marguerita and I endeavoured to verify the height of the rocky group of Macanao. It appeared under an angle of 0 degrees 16 minutes 35 seconds; which in a distance estimated at sixty miles would give the mica-slate group of Macanao the elevation of about 660 toises, a result which, in a zone where the terrestrial refractions are so unchanging, leads me to think that the island was less distant than we supposed. The dome of the Silla of Caracas, lying 62 degrees to the south-west, long fixed our attention. At those times when the coast is not loaded with vapours the Silla must be visible at sea, without reckoning the effects of refraction, at thirty-three leagues distance. During the 26th, and the three following days, the sea was covered with a bluish film which, when examined by a compound microscope, appeared formed of an innumerable quantity of filaments. We frequently find these filaments in the Gulf-stream, and the Channel of Bahama, as well as near the coast of Buenos Ayres. Some naturalists are of opinion that they are vestiges of the eggs of mollusca: but they appear to be more like fragments of fuci. The phosphorescence of sea-water seems however to be augmented by their presence, especially between 28 and 30 degrees of north latitude, which indicates an origin of some sort of animal nature.

On the 27th we slowly approached the island of Orchila. Like all the small islands in the vicinity of the fertile coast of the continent it has never been inhabited. I found the latitude of the northern cape 11 degrees 51 minutes 44 seconds and the longitude of the eastern cape 68 degrees 26 minutes 5 seconds (supposing Nueva Barcelona to be 67 degrees 4 minutes 48 seconds). Opposite the western cape there is a small rock against which the waves beat turbulently. Some angles taken with the sextant gave, for the length of the island from east to west, 8.4 miles (950 toises); and for the breadth scarcely three miles. The island of Orchila which, from its name, I figured to myself as a bare rock covered with lichens, was at that period beautifully verdant. The hills of gneiss were covered with grasses. It appears that the geological constitution of Orchila resembles, on a small scale, that of Marguerita. It consists of two groups of rocks joined by a neck of land; it is an isthmus covered with sand which seems to have issued from the floods by the successive lowering of the level of the sea. The rocks, like all those which are perpendicular and insulated in the middle of the sea, appear much more elevated than they really are, for they scarcely exceed from 80 to 90 toises. The Punta rasa stretches to the north-west and is lost, like a sandbank, below the waters. It is dangerous for navigators, and so is likewise the Mogote which, at the distance of two miles from the western cape, is surrounded by breakers. On a very near examination of these rocks we saw the strata of gneiss inclined towards the north-west and crossed by thick layers of quartz. The destruction of these layers has doubtless created the sands of the surrounding beach. Some clumps of trees shade the valleys, the summits of the hills are crowned with fan-leaved palm-trees; probably the palma de sombrero of the Llanos (Corypha tectorum). Rain is not abundant in these countries; but probably some springs might be found on the island of Orchila if sought for with the same care as in the mica-slate rocks of Punta Araya. When we recollect how many bare and rocky islands are inhabited and cultivated between the 17th and 26th degrees of latitude in the archipelago of the Lesser Antilles and Bahama islands, we are surprised to find those islands desert which are near to the coast of Cumana, Barcelona and Caracas. They would long have ceased to be so had they been under the dominion of any other government than that to which they belong. Nothing can engage men to circumscribe their industry within the narrow limits of a small island when a neighbouring continent offers them greater advantages.

We perceived at sunset the two points of the Roca de afuera, rising like towers in the midst of the ocean. A survey taken with the compass placed the most easterly of the points or roques at 0 degrees 19 minutes west of the western cape of Orchila. The clouds continued long accumulated over that island and showed its position from afar. The influence of a small tract of land in condensing the vapours suspended at an elevation of 800 toises is a very extraordinary phenomenon, although familiar to all mariners. From this accumulation of clouds the position of the lowest island may be recognized at a great distance.

On the 29th November we still saw very distinctly, at sunrise, the summit of the Silla of Caracas just rising above the horizon of the sea. At noon everything denoted a change of weather in the direction of the north: the atmosphere suddenly cooled to 12.6 degrees, while the sea maintained a temperature of 25.6 degrees at its surface. At the moment of the observation of noon the oscillations of the horizon, crossed by streaks or black bands of very variable size, produced changes of refraction from 3 to 4 degrees. The sea became rough in very calm weather and everything announced a stormy passage between Cayman Island and Cape St. Antonio. On the 30th the wind veered suddenly to north-north-east and the surge rose to a considerable height. Northward a darkish blue tint was observable on the sky, the rolling of our small vessel was violent and we perceived amidst the dashing of the waves two seas crossing each other, one the from north and the other from north-north-east. Waterspouts were formed at the distance of a mile and were carried rapidly from north-north-east to north-north-west. Whenever the waterspout drew near us we felt the wind grow sensibly cooler. Towards evening, owing to the carelessness of our American cook, our deck took fire; but fortunately it was soon extinguished. On the morning of the 1st of December the sea slowly calmed and the breeze became steady from north-east. On the 2nd of December we descried Cape Beata, in a spot where we had long observed the clouds gathered together. According to the observations of Acherner, which I obtained in the night, we were sixty-four miles distant. During the night there was a very curious optical phenomenon, which I shall not undertake to account for. At half-past midnight the wind blew feebly from the east; the thermometer rose to 23.2 degrees, the whalebone hygrometer was at 57 degrees. I had remained upon the deck to observe the culmination of some stars. The full-moon was high in the heavens. Suddenly, in the direction of the moon, 45 degrees before its passage over the meridian, a great arch was formed tinged with the prismatic colours, though not of a bright hue. The arch appeared higher than the moon; this iris-band was near 2 degrees broad, and its summit seemed to rise nearly from 80 to 85 degrees above the horizon of the sea. The sky was singularly pure; there was no appearance of rain; and what struck me most was that this phenomenon, which perfectly resembled a lunar rainbow, was not in the direction opposite to the moon. The arch remained stationary, or at least appeared to do so, during eight or ten minutes; and at the moment when I tried if it were possible to see it by reflection in the mirror of the sextant, it began to move and descend, crossing successively the Moon and Jupiter. It was 12 hours 54 minutes (mean time) when the summit of the arch sank below the horizon. This movement of an arch, coloured like the rainbow, filled with astonishment the sailors who were on watch on the deck. They alleged, as they do on the appearance of every extraordinary meteor, that it denoted wind. M. Arago examined the sketch of this arch in my journal; and he is of opinion that the image of the moon reflected in the waters could not have given a halo of such great dimensions. The rapidity of the movement is no small obstacle in the way of explanation of a phenomenon well worthy of attention.

On the 3rd of December we felt some uneasiness on account of the proximity of a small vessel supposed to be a pirate but which, as it drew near, we recognized to be the Balandra del Frayle (the sloop of the Monk). I was at a loss to conceive what so strange a denomination meant. The bark belonged to a Franciscan missionary, a rich priest of am Indian village in the savannahs (Llanos) of Barcelona, who had for several years carried on a very lucrative contraband trade with the Danish islands. M. Bonpland and several passengers saw in the night at the distance of a quarter of a mile, with the wind, a small flame on the surface of the ocean; it ran in the direction of south-west and lighted up the atmosphere. No shock of earthquake was felt and there was no change in the direction of the waves. Was it a phosphoric gleam produced by a great accumulation of mollusca in a state of putrefaction;

or did this flame issue from the depth of the sea, as is said to have been sometimes observable in latitudes agitated by volcanoes? The latter supposition appears to me devoid of all probability. The volcanic flame can only issue from the deep when the rocky bed of the ocean is already heaved up so that the flames and incandescent scoriae escape from the swelled and creviced part without traversing the waters.

At half-past ten in the morning of the 4th of December we were in the meridian of Cape Bacco (Punta Abacou) which I found in 76 degrees 7 minutes 50 seconds, or 9 degrees 3 minutes 2 seconds west of Nueva Barcelona. Having attained the parallel of 17 degrees, the fear of pirates made us prefer the direct passage across the bank of Vibora, better known by the name of the Pedro Shoals. This bank occupies more than two hundred and eighty square sea leagues and its configuration strikes the eye of the geologist by its resemblance to that of Jamaica, which is in its neighbourhood. It forms an island almost as large as Porto Rico.

From the 5th of December, the pilots believed they took successively the measurement at a distance of the island of Ranas (Morant Keys), Cape Portland and Pedro Keys. They may probably have been deceived in several of these distances, which were taken from the mast-head. I have elsewhere noted these measurements, not with the view of opposing them to those which have been made by able English navigators in these frequented latitudes, but merely to connect, in the same system of observations, the points I determined in the forests of the Orinoco and in the archipelago of the West Indies. The milky colour of the waters warned us that we were on the eastern part of the bank; the centigrade thermometer which at a distance from the bank and on the surface of the sea had for several days kept at 27 and 27.3 degrees (the air being at 21.2 degrees) sank suddenly to 25.7 degrees. The weather was bad from the 4th to the 6th of December: it rained fast; thunder rolled at a distance, and the gusts of wind from the north-north-east became more and more violent. We were during some part of the night in a critical position; we heard before us the noise of the breakers over which we had to pass, and we could ascertain their direction by the phosphoric gleam reflected from the foam of the sea. The scene resembled the Raudal of Garzita and other rapids which we had seen in the bed of the Orinoco. We succeeded in changing our course and in less than a quarter of an hour were out of danger. While we traversed the bank of the Vibora from south-south-east to north-north-west I repeatedly tried to ascertain the temperature of the water on the surface of the sea. The cooling was less sensible on the middle of the bank than on its edge, a circumstance which we attributed to the currents that there mingle waters from different latitudes. On the south of Pedro Keys the surface of the sea, at twenty-five fathoms deep, was 26.4 and at fifteen fathoms deep 26.2 degrees. The temperature of the sea on the east of the bank had been 26.8 degrees. Some American pilots affirm that among the Bahama Islands they often know, when seated in the cabin, that they are passing over sand-banks; they allege that the lights are surrounded with small coloured halos and that the air exhaled from the lungs is visibly condensed. The latter circumstance appears very doubtful; below 30 degrees of latitude the cooling produced by the waters of the bank is not sufficiently considerable to cause this phenomenon. During the time we passed on the bank of the Vibora the constitution of the air was quite different from what it had been when we quitted it. The rain was circumscribed by the limits of the bank of which we could distinguish the form from afar by the mass of vapour with which it was covered.

On the 9th of December, as we advanced towards the Cayman Islands,* the north-east wind again blew with violence. (* Christopher Columbus in 1503 named the Cayman Islands Penascales de las Tortugas on account of the sea-tortoises which he saw swimming in those latitudes.) I nevertheless obtained some altitudes of the sun at the moment when we believed ourselves, though twelve miles distant, in the meridian of the centre of the Great Cayman, which is covered with cocoa-trees.

The weather continued bad and the sea extremely rough. The wind at length fell as we neared Cape St. Antonio. I found the northern extremity of the cape 87 degrees 17 minutes 22 seconds, or 2 degrees 34 minutes 14 seconds eastward of the Morro of the Havannah: this is the longitude now marked on the best charts. We were at the distance of three miles from land but we were made aware of the proximity of the island of Cuba by a delicious aromatic odour. The sailors affirm that this odour is not perceived when they approach from Cape Catoche on the barren coast of Mexico. As the weather grew clearer the thermometer rose gradually in the shade to 27 degrees: we advanced rapidly northward, carried on by a current from south-south-east, the temperature of which rose at the surface of the water to 26.7 degrees; while out of the current it was 24.6 degrees. We anchored in the port of the Havannah on the 19th December after a passage of twenty-five days in continuous bad weather.

CHAPTER 3.29.
POLITICAL ESSAY ON THE ISLAND OF CUBA. THE HAVANNAH. HILLS OF GUANAVACOA, CONSIDERED IN THEIR GEOLOGICAL RELATIONS. VALLEY OF LOS GUINES, BATABANO, AND PORT OF TRINIDAD. THE KING AND QUEEN'S GARDENS.

Cuba owes its political importance to a variety of circumstances, among which may be enumerated the extent of its surface, the fertility of its soil, its naval establishments, and the nature of its population, of which three-fifths are free men. All these advantages are heightened by the admirable position of the Havannah. The northern part of the Caribbean Sea, known by the name of the Gulf of Mexico, forms a circular basin more than two hundred and fifty leagues in diameter: it is a Mediterranean with two outlets. The island of Cuba, or rather its coast between Cape St. Antonio and the town of Matanzas, situated at the opening of the old channel, closes the Gulf of Mexico on the south-east, leaving the ocean current known by the name of the Gulf Stream, no other outlet on the south than a strait between Cape St. Antonio and Cape Catoche; and no other on the north than the channel of Bahama, between Bahia-Honda and the shoals of Florida. Near the northern outlet, where the highways of so many nations may be said to cross each other, lies the fine port of the Havannah, fortified at once by nature and by art. The fleets which sail from this port and which are partly constructed of the cedrela and the mahogany of the island of Cuba, might, at the entrance of the Mexican Mediterranean, menace the opposite coast, as the fleets that sail from Cadiz command the Atlantic near the Pillars of Hercules. In the meridian of the Havannah the Gulf of Mexico, the old channel, and the channel of Bahama unite. The opposite direction of the currents and the violent agitations of the atmosphere at the setting-in of winter impart a peculiar character to these latitudes at the extreme limit of the equinoctial zone.

The island of Cuba is the largest of the Antilles.* (* Its area is little less in extent than that of England not including Wales.) Its long and narrow form gives it a vast development of coast and places it in proximity with Hayti and Jamaica, with the most southern province of the United States (Florida) and the most easterly province of the Mexican Confederation (Yucatan).* (* These places are brought into communication one with another by a voyage of ten or twelve days.) This circumstance claims serious attention when it is considered that Jamaica, St. Domingo, Cuba and the southern parts of the United States (from Louisiana to Virginia) contain nearly two million eight hundred thousand Africans. Since the separation of St. Domingo, the Floridas and New Spain from the mother-country, the island of Cuba is connected only by similarity of religion, language and manners with the neighbouring countries, which, during ages, were subject to the same laws.

Florida forms the last link in that long chain, the northern extremity of which reaches the basin of St. Lawrence and extends from the region of palm-trees to that of the most rigorous winter. The inhabitant of New England regards the increasing augmentation of the black population, the preponderance of the slave states and the predilection for the cultivation of colonial products as a public danger; and earnestly wishes that the strait of Florida, the present limit of the great American confederation, may never be passed but with the views of free trade, founded on equal rights. If he fears events which may place the Havannah under the dominion of a European power more formidable than Spain, he is not the less desirous that the political ties by which Louisiana, Pensacola and Saint Augustin of Florida were heretofore united to the island of Cuba may for ever be broken.

The extreme sterility of the soil, joined to the want of inhabitants and of cultivation, have at all times rendered the proximity of Florida of small importance to the trade of the Havannah; but the case is different on the coast of Mexico. The shores of that country, stretching in a semicircle from the frequented ports of Tampico, Vera Cruz, and Alvarado to

Cape Catoche, almost touch, by the peninsula of Yucatan, the western part of the island of Cuba. Commerce is extremely active between the Havannah and the port of Campeachy; and it increases, notwithstanding the new order of things in Mexico, because the trade, equally illicit with a more distant coast, that of Caracas or Columbia, employs but a small number of vessels. In such difficult times the supply of salt meat (tasajo) for the slaves is more easily obtained from Buenos Ayres and the plains of Merida than from those of Cumana, Barcelona and Caracas. The island of Cuba and the archipelago of the Philippines have for ages derived from New Spain the funds necessary for their internal administration and for keeping up their fortifications, arsenals and dockyards. The Havannah was the military port of the New World; and, till 1808, annually received 1,800,000 piastres from the Mexican treasury. At Madrid it was long the custom to consider the island of Cuba and the archipelago of the Philippines as dependencies on Mexico, situated at very unequal distances east and west of Vera Cruz and Acapulco, but linked to the Mexican metropolis (then a European colony) by all the ties of commerce, mutual aid and ancient sympathies. Increased internal wealth has rendered unnecessary the pecuniary succour formerly furnished to Cuba from the Mexican treasury. Of all the Spanish possessions that island has been most prosperous: the port of the Havannah has, since the troubles of St. Domingo, become one of the most important points of the commercial world. A fortunate concurrence of political circumstances, joined to the intelligence and commercial activity of the inhabitants, have preserved to the Havannah the uninterrupted enjoyment of free intercourse with foreign nations.

I twice visited this island, residing there on one occasion for three months, and on the other for six weeks; and I enjoyed the confidence of persons who, from their abilities and their position, were enabled to furnish me with the best information. In company with M. Bonpland I visited only the vicinity of the Havannah, the beautiful valley of Guines and the coast between Batabano and the port of Trinidad. After having succinctly described the aspect of this scenery and the singular modifications of a climate so different from that of the other islands, I will proceed to examine the general population of the Island of Cuba; its area calculated from the most accurate sketch of the coast; the objects of trade and the state of the public revenue.

The aspect of the Havannah, at the entrance of the port, is one of the gayest and most picturesque on the shore of equinoctial America north of the equator. This spot is celebrated by travellers of all nations. It boasts not the luxuriant vegetation that adorns the banks of the river Guayaquil nor the wild majesty of the rocky coast of Rio de Janeiro; but the grace which in those climates embellishes the scenes of cultivated nature is at the Havannah mingled with the majesty of vegetable forms and the organic vigour that characterizes the torrid zone. On entering the port of the Havannah you pass between the fortress of the Morro (Castillo de los Santos Reyes) and the fort of San Salvador de la Punta: the opening being only from one hundred and seventy to two hundred toises wide. Having passed this narrow entrance, leaving on the north the fine castle of San Carlos de la Cabana and the Casa Blanca, we reach a basin in the form of a trefoil of which the great axis, stretching from south-south-west to north-north-east, is two miles and one-fifth long. This basin communicates with three creeks, those of Regla, Guanavacoa and Atares; in this last there are some springs of fresh water. The town of the Havannah, surrounded by walls, forms a promontory bounded on the south by the arsenal and on the north by the fort of La Punta. After passing beyond some wrecks of vessels sunk in the shoals of La Luz, we no longer find eight or ten, but five or six fathoms of water. The castles of Santo Domingo de Atares and San Carlos del Principe defend the town on the westward; they are distant from the interior wall, on the land side, the one 660 toises, the other 1240. The intermediate space is filled by the suburbs (arrabales or barrios extra muros) of the Horcon, Jesu-Maria, Guadaloupe and Senor de la Salud, which from year to year encroach on the Field of Mars (Campo de Marte). The great edifices of the Havannah, the cathedral, the Casa del Govierno, the house of the commandant of the marine, the Correo or General Post Office and the factory of Tobacco are less remarkable for beauty than for solidity of structure. The streets are for the most part narrow and unpaved. Stones being brought from Vera Cruz, and very difficult of transport, the idea was conceived a short time before my voyage of joining great trunks of trees together, as is done in Germany and Russia, when dykes are constructed across marshy places. This project was soon abandoned and travellers newly arrived beheld with surprise fine trunks of mahogany sunk in the mud of the Havannah. At the time of my sojourn there few towns of Spanish America presented, owing to the want of a good police, a more unpleasant aspect. People walked in mud up to the knee; and the multitude of caleches or volantes (the characteristic equipage of the Havannah) of carts loaded with casks of sugar, and porters elbowing passengers, rendered walking most disagreeable. The smell of tasajo often poisons the houses and the winding streets. But it appears that of late the police has interposed and that a manifest improvement has taken place in the cleanliness of the streets; that the houses are more airy and that the Calle de los Mercadores presents a fine appearance. Here, as in the oldest towns of Europe. an ill-traced plan of streets can only be amended by slow degrees.

There are two fine public walks; one called the Alameda, between the hospital of Santa Paula and the theatre, and the other between the Castillo de la Punta and the Puerta de la Muralla, called the Paseo extra muros; the latter is deliciously cool and is frequented by carriages after sunset. It was begun by the Marquis de la Torre, governor of the island, who gave the first impulse to the improvement of the police and the municipal government. Don Luis de las Casas and the Count de Santa Clara enlarged the plantations. Near the Campo de Marte is the Botanical Garden which is well worthy to fix the attention of the government; and another place fitted to excite at once pity and indignation—the barracoon, in front of which the wretched slaves are exposed for sale. A marble statue of Charles III has been erected since my return to Europe, in the extra muros walk. This spot was at first destined for a monument to Christopher Columbus whose ashes, after the cession of the Spanish part of St. Domingo, were brought to the island of Cuba.*

(* Columbus lies buried in the cathedral of the Havannah, close to the wall near the high altar. On the tomb is the following inscription:

O restos y Imagen del grande Colon;
Mil siglos duran guardados en la Urna,
Y en remembranca de nuestra Nacion.

Oh relics and image of the great Colon (Columbus)
A thousand ages are encompassed in thy Urn,
And in the memory of our Nation.

His remains were first deposited at Valladolid and thence were removed to Seville. In 1536 the bodies of Columbus and of his son Diego (El Adelantado) were carried to St. Domingo and there interred in the cathedral; but they were afterwards removed to the place where they now repose.)

The same year the ashes of Fernando Cortez were transferred in Mexico from one church to another: thus, at the close of the eighteenth century, the remains of the two greatest men who promoted the conquest of America were interred in new sepulchres.

The most majestic palm-tree of its tribe, the palma real, imparts a peculiar character to the landscape in the vicinity of the Havannah; it is the Oreodoxa regia of our description of American palm-trees. Its tall trunk, slightly swelled towards the middle, grows to the height of 60 or 80 feet; the upper part is glossy, of a delicate green, newly formed by the closing and dilatation of the petioles, contrasts with the rest, which is whitish and fendilated. It appears like two columns, the one surmounting the other. The palma real of the island of Cuba has feathery leaves rising perpendicularly towards the sky, and curved only at the point. The form of this plant reminded us of the vadgiai palm-tree which covers the rocks in the cataracts of the Orinoco, balancing its long points over a mist of foam. Here, as in every place where the population is concentrated, vegetation diminishes. Those palm-trees round the Havannah and in the amphitheatre of Regla on which I delighted to gaze are disappearing by degrees. The marshy places which I saw covered with bamboos are cultivated and drained. Civilization advances; and the soil, gradually stripped of plants, scarcely offers any trace of its wild abundance. From the Punta to San Lazaro, from Cabana to Regla and from Regla to Atares the road is covered with houses, and those that surround the bay are of light and elegant construction. The plan of these houses is traced out by the owners, and they are ordered from the United States, like pieces of furniture. When the yellow fever rages at the Havannah the proprietors withdraw to those country houses and to the hills between Regla and Guanavacoa to breathe a purer air. In the coolness of night, when the boats cross the bay, and owing to the phosphorescence of the water, leave behind them long tracks of light, these romantic scenes afford charming and peaceful retreats for those who wish to withdraw from the tumult of a populous city. To judge of the progress of cultivation travellers should visit the small plots of maize and other alimentary plants, the rows of pine-apples (ananas) in the fields of Cruz de Piedra and the bishop's garden (Quinta del Obispo) which of late is become a delicious spot.

The town of the Havannah, properly so called, surrounded by walls, is only 900 toises long and 500 broad; yet more than 44,000 inhabitants, of whom 26,000 are negroes and mulattoes, are crowded together in this narrow space. A population nearly as considerable occupies the two great suburbs of Jesu-Maria and La Salud.* (* Salud signifies Health.) The latter place does not verify the name it bears; the temperature of the air is indeed lower than in the city but the streets might have been larger and better planned. Spanish engineers, who have been waging war for thirty years past with the inhabitants of the suburbs (arrabales), have convinced the government that the houses are too near the fortifications, and that the enemy might establish himself there with impunity. But the government has not courage to demolish the suburbs and disperse a population of 28,000 inhabitants collected in La Salud only. Since the great fire of 1802 that quarter has been considerably enlarged; barracks were at first constructed, but by degrees they have been converted into private houses. The defence of the Havannah on the west is of the highest importance: so long as the besieged are masters of the town, properly so called, and of the southern part of the bay, the Morro and La Cabana, they are impregnable because they can be provisioned by the Havannah, and the losses of the garrison repaired. I have heard well-informed French engineers observe that an enemy should begin his operations by taking the town, in order to bombard the Cabana, a strong fortress, but where the garrison, shut up in the casemates, could not long resist the insalubrity of the climate. The English took the Morro without being masters of the Havannah; but the Cabana and the Fort Number 4 which commands the Morro did not then exist. The most important works on the south and west are the Castillos de Atares y del Principe, and the battery of Santa Clara.

We employed the months of December, January and February in making observations in the vicinity of the Havannah and the fine plains of Guines. We experienced, in the family of Senor Cuesta (who then formed with Senor Santa Maria one of the greatest commercial houses in America) and in the house of Count O'Reilly, the most generous hospitality. We lived with the former and deposited our collections and instruments in the spacious hotel of Count O'Reilly, where the terraces favoured our astronomical observations. The longitude of the Havannah was at this period more than one fifth of a degree uncertain.* (* I also fixed, by direct observations, several positions in the interior of the island of Cuba: namely Rio Blanco, a plantation of Count Jaruco y Mopex; the Almirante, a plantation of the Countess Buenavista; San Antonio de Beitia; the village of Managua; San Antonio de Bareto; and the Fondadero, near the town of San Antonio de los Banos.). It had been fixed by M. Espinosa, the learned director of the Deposito hidrografico of Madrid, at 5 degrees 38 minutes 11 seconds, in a table of positions which he communicated to me on leaving Madrid. M. de Churruca fixed the Morro at 5 hours 39 minutes 1 second. I met at the Havannah with one of the most able officers of the Spanish navy, Captain Don Dionisio Galeano, who had taken a survey of the coast of the strait of Magellan. We made observations together on a series of eclipses of the satellites of Jupiter, of which the mean result gave 5 hours 38 minutes 50 seconds. M. Oltmanns deduced in 1805 the whole of those observations which I marked for the Morro, at 5 hours 38 minutes 52.5 seconds—84 degrees 43 minutes 7.5 seconds west of the meridian of Paris. This longitude was confirmed by fifteen occultations of stars observed from 1809 to 1811 and calculated by M. Ferrer: that excellent observer fixes the definitive result at 5 degrees 38 minutes 50.9 seconds. With respect to the magnetic dip I found it by the compass of Borda (December 1800) 53 degrees 22 minutes of the old sexagesimal division: twenty-two years before, according to the very accurate observations made by Captain Sabine in his memorable voyage to the coasts of Africa, America and Spitzbergen, the dip was only 51 degrees 55 minutes; it had therefore diminished 1 degree 27 minutes.

The island of Cuba being surrounded with shoals and breakers along more than two-thirds of its length, and as ships keep out beyond those dangers, the real shape of the island was for a long time unknown. Its breadth, especially between the Havannah and the port of Batabano, has been exaggerated; and it is only since the Deposito hidrografico of Madrid published the observations of captain Don Jose del Rio, and lieutenant Don Ventura de Barcaiztegui, that the area of the island of Cuba could be calculated with any accuracy. Wishing to furnish in this work the most accurate result that can be obtained in the present state of our astronomical knowledge, I engaged M. Bauza to calculate the area. He found, in June, 1835, the surface of the island of Cuba, without the Isla dos Pinos, to be 3520 square sea leagues, and with that island 3615. From this calculation, which has been twice repeated, it results that the island of Cuba is one-seventh less than has hitherto been believed; that it is 32/100 larger than Hayti, or San Domingo; that its surface equals that of Portugal, and within one-eighth that of England without Wales; and that if the whole archipelago of the Antilles presents as great an area as the half of Spain, the island of Cuba alone almost equals in surface the other Great and Small Antilles. Its greatest length, from Cape San Antonio to Point Maysi (in a direction from west-south-west to east-north-east and from west-north-west to east-south-east) is 227 leagues; and its greatest breadth (in the direction north and south), from Point Maternillo to the mouth of the Magdalena, near Peak Tarquino, is 37 leagues. The mean breadth of the island, on four-fifths of its length, between the Havannah and Puerto Principe, is 15 leagues. In the best cultivated part, between the Havannah and Batabano, the isthmus is only eight sea leagues. Among the great islands of the globe, that of Java most resembles the island of Cuba in its form and area (4170 square leagues). Cuba has a circumference of coast of 520 leagues, of which 280 belong to the south shore, between Cape San Antonio and Punta Maysi.

The island of Cuba, over more than four-fifths of its surface, is composed of low lands. The soil is covered with secondary and tertiary formations, formed by some rocks of gneiss-granite, syenite and euphotide. The knowledge obtained hitherto of the geologic configuration of the country, is as unsatisfactory as what is known respecting the relative age and nature of the soil. It is only ascertained that the highest group of mountains lies at the south-eastern extremity of the island, between Cape Cruz, Punta Maysi, and Holguin. This mountainous part, called the Sierra or Las Montanas del Cobre (the Copper Mountains), situated north-west of the town of Santiago de Cuba, appears to be about 1200 toises in height. If this calculation be correct, the summits of the Sierra would command those of the Blue Mountains of Jamaica, and the peaks of La Selle and La Hotte in the island of San Domingo. The Sierra of Tarquino, fifty miles west of the town of Cuba, belongs to the same group as the Copper Mountains. The island is crossed from east-south-east to west-north-west by a chain of hills, which approach the southern coast between the meridians of La Ciudad de Puerto Principe and the Villa Clara; while, further to the westward towards Alvarez and Matanzas, they stretch in the direction of the northern coast. Proceeding from the mouth of the Rio Guaurabo to the Villa de la Trinidad, I saw on the north-west, the Lomas de San Juan, which form needles or horns more than 300 toises high, with their declivities sloping regularly to the south. This calcareous group presents a majestic aspect, as seen from the anchorage near the Cayo de Piedras. Xagua and Batabano are low coasts; and I believe that, in general, west of the meridian of Matanzas, there is no hill more than 200 toises high, with the exception of the Pan de Guaixabon. The land in the interior of the island is gently undulated, as in England; and it rises only from 45 to 50 toises above the level of the sea. The objects most visible at a distance, and most celebrated by navigators, are the Pan de Matanzas, a truncated cone which has the form of a small monument; the Arcos de Canasi, which appear between Puerto Escondido and Jaruco, like small segments of a circle; the Mesa de Mariel, the Tetas de Managua, and the Pan de Guaixabon. This gradual slope of the limestone formations of the island of Cuba towards the north and west indicates the submarine connection of those rocks with the equally low lands of the Bahama Islands, Florida and Yucatan.

Intellectual cultivation and improvement were so long restricted to the Havannah and the neighbouring districts, that we cannot be surprised at the ignorance prevailing among the inhabitants respecting the geologic formation of the Copper Mountains. Don Francisco Ramirez, a traveller versed in chemical and mineralogical science, informed me that the western part of the island is granitic, and that he there observed gneiss and primitive slate. Probably the alluvial deposits of auriferous sand which were explored with much ardour* at the beginning of the conquest, to the great misfortune of the natives came from those granitic formations (* At Cubanacan, that is, in the interior of the island, near Jagua and Trinidad, where the auriferous sands have been washed by the waters as far as the limestone soil. Martyr d'Anghiera, the most intelligent writer on the Conquest, says: "Cuba is richer in gold than Hispaniola (San Domingo); and at the moment I am writing, 180,000 castillanos of ore have been collected at Cuba." Herrera estimates the tax called King's-fifth (quinto del Rey), in the island of Cuba, at 6000 pesos, which indicates an annual product of 2000 marks of gold, at 22 carats; and consequently purer than the gold of Sibao in San Domingo. In 1804 the mines of Mexico altogether produced 7000 marks of gold; and those of Peru 3400. It is difficult, in these calculations, to distinguish between the gold sent to Spain by the first Conquistadores, that obtained by washings, and that which had been accumulated for ages in the hands of the natives, who were pillaged at will. Supposing that in the two islands of Cuba and San Domingo (in Cubanacan and Cibao) the product of the washings was 3000 marks of gold, we find a quantity three times less than the gold furnished annually (1790 to 1805) by the small province of Choco. In this supposition of ancient wealth there is nothing improbable; and if we are surprised at the scanty produce of the gold-washings attempted in our days at Cuba and San Domingo, which were heretofore so prolific, it must be recollected that at Brazil also the product of the gold-washings has fallen, from 1760 to 1820, from 6600 gold kilogrammes to less than 595. Lumps of gold weighing several pounds, found in our days in Florida and North and South Carolina, prove the primitive wealth of the whole basin of the Antilles from the island of Cuba to the Appalachian chain. It is also natural that the product of the gold-washings should diminish with greater rapidity than that of the subterraneous working of the veins. The metals not being renewed in the clefts of the veins (by sublimation) now accumulate in alluvial soil by the course of the rivers where the table-lands are higher than the level of the surrounding running waters. But in rocks with metalliferous veins the miner does not at once know all he has to work. He may chance to lengthen the labours, to go deep, and to cross other accompanying veins. Alluvial soils are generally of small depth where they are auriferous; they most frequently rest upon sterile rocks. Their superficial position and uniformity of composition help to the knowledge of their limits, and wherever workmen can be collected, and where the waters for the washings abound, accelerate the total working of the auriferous clay. These considerations, suggested by the history of the Conquest, and by the science of mining, may throw some light on the problem of the metallic wealth of Hayti. In that island, as well as at Brazil, it would be more profitable to attempt subterraneous workings (on veins) in primitive and intermediary soils than to renew the gold-washings which were abandoned in the ages of barbarism, rapine and carnage.); traces of that sand are still found in the rivers Holguin and Escambray, known in general in the vicinity of Villa-Clara, Santo Espiritu, Puerto del Principe de Bayamo and the Bahia de Nipe. The abundance of copper mentioned by the Conquistadores of the sixteenth century, at a period when the Spaniards were more attentive than they have been in latter times to the natural productions of America, may possibly be attributed to the formations of amphibolic slate, transition clay-slate mixed with diorite, and to euphotides analogous to those I found in the mountains of Guanabacoa.

The central and western parts of the island contain two formations of compact limestone; one of clayey sandstone and another of gypsum. The former has, in its aspect and composition, some resemblance to the Jura formation. It is white, or of a clear ochre-yellow, with a dull fracture, sometimes conchoidal, sometimes smooth; divided into thin layers, furnishing some balls of pyromac silex, often hollow (at Rio Canimar two leagues east of Matanzas), and petrifications of pecten, cardites, terebratules and madrepores.* (* I saw neither gryphites nor ammonites of Jura limestone nor the nummulites and cerites of coarse limestone.) I found no oolitic beds, but porous beds almost bulbous, between the Potrero del Conde de Mopox, and the port of Batabano, resembling the spongy beds of Jura limestone in Franconia, near Dondorf, Pegnitz, and Tumbach. Yellowish cavernous strata, with cavities from three to four inches in diameter, alternate with strata altogether compact,* and poorer in petrifications. (* The western part of the island has no deep ravines; and we recognize this alternation in travelling from the Havannah to Batabano, the deepest beds (inclined from 30 to 40 degrees north-east) appear as we advance.) The chain of hills that borders the plain of Guines on the north and is linked with the Lomas de Camua, and the Tetas de Managua, belongs to the latter variety, which is reddish white, and almost of lithographic nature, like the Jura limestone of Pappenheim. The compact and cavernous beds contain nests of brown ochreous iron; possibly the red earth (tierra colorada) so much sought for by the coffee planters (haciendados) owes its origin to the decomposition of some superficial beds of oxidated iron, mixed with silex and clay, or to a reddish sandstone* (* Sandstone and ferruginous sand; iron-sand?) superposed on limestone. The whole of this formation, which I shall designate by the name of the limestone of Guines, to distinguish it from another much more recent, forms, near Trinidad, in the Lomas of St. Juan, steep declivities, resembling the mountains of limestone of Caripe, in the vicinity of Cumana. They also contain great caverns, near Matanzas and Jaruco, where I have not heard that any fossil bones have been found. The frequency of caverns in which the pluvial waters accumulate, and where small rivers disappear, sometimes causes a sinking of the earth. I am of opinion that the gypsum of the island of Cuba belongs not to tertiary but to secondary soil; it is worked in several places on the east of Matanzas, at San Antonio de los Banos, where it contains sulphur, and at the Cayos, opposite San Juan de los Remedios. We must not confound with this limestone of Guines, sometimes porous, sometimes compact, another formation so recent that it seems to augment in our days. I allude to the calcareous agglomerates, which I saw in the islands of Cayos that border the coast between the Batabano and the bay of Xagua, principally south of the Cienega de Zapata, Cayo Buenito, Cayo Flamenco and Cayo de Piedras. The soundings prove that they are rocks rising abruptly from a bottom of between twenty and thirty fathoms. Some are at the water's edge, others one-fourth or one-fifth of a toise above the surface of the sea. Angular fragments of madrepores, and cellularia from two to three cubic inches, are found cemented by grains of quartzose sand. The inequalities of the rocks are covered by mould, in which, by help of a microscope, we only distinguish the detritus of shells and corals. This tertiary formation no doubt belongs to that of the coast of Cumana, Carthagena, and the Great Land of Guadaloupe, noticed in my geognostic table of South America.* (* M. Moreau de Jonnes has well distinguished, in his Histoire physique des Antilles Francoises, between the Roche a ravets of Martinique and Hayti, which is porous, filled with terebratulites, and other vestiges of sea-shells, somewhat analogous to the limestone of Guines and the calcareous pelagic sediment called at Guadaloupe Platine, or Maconne bon Dieu. In the cayos of the island of Cuba, or Jardinillos del Rey y del Reyna, the whole coral rock lying above the surface of the water appeared to me to be fragmentary, that is, composed of broken blocks. It is, however, probable, that in the depth it reposes on masses of polypi still living.) MM. Chamiso and Guiamard have recently thrown great light on the formation of the coral islands in the Pacific. At the foot of the Castillo de in Punta, near the Havannah, on shelves of cavernous rocks,* covered with verdant sea-weeds and living polypi, we find enormous masses of madrepores and other lithophyte corals set in the texture of those shelves. (* The surface of these shelves, blackened and excavated by the waters, presents ramifications like the cauliflower, as they are observed on the currents of lava. Is the change of colour produced by the waters owing to the manganese which we recognize by some dendrites? The sea, entering into the clefts of the rocks, and in a cavern at the foot of the Castillo del Morro, compresses the air and makes it issue with a tremendous noise. This noise explains the phenomena of the baxos roncadores (snoring bocabeoos), so well known to navigators who cross from Jamaica to the mouth of Rio San Juan of Nicaragua, or to the island of San Andres.) We are at first tempted to admit that the whole of this limestone rock, which constitutes the principal portion of the island of Cuba, may be traced to an uninterrupted operation of nature—to the action of productive organic forces—an action which continues in our days in the bosom of the ocean; but this apparent novelty of limestone formations soon vanishes when we quit the shore, and recollect the series of coral rocks which contain the formations of different ages, the muschelkalk, the Jura limestone and coarse limestone. The same coral rocks as those of the Castillo and La Punta are found in the lofty inland mountains, accompanied with petrifications of bivalve shells, very different from those now seen on the coasts of the Antilles. Without positively assigning a determinate place in the table of formations to the limestone of Guines, which is that of the Castillo and La Punta, I have no doubt of the relative antiquity of that rock with respect to the calcareous agglomerate of the Cayos, situated south of Batabano, and east of the island of Pinos. The globe has undergone great revolutions between the periods when these two soils were formed; the one containing the great caverns of Matanzas, the other daily augmenting by the agglutination of fragments of coral and quartzose sand. On the south of the island of Cuba, the latter soil seems to repose sometimes on the Jura limestone of Guines, as in the Jardinillos, and sometimes (towards Cape Cruz) immediately over primitive rocks. In the lesser Antilles the corals are covered with volcanic productions. Several of the Cayos of the island of Cuba contain fresh water; and I found this water very good in the middle of the Cayo de Piedras. When we reflect on the extreme smallness of these islands we can scarcely believe that the fresh-water wells are filled with rain-water not evaporated. Do they prove a submarine communication between the limestone of the coast with the limestone serving as the basis of lithophyte polypi, and is the fresh water of Cuba raised up by hydrostatic pressure across the coral rocks of Cayos, as it is in the bay of Xagua, where, in the middle of the sea, it forms springs frequented by the lamantins?

The secondary formations on the east of the Havannah are pierced in a singular manner by syenitic and euphotide rocks united in groups. The southern bottom of the bay as well as the northern part (the hills of the Morro and the Cabana) are of Jura limestone; but on the eastern bank of the two Ensenadas de Regla and Guanabacoa, the whole is transition soil. Going from north to south, and first near Marimelena, we find syenite consisting of a great quantity of hornblende, partly decomposed, a little quartz, and a reddish-white feldspar seldom crystallized. This fine syenite, the strata of which incline to the north-west, alternates twice with serpentine. The layers of intercalated serpentine are three toises thick. Farther south, towards Regla and Guanabacoa, the syenite disappears, and the whole soil is covered with serpentine, rising in hills from thirty to forty toises high, and running from east to west. This rock is much fendillated, externally of a bluish-grey, covered with dendrites of manganese, and internally of a leek and asparagus-green, crossed by small veins of asbestos. It contains no garnet or amphibole, but metalloid diallage disseminated in the mass. The serpentine is sometimes of an esquillous, sometimes of a conchoidal fracture: this was the first time I had found metalloid diallage within the tropics. Several blocks of serpentine have magnetic poles; others are of such a homogeneous texture, and have such a glossiness, that at a distance they may be taken for pechstein (resinite). It were to be wished that these fine masses were employed in the arts as they are in several parts of Germany. In approaching Guanabacoa we find serpentine crossed by veins between twelve and fourteen inches thick, and filled with fibrous quartz, amethyst, and fine mammelonnes, and stalactiforme chalcedonies; it is possible that chrysoprase may also one day be found. Some copper pyrites appear among these veins accompanied, it is said, by silvery-grey copper. I found no traces of this grey copper: it is probably the metalloid diallage that has given the Cerro de Guanabacoa the reputation of riches in gold and silver which it has enjoyed for ages. In some places petroleum flows* from rents in the serpentine. (* Does there exist in the Bay of the Havannah any other source of petroleum than that of Guanabacoa, or must it be admitted that the betun liquido, which in 1508 was employed by Sebastian de Ocampo for the caulking of ships, is dried up? That spring, however, fixed the attention of Ocampo on the port of the Havannah, where he gave it the name of Puerto de Carenas. It is said that abundant springs of petroleum are also found in the eastern part of the island (Manantialis de betun y chapapote) between Holguin and Mayari, and on the coast of Santiago de Cuba.) Springs of water are frequent; they contain a little sulphuretted hydrogen, and deposit oxide of iron. The Baths of Bareto are agreeable, but of nearly the same temperature as the atmosphere. The geologic constitution of this group of serpentine rocks, from its insulated position, its veins, its connection with syenite and the fact of its rising up across shell-formations, merits particular attention. Feldspar with a basis of souda (compact feldspar) forms, with diallage, the euphotide and serpentine; with pyroxene, dolerite and basalt; and with garnet, eclogyte. These five rocks, dispersed over the whole globe, charged with oxidulated and titanious iron, are probably of similar origin. It is easy to distinguish two formations in the euphotide; one is destitute of amphibole, even when it alternates with amphibolic rocks (Joria in Piedmont, Regla in the island of Cuba) rich in pure serpentine, in metalloid diallage and sometimes in jasper (Tuscany, Saxony); the other, strongly charged with amphibole, often passing to diorite,* has no jasper in layers, and sometimes contains rich veins of copper; (Silesia, Mussinet in Piedmont, the Pyrenees, Parapara in Venezuela, Copper Mountains of North America). (* On a serpentine that flows like a penombre, veins of greenstone (diorite) near Lake Clunie in Perthshire. See MacCulloch in Edinburgh Journal of Science 1824 July pages 3 to 16. On a vein of serpentine, and the alterations it produces on the banks of Carity, near West-Balloch in Forfarshire see Charles Lyell l.c. volume 3 page 43.) It is the latter formation of euphotide which, by its mixture with diorite, is itself linked with hyperthenite, in which real beds of serpentine are sometimes developed in Scotland and in Norway. No volcanic rocks of a more recent period have hitherto been discovered in the island of Cuba; for instance, neither trachytes, dolerites, nor basalts. I know not whether they are found in the rest of the Great Antilles, of which the geologic constitution differs essentially from that of the series of calcareous and volcanic islands which stretch from Trinidad to the Virgin Islands. Earthquakes, which are in general less fatal at Cuba than at Porto Rico and Hayti, are most felt in the eastern part, between Cape Maysi, Santiago de Cuba and La Ciudad de Puerto Principe. Perhaps towards those regions the action of the crevice extends laterally, which is believed to cross the neck of granitic land between Port-au-Prince and Cape Tiburon and on which whole mountains were overthrown in 1770.

The cavernous texture of the limestone formations (soboruco) just described, the great inclination of the shelvings, the smallness of the island, the nakedness of the plains and the proximity of the mountains that form a lofty chain on the southern coast, may be considered as among the principal causes of the want of rivers and the drought which is felt, especially in the western part of Cuba. In this respect, Hayti, Jamaica, and several of the Lesser Antilles, which contain volcanic heights covered with forests, are more favoured by nature. The lands most celebrated for their fertility are the districts of Xagua, Trinidad, Matanzas and Mariel. The valley of Guines owes its reputation to artificial irrigation (sanjas de riego). Notwithstanding the want of great rivers and the unequal fertility of the soil, the island of Cuba, by its undulated surface, its continually renewed verdure, and the distribution of its vegetable forms, presents at every step the most varied and beautiful landscape. Two trees with large, tough, and glossy leaves, the Mammea and the Calophyllum calaba, five species of palm-trees (the palma real, or Oreodoxa regia, the common cocoa-tree, the Cocos crispa, the Corypha miraguama and the C. maritima), and small shrubs constantly loaded with flowers, decorate the hills and the savannahs. The Cecropia peltata marks the humid spots. It would seem as if the whole island had been originally a forest of palm, lemon, and wild orange trees. The latter, which bear a small fruit, are probably anterior to the arrival of Europeans,* who transported thither the agrumi of the gardens; they rarely exceed the height of from ten to fifteen feet. (* The best informed inhabitants of the island assert that the cultivated orange-trees brought from Asia preserve the size and all the properties of their fruits when they become wild. The Brazilians affirm that the small bitter orange which bears the name of loranja do terra and is found wild, far from the habitations of man, is of American origin. Caldcleugh, Travels in South America.) The lemon and orange trees are most frequently separate; and the new planters, in clearing the ground by fire, distinguish the quality of the soil according as it is covered with one or other of those groups of social plants; they prefer the soil of the naranjal to that which produces the small lemon. In a country where the making of sugar is not sufficiently improved to admit of the employment of any other fuel than the bagasse (dried sugar-cane) the progressive destruction of the small woods is a positive calamity. The aridity of the soil augments in proportion as it is stripped of the trees that sheltered it from the heat of the sun; for the leaves, emitting heat under a sky always serene, occasion, as the air cools, a precipitation of aqueous vapours.

83

Among the few rivers worthy of attention, the Rio Guines may be noticed, the Rio Armendaris or Chorrera, of which the waters are led to the Havannah by the Sanja de Antoneli; the Rio Canto on the north of the town of Bayamo; the Rio Maximo which rises on the east of Puerto Principe; the Rio Sagua Grande near Villa Clara; the Rio de las Palmas which issues opposite Cayo Galiado; the small rivers of Jaruco and Santa Cruz between Guanabo and Matanzas, navigable at the distance of some miles from their mouths and favourable for the shipment of sugar-casks; the Rio San Antonio which, like many others, is engulfed in the caverns of limestone rocks; the Rio Guaurabo west of the port of Trinidad; and the Rio Galafre in the fertile district of Filipinas, which throws itself into the Laguna de Cortez. The most abundant springs rise on the southern coast where, from Xagua to Punta de Sabina, over a length of forty-six leagues, the soil is extremely marshy. So great is the abundance of the waters which filter by the clefts of the stratified rock that, from the effect of an hydrostatic pressure, fresh water springs far from the coast, and amidst salt water. The jurisdiction of the Havannah is not the most fertile part of the island; and the few sugar-plantations that existed in the vicinity of the capital are now converted into farms for cattle (potreros) and fields of maize and forage, of which the profits are considerable. The agriculturists of the island of Cuba distinguish two kinds of earth, often mixed together like the squares of a draught-board, black earth (negra o prieta), clayey and full of moisture, and red earth (bermeja), more silicious and containing oxide of iron. The tierra negra is generally preferred (on account of its best preserving humidity) for the cultivation of the sugarcane, and the tierra bermeja for coffee; but many sugar plantations are established on the red soil.

The climate of the Havannah is in accordance with the extreme limits of the torrid zone: it is a tropical climate, in which a more unequal distribution of heat at different parts of the year denotes the passage to the climates of the temperate zone. Calcutta (latitude 22 degrees 34 minutes north), Canton (latitude 23 degrees 8 minutes north), Macao (latitude 22 degrees 12 minutes north), the Havannah (latitude 23 degrees 9 minutes north) and Rio Janeiro (latitude 22 degrees 54 minutes south) are places which, from their position at the level of the ocean near the tropics of Cancer and Capricorn, consequently at an equal distance from the equator, afford great facilities for the study of meteorology. This study can only advance by the determination of certain numerical elements which are the indispensable basis of the laws we seek to discover. The aspect of vegetation being identical near the limits of the torrid zone and at the equator, we are accustomed to confound vaguely the climates of two zones comprised between 0 and 10 degrees, and between 15 and 23 degrees of latitude. The region of palm-trees, bananas and arborescent gramina extends far beyond the two tropics: but it would be dangerous to apply what has been observed at the extremity of the tropical zone to what may take place in the plains near the equator. In order to rectify those errors it is important that the mean temperature of the year and months be well known, as also the thermometric oscillations in different seasons at the parallel of the Havannah; and to prove by an exact comparison with other points alike distant from the equator, for instance, with Rio Janeiro and Macao, that the lowering of temperature observed in the island of Cuba is owing to the irruption and the stream of layers of cold air, borne from the temperate zones towards the tropics of Cancer and Capricorn. The mean temperature of the Havannah, according to four years of good observations, is 25.7 degrees (20.6 degrees R.), only 2 degrees centigrade above that of the regions of America nearest the equator. The proximity of the sea raises the mean temperature of the year on the coast; but in the interior of the island, when the north winds penetrate with the same force, and where the soil rises to the height of forty toises, the mean temperature attains only 23 degrees (18.4 degrees R.) and does not exceed that of Cairo and Lower Egypt. The difference between the mean temperature of the hottest and coldest months rises to 12 degrees in the interior of the island; at the Havannah and on the coast, to 8 degrees; at Cumana, to scarcely 3 degrees. The hottest months, July and August, attain 28.8 degrees, at the island of Cuba, perhaps 29.5 degrees of mean temperature, as at the equator. The coldest months are December and January; their mean temperature in the interior of the island, is 17 degrees; at the Havannah, 21 degrees, that is, 5 to 8 degrees below the same months at the equator, yet still 3 degrees above the hottest month at Paris.

It will be interesting to compare the climate of the Havannah with that of Macao and Rio Janeiro; two places, one of which is near the limit of the northern torrid zone, on the eastern coast of Asia; and the other on the eastern coast of America, towards the extremity of the southern torrid zone.

The climate of the Havannah, notwithstanding the frequency of the north and north-west winds, is hotter than that of Macao and Rio Janeiro. The former partakes of the cold which, owing to the frequency of the west winds, is felt in winter along all the eastern coast of a great continent. The proximity of spaces of land covered with mountains and table-lands renders the distribution of heat in different months of the year more unequal at Macao and Canton than in an island bounded on the west and north by the hot waters of the Gulf-stream. The winters are therefore much colder at Canton and Macao than at the Havannah: yet the latitude of Macao is 1 degree more southerly than that of the Havannah; and the latter town and Canton are, within nearly a minute, on the same parallel. The thermometer at Canton has sometimes almost reached the point zero; and by the effect of reflection, ice has been found on the terraces of houses. Although this great cold never lasts more than one day, the English merchants residing at Canton like to make chimney-fires in their apartments from November to January; while at the Havannah, the artificial warmth even of a brazero is not required. Hail is frequent and the hail-stones are extremely large in the Asiatic climate of Canton and Macao, while it is scarcely seen once in fifteen years at the Havannah. In these three places the thermometer sometimes keeps up for several hours between 0 and 4 degrees (centigrade); and yet (a circumstance which appears to be very remarkable) snow has never been seen to fall; and notwithstanding the great lowering of the temperature, the bananas and the palm-trees are as beautiful around Canton, Macao and the Havannah as in the plains nearest the equator.

In the island of Cuba the lowering of the temperature lasts only during intervals of such short duration that in general neither the banana, the sugar-cane nor other productions of the torrid zone suffer much. We know how well plants of vigorous organization resist temporary cold, and that the orange trees of Genoa survive the fall of snow and endure cold which does not more than exceed 6 or 7 degrees below freezing-point. As the vegetation of the island of Cuba bears the character of the vegetation of the regions near the equator, we are surprised to find even in the plains a vegetable form of the temperate climates and mountains of the equatorial part of Mexico. I have often directed the attention of botanists to this extraordinary phenomenon in the geography of plants. The pine (Pinus occidentalis) is not found in the Lesser Antilles; not even in Jamaica (between 17 3/4 and 18 1/2 degrees of latitude). It is only seen further north, in the mountains of San Domingo, and in all that part of the island of Cuba situated between 20 and 23 degrees of latitude. It attains a height of from sixty to seventy feet; and it is remarkable that the cahoba* (mahogany (* Swieteinia Mahogani, Linn.)) and the pine vegetate at the island of Pinos in the same plains. We also find pines in the south eastern part of the island of Cuba, on the declivity of the Copper Mountains where the soil is barren and sandy. The interior table-land of Mexico is covered with the same species of coniferous plants; at least the specimens brought by M. Bonpland and myself from Acaguisotla, Nevado de Toluca and Cofre de Perote do not appear to differ specifically from the Pinus occidentalis of the West India Islands described by Schwartz. Now those pines which we see at sea level in the island of Cuba, in 20 and 22 degrees of latitude, and which belong only to the southern part of that island, do not descend on the Mexican continent between the parallels of 17 1/2 and 19 1/2 degrees, below the elevation of 500 toises. I even observed that, on the road from Perote to Xalapa in the eastern mountains opposite to the island of Cuba, the limit of the pines is 935 toises; while in the western mountains, between Chilpanzingo and Acapulco, near Quasiniquilapa, two degrees further south, it is 580 toises and perhaps on some points 450. These anomalies of stations are very rare in the torrid zone and are probably less connected with the temperature than with the nature of the soil. In the system of the migration of plants we must suppose that the Pinus occidentalis of Cuba came from Yucatan before the opening of the channel between Cape Catoche and Cape San Antonio, and not from the United States, so rich in coniferous plants; for in Florida the species of which we have here traced the botanical geography has not been discovered.

About the end of April, M. Bonpland and myself, having completed the observations we proposed to make at the northern extremity of the torrid zone, were on the point of proceeding to Vera Cruz with the squadron of Admiral Ariztizabal; but being misled by false intelligence respecting the expedition of Captain Baudin, we were induced to relinquish the project of passing through Mexico on our way to the Philippine Islands. The public journals announced that two French sloops, the Geographe and Naturaliste, had sailed for Cape Horn; that they were to proceed along the coasts of Chili and Peru, and thence to New Holland. This intelligence revived in my mind all the projects I had formed during my stay in Paris, when I solicited the Directory to hasten the departure of Captain Baudin. On leaving Spain, I had promised to rejoin the expedition wherever I could reach it. M. Bonpland and I resolved instantly to divide our herbals into three portions, to avoid exposing to the risks of a long voyage the objects we had obtained with so much difficulty on the banks of the Orinoco, the Atabapo and the Rio Negro. We sent one collection by way of England to Germany, another by way of Cadiz to France, and a third remained at the Havannah. We had reason to congratulate ourselves on this foresight: each collection contained nearly the same species, and no precautions were neglected to have the cases, if taken by English or French vessels, remitted to Sir Joseph Banks or to the professors of natural history at the Museum at Paris. It happened fortunately that the manuscripts which I at first intended to send with the collection to Cadiz were not intrusted to our much esteemed friend and fellow traveller, Fray Juan Gonzales, of the order of the Observance of St. Francis, who had followed us to the Havannah with the view of returning to Spain. He left the island of Cuba soon after us, but the vessel in which he sailed foundered on the coast of Africa, and the cargo and crew were all lost. By this event we lost some of the duplicates of our herbals, and what was more important, all the insects which M. Bonpland had with great difficulty collected during our voyage to the Orinoco and the Rio Negro. By a singular fatality, we remained two years in the Spanish colonies without receiving a single letter from Europe; and those which arrived in the three following years made no mention of what we had transmitted. The reader may imagine my uneasiness for the fate of a journal which contained astronomical observations and barometrical measurements, of which I had not made any copy. After having visited New Grenada, Peru and Mexico, and just when I was preparing to leave the New Continent, I happened, at a public library of Philadelphia, to cast my eyes on a scientific Publication, in which I found these words: "Arrival of M. de Humboldt's manuscripts at his brother's house in Paris, by way of Spain!" I could scarcely suppress an exclamation of joy.

While M. Bonpland laboured day and night to divide and put our collections in order, a thousand obstacles arose to impede our departure. There was no vessel in the port of the Havannah that would convey us to Porto Bello or Carthagena. The persons I consulted seemed to take pleasure in exaggerating the difficulties of the passage of the isthmus, and the dangerous voyage from Panama to Guyaquil, and from Guyaquil to Lima and Valparaiso. Not being able to find a passage in any neutral vessel, I freighted a Catalonian sloop, lying at Batabano, which was to be at my disposal to take me either to Porto Bello or Carthagena, according as the gales of Saint Martha might permit.* (* The gales of Saint Martha blow with great violence at that season below latitude 12 degrees.) The prosperous state of commerce at the Havannah and the multiplied connections of that city with the ports of the Pacific would facilitate for me the means of procuring funds for several years. General Don Gonzalo O'Farrill resided at that time in my native country as minister of the court of Spain. I could exchange my revenues in Prussia for a part of his at the island of Cuba; and the family of Don Ygnacio O'Farrill y Herera, brother of the general, concurred kindly in all that could favour my new projects. On the 6th of March the vessel I had freighted was ready to receive us. The road to Batabano led us once more by Guines to the plantation of Rio Blanco, the property of Count Jaruco y Mopox.

The road from Rio Blanco to Batabano runs across an uncultivated country, half covered with forests; in the open spots the indigo plant and the cotton-tree grow wild. As the capsule of the Gossypium opens at the season when the northern storms are most frequent, the down that envelops the seed is swept from one side to the other; and the gathering of the cotton, which is of a very fine quality, suffers greatly. Several of our friends, among whom was Senor de Mendoza, captain of the port of Valparaiso, and brother to the celebrated astronomer who resided so long in London, accompanied us to Potrero de Mopox. In herborizing further southward, we found a new palm-tree with fan-leaves (Corypha maritima), having a free thread between the interstices of the folioles. This Corypha covers a part of the southern coast and takes the place of the majestic palma real and the Cocos crispa of the northern coast. Porous limestone (of the Jura formation) appeared from time to time in the plain.

Batabano was then a poor village and its church had been completed only a few years previously. The Sienega begins at the distance of half a league from the village; it is a tract of marshy soil, extending from the Laguna de Cortez as far as the mouth of the Rio Xagua, on a length of sixty leagues from west to east. At Batabano it is believed that in those regions the sea continues to gain upon the land, and that the oceanic irruption was particularly remarkable at the period of the great upheaving which took place at the end of the eighteenth century, when the tobacco mills disappeared, and the Rio Chorrera changed its course. Nothing can be more gloomy than the aspect of these marshes around Batabano. Not a shrub breaks the monotony of the prospect: a few stunted trunks of palm-trees rise like broken masts, amidst great tufts of Junceae and Irides. As we stayed only one night at Batabano, I regretted much that I was unable to obtain precise information relative to the two species of crocodiles which infest the Sienega. The inhabitants give to one of these animals the name of cayman, to the other that of crocodile; or, as they say commonly in Spain, of cocodrilo. They assured us that the latter has most agility, and measures most in height: his snout is more pointed than that of the cayman, and they are never found together. The crocodile is very courageous and is said to climb into boats when he can find a support for his tail. He frequently wanders to the distance of a league from the Rio Cauto and the marshy coast of Xagua to devour the pigs on the islands. This animal is sometimes fifteen feet long, and will, it is said, pursue a man on horseback, like the wolves in Europe; while the animals exclusively called caymans at Batabano are so timid that people bathe without apprehension in places where they live in bands. These peculiarities, and the name of cocodrilo, given at the island of Cuba, to the most dangerous of the carnivorous reptiles, appear to me to indicate a different species from the great animals of the Orinoco, Rio Magdalena and Saint Domingo. In other parts of the Spanish American continent the settlers, deceived by the exaggerated accounts of the ferocity of crocodiles in Egypt, allege that the real crocodile is only found in the Nile. Zoologists have, however, ascertained that there are in America caymans or alligators with obtuse snouts, and legs not indented, and crocodiles with pointed snouts and indented legs; and in the old continent, both crocodiles and gaviales. The Crocodilus acutus of San Domingo, in which I cannot hitherto specifically distinguish the crocodiles of the great rivers of the Orinoco and the Magdalena, has, according to Cuvier, so great a resemblance to the crocodile of the Nile,* that it required a minute examination to prove that the rule laid down by Buffon relative to the distribution of species between the tropical regions of the two continents was correct (* This striking analogy was ascertained by M. Geoffroy de Saint Hilaire in 1803 when General Rochambeau sent a crocodile from San Domingo to the Museum of Natural History at Paris. M. Bonpland and myself had made drawings and detailed descriptions in 1801 and 1802 of the same species which inhabit the great rivers of South America, during our passage on the Apure, the Orinoco and the Magdalena. We committed the mistake so common to travellers, of not sending them at once to Europe, together with some young specimens.)

On my second visit to the Havannah, in 1804, I could not return to the Sienega of Batabano; and therefore I had the two species, called caymans and crocodiles by the inhabitants, brought to me, at a great expense. Two crocodiles arrived alive; the oldest was four feet three inches long; they had been caught with great difficulty and were conveyed, muzzled and bound, on a mule, for they were exceedingly vigorous and fierce. In order to observe their habits and movements,* we placed them in a great hall, where, by climbing on a very high piece of furniture, we could see them attack great dogs. (* M. Descourtils, who knows the habits of the crocodile better than any other author who has written on that reptile, saw, like Dampier and myself, the Crocodilus acutus often touch his tail with his mouth.) Having seen much of crocodiles during six months, on the Orinoco, the Rio Apure and the Magdalena, we were glad to have another opportunity of observing their habits before our return to Europe. The animals sent to us from Batabano had the snout nearly as sharp as the crocodiles of the Orinoco and the Magdalena (Crocodilus acutus, Cuv.); their colour was dark-green on the back, and white below the belly, with yellow spots on the flanks. I counted, as in all the real crocodiles, thirty-eight teeth in the upper jaw, and thirty in the lower; in the former, the tenth and ninth; and in the latter, the first and fourth, were the largest. In the description made by M. Bonpland and myself on the spot, we have expressly marked that the lower fourth tooth rises over the upper jaw. The posterior extremities were palmated. These crocodiles of Batabano appeared to us to be specifically identical with the Crocodilus acutus. It is true that the accounts we heard of their habits did not quite agree with what we had ourselves observed on the Orinoco; but carnivorous reptiles of the same species are milder and more timid, or fiercer and more courageous, in the same river, according to the nature of the localities. The animal called the cayman, at Batabano, died on the way, and was not brought to us, so that we could make no comparison of the two species.* (* The four bags filled with musk (bolzas del almizcle) are, in the crocodile of Batabano, exactly in the same position as in that of the Rio Magdalena, beneath the lower jaw and near the anus. I was much surprised at not perceiving the smell of musk at the Havannah, three days after the death of the animal, in a temperature of 30 degrees, while at Mompox, on the banks of the Magdalena, living crocodiles infected our apartment. I have since found that Dampier also remarked an absence of smell in the crocodile of Cuba where the caymans spread a very strong smell of musk.) I have no doubt that the crocodile with a sharp snout, and the alligator or cayman with a snout like a pike,* (* Crocodilus acutus of San Domingo. Alligator lucius of Florida and the Mississippi.) inhabit together, but in distinct bands, the marshy coast between Xagua, the Surgidero of Batabano, and the island of Pinos. In that island Dampier was struck with the great difference between the caymans and the American crocodiles. After having described, though not always with perfect correctness, several of the characteristics which distinguish crocodiles from caymans, he traces the geographical distribution of those enormous saurians. "In the bay of Campeachy," he says, "I saw only caymans or alligators; at the island of Great Cayman, there are crocodiles and no alligators; at the island of Pinos, and in the innumerable creeks of the coast of Cuba, there are both crocodiles and caymans."* (* Dampier's Voyages and Descriptions, 1599.) To these valuable observations of Dampier I may add that the real crocodile (Crocodilus acutus) is found in the West India Islands nearest the mainland, for instance, at the island of Trinidad; at Marguerita; and also, probably, at Curacao, notwithstanding the want of fresh water. It is observed, further south, in the Neveri, the Rio Magdalena, the Apure and the Orinoco, as far as the confluence of the Cassiquiare with the Rio Negro (latitude 2 degrees 2 minutes), consequently more than four hundred leagues from Batabano. It would be interesting to verify on the eastern coast of Mexico and Guatimala, between the Mississippi and the Rio Chagres (in the isthmus of Panama), the limit of the different species of carnivorous reptiles.

We set sail on the 9th of March, somewhat incommoded by the extreme smallness of our vessel, which afforded us no sleeping-place but upon deck. The cabin (camera de pozo) received no air or light but from above; it was merely a hold for provisions, and it was with difficulty that we could place our instruments in it. The thermometer kept up constantly at 32 and 33 degrees (centesimal.) Luckily these inconveniences lasted only twenty days. Our several voyages in the canoes of the Orinoco, and a passage in an American vessel laden with several thousand arrobas of salt meat dried in the sun had rendered us not very fastidious.

The gulf of Batabano, bounded by a low and marshy coast, looks like a vast desert. The fishing birds, which are generally at their post whilst the small land birds, and the indolent vultures (Vultur aura.) are at roost, are seen only in small numbers. The sea is of a greenish-brown hue, as in some of the lakes of Switzerland; while the air, owing to its extreme purity, had, at the moment the sun appeared above the horizon, a cold tint of pale blue, similar to that which landscape painters observe at the same hour in the south of Italy, and which makes distant objects stand out in strong relief. Our sloop was the only vessel in the gulf; for the roadstead of Batabano is scarcely visited except by smugglers, or, as they are here politely called, the traders (los tratantes). The projected canal of Guines will render

Batabano an important point of communication between the island of Cuba and the coast of Venezuela. The port is within a bay bounded by Punta Gorda on the east, and by Punta de Salinas on the west: but this bay is itself only the upper or concave end of a great gulf measuring nearly fourteen leagues from south to north, and along an extent of fifty leagues (between the Laguna de Cortez and the Cayo de Piedras) inclosed by an incalculable number of flats and chains of rocks. One great island only, of which the superficies is more than four times the dimensions of that of Martinique, with mountains crowned with majestic pines, rises amidst this labyrinth. This is the island of Pinos, called by Columbus El Evangelista, and by some mariners of the sixteenth century, the Isla de Santa Maria. It is celebrated for its mahogany (Swietenia mahogoni) which is an important article of commerce. We sailed east-south-east, taking the passage of Don Cristoval, to reach the rocky island of Cayo de Piedras, and to clear the archipelago, which the Spanish pilots, in the early times of the conquest, designated by the names of Gardens and Bowers (Jardines y Jardinillos). The Queen's Gardens, properly so called, are nearer Cape Cruz, and are separated from the archipelago by an open sea thirty-five leagues broad. Columbus gave them the name they bear, in 1494, when, on his second voyage, he struggled during fifty-eight days with the winds and currents between the island of Pinos and the eastern cape of Cuba. He describes the islands of this archipelago as verdant, full of trees and pleasant* (verdes, llenos de arboledas, y graciosos). (* There exists great geographical confusion, even at the Havannah, in reference to the ancient denominations of the Jardines del Rey and Jardines de la Reyna. In the description of the island of Cuba, given in the Mercurio Americano, and in the Historia Natural de la Isla de Cuba, published at the Havannah by Don Antonio Lopez Gomez, the two groups are placed on the southern coast of the island. Lopez says that the Jardines del Rey extend from the Laguna de Cortez to Bahia de Xagua; but it is historically certain that the governor Diego Velasquez gave his name to the western part of the chain of rocks of the Old Channel, between Cayo Frances and Le Monillo, on the northern coast of the island of Cuba. The Jardines de la Reyna, situated between Cabo Cruz and the port of the Trinity, are in no manner connected with the Jardines and Jardinillos of the Isla de Pinos. Between the two groups of the chain of rocks are the flats (placeres) of La Paz and Xagua.)

A part of these so-styled gardens is indeed beautiful; the voyager sees the scene change every moment, and the verdure of some of the islands appears the more lovely from its contrast with chains of rocks, displaying only white and barren sands. The surface of these sands, heated by the rays of the sun, seems to be undulating like the surface of a liquid. The contact of layers of air of unequal temperature produces the most varied phenomena of suspension and mirage from ten in the morning till four in the afternoon. Even in those desert places the sun animates the landscape, and gives mobility to the sandy plain, to the trunks of trees, and to the rocks that project into the sea like promontories. When the sun appears these inert masses seem suspended in air; and on the neighbouring beach the sands present the appearance of a sheet of water gently agitated by the winds. A train of clouds suffices to seat the trunks of trees and the suspended rocks again on the soil; to render the undulating surface of the plains motionless; and to dissipate the charm which the Arabian, Persian, and Hindoo poets have celebrated as "the sweet illusions of the solitary desert."

We doubled Cape Matahambre very slowly. The chronometer of Louis Berthoud having kept time accurately at the Havannah, I availed myself of this occasion to determine, on this and the following days, the positions of Cayo de Don Cristoval, Cayo Flamenco, Cayo de Diego Perez and Cayo de Piedras. I also employed myself in examining the influence which the changes at the bottom of the sea produce on its temperature at the surface. Sheltered by so many islands, the surface is calm as a lake of fresh water, and the layers of different depths being distinct and separate, the smallest change indicated by the lead acts on the thermometer. I was surprised to see that on the east of the little Cayo de Don Cristoval the high banks are only distinguished by the milky colour of the water, like the bank of Vibora, south of Jamaica, and many other banks, the existence of which I ascertained by means of the thermometer. The bottom of the rock of Batabano is a sand composed of coral detritus; it nourishes sea-weeds which scarcely ever appear on the surface: the water, as I have already observed, is greenish; and the absence of the milky tint is, no doubt, owing to the perfect calm which pervades those regions. Whenever the agitation is propagated to a certain depth, a very fine sand, or a mass of calcareous particles suspended in the water, renders it troubled and milky. There are shallows, however, which are distinguished neither by the colour nor by the low temperature of the waters; and I believe that phenomenon depends on the nature of a hard and rocky bottom, destitute of sand and corals; on the form and declivity of the shelvings; the swiftness of the currents; and the absence of the propagation of motion towards the lower layers of the water. The cold frequently indicated by the thermometer, at the surface of the high banks, must be traced to the molecules of water which, owing to the rays of heat and the nocturnal cooling, fall from the surface to the bottom, and are stopped in their fall by the high banks; and also to the mingling of the layers of very deep water that rise on the shelvings of the banks as on an inclined plane, to mix with the layers of the surface.

Notwithstanding the small size of our bark and the boasted skill of our pilot, we often ran aground. The bottom being soft, there was no danger; but, nevertheless, at sunset, near the pass of Don Cristoval, we preferred to lie at anchor. The first part of the night was beautifully serene: we saw an incalculable number of falling-stars, all following one direction, opposite to that from whence the wind blew in the low regions of the atmosphere. The most absolute solitude prevails in this spot, which, in the time of Columbus, was inhabited and frequented by great numbers of fishermen. The inhabitants of Cuba then employed a small fish to take the great sea turtles; they fastened a long cord to the tail of the reves (the name given by the Spaniards to that species of Echeneis*). (* To the sucet or guaican of the natives of Cuba the Spaniards have given the characteristic name of reves, that is, placed on its back, or reversed. In fact, at first sight, the position of the back and the abdomen is confounded. Anghiera says: Nostrates reversum appellant, quia versus venatur. I examined a remora of the South Sea during the passage from Lima to Acapulco. As he lived a long time out of the water, I tried experiments on the weight he could carry before the blades of the disk loosened from the plank to which the animal was fixed; but I lost that part of my journal. It is doubtless the fear of danger that causes the remora not to loose his hold when he feels that he is pulled by a cord or by the hand of man. The sucet spoken of by Columbus and Martin d'Anghiera was probably the Echeneis naucrates and not the Echeneis remora.) The fisher-fish, formerly employed by the Cubans by means of the flattened disc on his head, furnished with suckers, fixed himself on the shell of the sea-turtle, which is so common in the narrow and winding channels of the Jardinillos. "The reves," says Christopher Columbus, "will sooner suffer himself to be cut in pieces than let go the body to which he adheres." The Indians drew to the shore by the same cord the fisher-fish and the turtle. When Gomara and the learned secretary of the emperor Charles V, Peter Martyr d'Anghiera, promulgated in Europe this fact which they had learnt from the companions of Columbus, it was received as a traveller's tale. There is indeed an air of the marvellous in the recital of d'Anghiera, which begins in these words: Non aliter ac nos canibus gallicis per aequora campi lepores insectamur, incolae [Cubae insulae] venatorio pisce pisces alios capiebant. (Exactly as we follow hares with greyhounds in the fields, so do the natives [of Cuba] take fishes with other fish trained for that purpose). We now know, from the united testimony of Rogers, Dampier and Commerson, that the artifice resorted to in the Jardinillos to catch turtles is employed by the inhabitants of the eastern coast of Africa, near Cape Natal, at Mozambique and at Madagascar. In Egypt, at San Domingo and in the lakes of the valley of Mexico, the method practised for catching ducks was as follows: men, whose heads were covered with great calabashes pierced with holes, hid themselves in the water, and seized the birds by the feet. The Chinese, from the remotest antiquity, have employed the cormorant, a bird of the pelican family, for fishing on the coast: rings are fixed round the bird's neck to prevent him from swallowing his prey and fishing for himself. In the lowest degree of civilization, the sagacity of man is displayed in the stratagems of hunting and fishing: nations who probably never had any communication with each other furnish the most striking analogies in the means they employ in exercising their empire over animals.

Three days elapsed before we could emerge from the labyrinth of Jardines and Jardinillos. At night we lay at anchor; and in the day we visited those islands or chains of rocks which were most easily accessible. As we advanced eastward the sea became less calm and the position of the shoals was marked by water of a milky colour. On the boundary of a sort of gulf between Cayo Flamenco and Cayo de Piedras we found that the temperature of the sea, at its surface, augmented suddenly from 23.5 to 25.8 degrees centigrade. The geologic constitution of the rocky islets that rise around the island of Pinos fixed my attention the more earnestly as I had always rather doubted of the existence of those huge masses of coral which are said to rise from the abyss of the Pacific to the surface of the water. It appeared to me more probable that these enormous masses had some primitive or volcanic rock for a basis, to which they adhered at small depths. The formation, partly compact and lithographic, partly bulbous, of the limestone of Guines, had followed us as far as Batabano. It is somewhat analogous to Jura limestone; and, judging from their external aspect, the Cayman Islands are composed of the same rock. If the mountains of the island of Pinos, which present at the same time (as it is said by the first historians of the conquest) the pineta and palmeta, be visible at the distance of twenty sea leagues, they must attain a height of more than five hundred toises: I have been assured that they also are formed of a limestone altogether similar to that of Guines. From these facts I expected to find the same rock (Jura limestone) in the Jardinillos: but I saw, in the chain of rocks that rises generally five to six inches above the surface of the water, only a fragmentary rock, in which angular pieces of madrepores are cemented by quartzose sand. Sometimes the fragments form a mass of from one to two cubic feet and the grains of quartz so disappear that in several layers one might imagine that the polypi have remained on the spot. The total mass of this chain of rocks appears to me a limestone agglomerate, somewhat analogous to the earthy limestone of the peninsula of Araya, near Cumana, but of much more recent formation. The inequalities of this coral rock are covered by a detritus of shells and madrepores. Whatever rises above the surface of the water is composed of broken pieces, cemented by carbonate of lime, in which grains of quartzose sand are set. Whether rocks formed by polypi still living are found at great depth below this fragmentary rock of coral or whether these polypi are raised on the Jura formation are questions which I am unable to answer. Pilots believe that the sea diminishes in these latitudes, because they see the chain of rocks augment and rise, either by the earth which the waves heave up, or by successive agglutinations. It is not impossible that the enlarging of the channel of Bahama, by which the waters of the Gulf-stream issue, may cause, in the lapse of ages, a slight lowering of the waters south of Cuba, and especially in the gulf of Mexico, the centre of the great current which runs along the shores of the United States, and casts the fruits of tropical plants on the coast of Norway.* (* "The Gulf-stream, between the Bahamas and Florida, is very little wider than Behring's Strait; and yet the water rushing through this passage is of sufficient force and quantity to put the whole Northern Atlantic in motion, and to make its influence be felt in the distant strait of Gibraltar and on the more distant coast of Africa." Quarterly Review February 1818.) The configuration of the coast, the direction, the force and the duration of certain winds and currents, the changes which the barometric heights undergo through the variable predominance of those winds, are causes, the concurrence of which may alter, in a long space of time, and in circumscribed limits of extent and height, the equilibrium of the seas.* (* I do not pretend to explain, by the same causes, the great phenomena of the coast of Sweden, where the sea has, on some points, the appearance of a very unequal lowering of from three to five feet in one hundred years. The great geologist, Leopold von Buch, has imparted new interest to these observations by examining whether it be not rather some parts of the continent of Scandinavia which insensibly heaves up. An analogous supposition was entertained by the inhabitants of Dutch Guiana.) When the coast is so low that the level of the soil, at a league within the island, does not change to extent of a few inches, these swellings and diminution of the waters strike the imagination of the inhabitants.

The Cayo bonito (Pretty Rock), which we first visited, fully merits its name from the richness of its vegetation. Everything denotes that it has been long above the surface of the ocean; and the central part of the Cayo is not more depressed than the banks. On a layer of sand and land shells, five to six inches thick, covered by a fragmentary madreporic rock, rises a forest of mangroves (Rhizophora). From their form and foliage they might at a distance be mistaken for laurel trees. The Avicennia, the Batis, some small Euphorbia and grasses, by the intertwining of their roots, fix the moving sands. But the characteristic distinction of the Flora of these coral islands is the magnificent Tournefortia gnaphalioides of Jacquin, with silvered leaves, which we found here for the first time. This is a social plant and is a shrub from four feet and a half to five feet high. Its flowers emit an agreeable perfume; and it is the ornament of Cayo Flamenco, Cayo Piedras and perhaps of the greater part of the low lands of the Jardinillos. While we were employed in herborizing,* our sailors were searching among the rocks for lobsters. (* We gathered Cenchrus myosuroides, Euphorbia buxifolia, Batis maritima, Iresine obtusifolia, Tournefortia gnaphalioides, Diomedea glabrata, Cakile cubensis, Dolichos miniatus, Parthenium hysterophorus, etc. The last-named plant, which we had previously found in the valley of Caracas and on the temperate table-lands of Mexico, between 470 and 900 toises high, covers the fields of the island of Cuba. It is used by the inhabitants for aromatic baths, and to drive away the fleas which are so numerous in tropical climates. At Cumana the leaves of several species of cassia are employed, on account of their smell, against those annoying insects.) Disappointed at not finding them, they avenged themselves by climbing on the mangroves and making a dreadful slaughter of the young alcatras, grouped in pairs in their nests. This name is given, in Spanish America, to the brown swan-tailed pelican of Buffon. With the want of foresight peculiar to the great pelagic birds, the alcatra builds his nest where several branches of trees unite together. We counted four or five nests on the same trunk of a mangrove. The young birds defended themselves valiantly with their enormous beaks, which are six or seven inches long; the old ones hovered over our heads, making hoarse and plaintive cries. Blood streamed from the tops of the trees, for the sailors were armed with great sticks and cutlasses (machetes). In vain we reproved them for this cruelty. Condemned to long obedience in the solitude of the seas, this class of men feel pleasure in exercising a cruel tyranny over animals when occasion offers. The ground was covered with wounded birds struggling in death. At our arrival a profound calm prevailed in this secluded spot; now, everything seemed to say: Man has passed this way.

The sky was veiled with reddish vapours, which however dispersed in the direction of south-west; we hoped, but in vain, to discern the heights of the island of Pinos. Those spots have a charm in which most parts of the New World are wanting. They are associated with recollections of the greatest names of the Spanish monarchy—those of Christopher Columbus and of Hernan Cortez. It was on the southern coast of the island of Cuba, between the bay of Xagua and the island of Pinos, that the great Spanish Admiral, in his second voyage, saw, with astonishment, "that mysterious king who spoke to his subjects only by signs, and that group of men who wore long white tunics, like the monks of La Merced, whilst the rest of the people were naked." "Columbus in his fourth voyage found in the Jardinillos, great boats filled with Mexican Indians, and laden with the rich productions and merchandise of Yucatan." Misled by his ardent imagination, he thought he had heard from those navigators, "that they came from a country where the men were mounted on horses,* and wore crowns of gold on their heads." (* Compare the Lettera rarissima di Christoforo Colombo, di 7 di Julio, 1503; with the letter of Herrera, dated December 1. Nothing can be more touching and pathetic than the expression of melancholy which prevails in the letter of Columbus, written at Jamaica, and addressed to King Ferdinand and Queen Isabella. I recommend to the notice of those who wish to understand the character of that extraordinary man, the recital of the nocturnal vision, in which he imagined that he heard a celestial voice, in the midst of a tempest, encouraging him by these words: Iddio maravigliosamente fece sonar tuo nome nella terra. Le Indie que sono pa te del mondo cosi ricca, te le ha date per tue; tu le hai repartite dove ti e piaciuto, e ti dette potenzia per farlo. Delli ligamenti del mare Oceano che erano serrati con catene cosi forte, ti dono le chiave, etc. [God marvellously makes thy name resound throughout the world. The Indies, which are so rich a portion of the world, he gives to thee for thyself; thou mayest distribute them in the way thou pleasest, and God gives thee power to do so. Of the shores of the Atlantic, which were closed by such strong chains, he gives thee the key.] This fragment has been handed down to us only in an ancient Italian tradition; for the Spanish original mentioned in the Biblioteca Nautica of Don Antonio Leon has not hitherto been found. I may add a few more lines, characterized by great simplicity, written by the discoverer of the New World: "Your Highness," says Columbus, "may believe me, the globe of the earth is far from being so great as the vulgar admit. I was seven years at your royal court, and during seven years was told that my enterprise was a folly. Now that I have opened the way, tailors and shoemakers ask the privilege of going to discover new lands. Persecuted, forgotten as I am, I never think of Hispaniola and Paria without my eyes being filled with tears. I was twenty years in the service of your Highness; I have not a hair that is not white; and my body is enfeebled. Heaven and earth now mourn for me; all who have pity, truth, and justice, mourn for me (pianga adesso il cielo e pianga per me la terra; pianga per me chi ha carita, verita, giustizia)." Lettera rarissima pages 13, 19, 34, 37.) "Catayo (China), the empire of the Great Khan, and the mouth of the Ganges," appeared to him so near, that he hoped soon to employ two Arabian interpreters, whom he had embarked at Cadiz, in going to America. Other remembrances of the island of Pinos, and the surrounding Gardens, are connected with the conquest of Mexico. When Hernan Cortes was preparing his great expedition, he was wrecked with his Nave Capitana on one of the flats of the Jardinillos. For the space of five days he was believed to be lost, and the valiant Pedro de Alvarado sent (in November 1518) from the port of Carenas* (the Havannah) three vessels in search of him. (* At that period there were two settlements, one at Puerto de Carenas in the ancient Indian province of the Havannah, and the other—the most considerable—in the Villa de San Cristoval de Cuba. These settlements were only united in 1519 when the Puerto de Carenas took the name of San Cristoval de la Habana. "Cortes," says Herrera, "paso a la Villa de San Cristoval que a la sazon estaba en la costa del sur, y despues se paso a la Habana." [Cortes proceeded to the town of San Cristoval, which at that time was on the sea-coast, and afterwards he repaired to the Havannah.]) In February, 1519, Cortes assembled his whole fleet near cape San Antonio, probably on the spot which still bears the name of Ensenada de Cortes, west of Batabano and opposite to the island of Pinos. From thence, believing he should better escape the snares laid for him by the governor, Velasquez, he passed almost clandestinely to the coast of Mexico. Strange vicissitude of events! the empire of Montezuma was shaken by a handful of men who, from the western extremity of the island of Cuba, landed on the coast of Yucatan; and in our days, three centuries later, Yucatan, now a part of the new confederation of the free states of Mexico, has nearly menaced with conquest the western coast of Cuba.

On the morning of the 11th March we visited Cayo Flamenco. I found the latitude 21 degrees 59 minutes 39 seconds. The centre of this island is depressed and only fourteen inches above the surface of the sea. The water here is brackish while in other cayos is quite fresh. The mariners of Cuba attribute this freshness of the water to the action of the sands in filtering sea-water, the same cause which is assigned for the freshness of the lagunes of Venice. But this supposition is not justified by any chemical analogy. The cayos are composed of rocks, and not of sands, and their smallness renders it extremely improbable that the pluvial waters should unite in a permanent lake. Perhaps the fresh water of this chain of rocks comes from the neighbouring coast, from the mountains of Cuba, by the effect of hydrostatic pressure. This would prove a prolongation of the strata of Jura limestone below the sea and a superposition of coral rock on that limestone.* (* Eruptions of fresh water in the sea, near Baiae, Syracuse and Aradus (in Phenicia) were known to the ancients. Strabo lib. 16 page 754. The coral islands that surround Radak, especially the low island of Otdia, furnish also fresh water. Chamisso in Kotzebue's Entdekkungs-Reise volume 3 page 108.)

It is too general a prejudice to consider every source of fresh or salt water to be merely a local phenomenon: currents of water circulate in the interior of lands between strata of rocks of a particular density or nature, at immense distances, like the floods that furrow the surface of the globe. The learned engineer, Don Francisco Le Maur, informed me that in the bay of Xagua, half a degree east of the Jardinillos, there issue in the middle of the sea, springs of fresh water, two leagues and a half from the coast. These springs gush up with such force that they cause an agitation of the water often dangerous for small canoes. Vessels that are not going to Xagua sometimes take in water from these ocean springs and the water is fresher and colder in proportion to the depth whence it is drawn. The manatees, guided by instinct, have discovered this region of fresh waters; and the fishermen who like the flesh of these herbivorous animals,* find them in abundance in the open sea. (* Possibly they subsist upon sea-weed in the ocean, as we saw them feed, on the banks of the Apure and the Orinoco, on several species of Panicum and Oplismenus (camalote?). It appears common enough, on the coast of Tabasco and Honduras, at the mouths of rivers, to find the manatees swimming in the sea, as crocodiles do sometimes. Dampier distinguishes between the fresh-water and the salt-water manatee. (Voyages and Descr. volume 2) Among the Cayos de las doce leguas, east of Xagua, some islands bear the name of Meganos del Manati.)

Half a mile east of Cayo Flamenco we passed close to two rocks on which the waves break furiously. They are the Piedras de Diego Perez (latitude 21 degrees 58 minutes 10 seconds.) The temperature of the sea at its surface lowers at this point to 22.6 degrees centigrade, the depth of the water being only about one fathom. In the evening we went on shore at Cayo de Piedras; two rocks connected together by breakers and lying in the direction of north-north-west to south-south-east. On these rocks which form the eastern extremity of the Jardinillos many vessels are lost, and they are almost destitute of shrubs because shipwrecked crews cut them to make fire-signals. The Cayo de Piedras is extremely precipitous on the side near the sea; and towards the middle there is a small basin of fresh water. We found a block of madrepore in the rock, measuring upwards of three cubic feet. Doubtless this limestone formation, which at a distance resembles Jura limestone, is a fragmentary rock. It would be well if this chain of cayos which surrounds the island of Cuba were examined by geologists with the view of determining what may be attributed to the animals which still work at the bottom of the sea, and what belongs to the real tertiary formations, the age of which may be traced back to the date of the coarse limestone abounding in remains of lithophite coral. In general, that which rises above the waters is only breccia, or aggregate of madreporic fragments cemented by carbonate of lime, broken shells, and sand. It is important to examine, in each of the cayos, on what this breccia reposes; whether it covers edifices of mollusca still living, or those secondary and tertiary rocks, which judging from the remains of coral they contain, seem to be the product of our days. The gypsum of the cayos opposite San Juan de los Remedios, on the northern coast of the island of Cuba, merits great attention. Its age is doubtless more remote than historic times, and no geologist will believe that it is the work of the mollusca of our seas.

From the Cayo de Piedras we could faintly discern in the direction of east-north-east the lofty mountains that rise beyond the bay of Xagua. During the night we again lay at anchor; and next day (12th March), having passed between the northern cape of the Cayo de Piedras and the island of Cuba, we entered a sea free from breakers. Its blue colour (a dark indigo tint) and the heightening of the temperature proved how much the depth of the water had augmented. We tried, under favour of the variable winds on sea and shore, to steer eastward as far as the port of La Trinidad so that we might be less opposed by the north-east winds which then prevail in the open sea, in making the passage to Carthagena, of which the meridian falls between Santiago de Cuba and the bay of Guantanamo. Having passed the marshy coast of Camareos,* (* Here the celebrated philanthropist Bartolomeo de las Casas obtained in 1514 from his friend Velasquez, the governor, a good repartimiento de Indios (grant of land so called). But this he renounced in the same year, from scruples of conscience, during a short stay at Jamaica.) we arrived (latitude 21 degrees 50 minutes) in the meridian of the entrance of the Bahia de Xagua. The longitude the chronometer gave me at this point was almost identical with that since published (in 1821) in the map of the Deposito hidrografico of Madrid.

The port of Xagua is one of the finest but least frequented of the island. "There cannot be another such in the world," is the remark of the Coronista major (Antonio de Herrera). The surveys and plans of defence made by M. Le Maur, at the time of the commission of Count Jaruco, prove that the anchorage of Xagua merits the celebrity it acquired even in the first years of the conquest. The town consists merely of a small group of houses and a fort (castillito.) On the east of Xagua, the mountains (Cerros de San Juan) near the coast, assume an aspect more and more majestic; not from their height, which does not seem to exceed three hundred toises, but from their steepness and general form. The coast, I was told, is so steep that a frigate may approach the mouth of the Rio Guaurabo. When the temperature of the air diminished at night to 23 degrees and the wind blew from the land it brought that delicious odour of flowers and honey which characterizes the shores of the island of Cuba.* (* Cuban wax, which is a very important object of trade, is produced by the bees of Europe (the species Apis, Latr.). Columbus says expressly that in his time the inhabitants of Cuba did not collect wax. The great loaf of that substance which he found in the island in his first voyage, and presented to King Ferdinand in the celebrated audience of Barcelona, was afterwards ascertained to have been brought thither by Mexican barques from Yucatan. It is curious that the wax of melipones was the first production of Mexico that fell into the hands of the Spaniards, in the month of November, 1492.) We sailed along the coast keeping two or three miles distant from land. On the 13th March a little before sunset we were opposite the mouth of the Rio San Juan, so much dreaded by navigators on account of the innumerable quantity of mosquitos and zancudos which fill the atmosphere. It is like the opening of a ravine, in which vessels of heavy burden might enter, but that a shoal (placer) obstructs the passage. Some horary angles gave me the longitude 82 degrees 40 minutes 50 seconds for this port which is frequented by the smugglers of Jamaica and the corsairs of Providence Island. The mountains that command the port scarcely rise to 230 toises. I passed a great part of the night on deck. The coast was dreary and desolate. Not a light announced a fisherman's hut. There is no village between Batabano and Trinidad, a distance of fifty leagues; scarcely are there more than two or three corrales or farm yards, containing hogs or cows. Yet, in the time of Columbus, this territory was inhabited along the shore. When the ground is dug to make wells, or when torrents furrow the surface of the earth in floods, stone hatchets and copper utensils* are often discovered; these are remains of the ancient inhabitants of America. (* Doubtless the copper of Cuba. The abundance of this metal in its native state would naturally induce the Indians of Cuba and Hayti to melt it. Columbus says that there were masses of native copper at Hayti, of the weight of six arrobas; and that the boats of Yucatan, which he met with on the eastern coast of Cuba, carried, among other Mexican merchandize, crucibles to melt copper.)

At sunrise I requested the captain to heave the lead. There was no bottom to be found at sixty fathoms; and the ocean was warmer at its surface than anywhere else; it was at 26.8 degrees; the temperature exceeded 4.2 degrees that which we had found near the breakers of Diego Perez. At the distance of half a mile from the coast, the sea water was not more than 2.5 degrees; we had no opportunity of sounding but the depth of the water had no doubt diminished. On the 14th of March we entered the Rio Guaurabo, one of the two ports of Trinidad de Cuba, to put on shore the practico, or pilot of Batabano, who had steered us across the flats of the Jardinillos, though not without causing us to run aground several times. We also hoped to find a packet-boat (correo maritimo) in this port, which would take us to Carthagena. I landed towards the evening, and placed Borda's azimuth compass and the artificial horizon on the shore for the purpose of observing the passage of some stars by the meridian; but we had scarcely begun our preparations when a party of small traders of the class called pulperos, who had dined on board a foreign ship recently arrived, invited us to accompany them to the town. These good people requested us mount two by two on the same horse; and, as the heat was excessive, we accepted their offer. The distance from the mouth of the Rio Guaurabo to Trinidad is nearly four miles in a north-west direction. The road runs across a plain which seems as if it had been levelled by a long sojourn of the waters. It is covered with vegetation, to which the miraguama, a palm-tree with silvered leaves (which we saw here for the first time), gives a peculiar character.* (* Corypha miraguama. Probably the same species which struck Messrs. John and William Fraser (father and son) in the vicinity of Matanzas. Those two botanists, who introduced a great number of valuable plants to the gardens of Europe, were shipwrecked on their voyage to the Havannah from the United States, and saved themselves with difficulty on the cayos at the entrance of the Old Channel, a few weeks before my departure for Carthagena.) This fertile soil, although of tierra colorada, requires only to be tilled and it would yield fruitful harvests. A very picturesque view opens westward on the Lomas of San Juan, a chain of calcareous mountains from 1800 to 2000 toises high and very steep towards the south. Their bare and barren summits form sometimes round blocks; and here and there rise up in points like horns,* a little inclined. (* Wherever the rock is visible I perceived compact limestone, whitish-grey, partly porous and partly with a smooth fracture, as in the Jura formation.) Notwithstanding the great lowering of the temperature during the season of the Nortes or north winds, snow never falls; and only a hoar-frost (escarcha) is seen on these mountains, as on those of Santiago. This absence of snow is difficult to be explained. In emerging from the forest we perceived a curtain of hills of which the southern slope is covered with houses; this is the town of Trinidad, founded in 1514, by the governor Diego Velasquez, on account of the rich mines of gold which were said to have been discovered in the little valley of Rio Arimao.* (* This river flows towards the east into the Bahia de Xagua.) The streets of Trinidad have all a rapid descent: there, as in most parts of Spanish America, it is complained that the Couquistadores chose very injudiciously the sites for new towns.* (* It is questionable whether the town founded by Velasquez was not situated in the plain and nearer the ports of Casilda and Guaurabo. It has been suggested that the fear of the French, Portuguese and English freebooters led to the selection, even in inland places, of sites on the declivity of mountains, whence, as from a watch-tower, the approach of the enemy could be discerned; but it seems to me that these fears could have had no existence prior to the government of Hernando de Soto. The Havannah was sacked for the first time by French corsairs in 1539.) At the northern extremity is the church of Nuestra Senora de la Popa, a celebrated place of pilgrimage. This point I found to be 700 feet above the level of the sea; it commands a magnificent view of the ocean, the two ports (Puerto Casilda and Boca Guaurabo), a forest of palm-trees and the group of the lofty mountains of San Juan. We were received at the town of Trinidad with the kindest hospitality by Senor Munoz, the Superintendent of the Real Hacienda. I made observations during a great part of the night and found the latitude near the cathedral by the Spica Virginis, alpha of the Centaur, and beta of the Southern Cross, under circumstances not equally favourable, to be 21 degrees 48 minutes 20 seconds. My chronometric longitude was 82 degrees 21 minutes 7 seconds. I was informed at my second visit to the Havannah, in returning from Mexico, that this longitude was nearly identical with that obtained by the captain of a frigate, Don Jose del Rio, who had long resided on that spot; but that he marked the latitude of the town at 21 degrees 42 minutes 40 seconds.

The Lieutenant-Governor (Teniente Governadore) of Trinidad, whose jurisdiction then extended to Villa Clara, Principe and Santo Espiritu, was nephew to the celebrated astronomer Don Antonio Ulloa. He gave us a grand entertainment, at which we met some French emigrants from San Domingo who had brought their talents and industry to Spanish America. The exportation of the sugar of Trinidad, by the registers of the custom-house, did not then exceed 4000 chests.

The advantage of having two ports is often discussed at Trinidad. The distance of the town from Puerto de Casilda and Puerto Guaurabo is nearly equal; yet the expense of transport is greatest in the former port. The Boca del Rio Guaurabo, defended by a new battery, furnishes safe anchorage, although less sheltered than that of Puerto Casilda. Vessels that draw little water or are lightened to pass the bar, can go up the river and approach the town within a mile. The packet-boats (correos) that touch at Trinidad de Cuba prefer, in general, the Rio Guaurabo, where they find safe anchorage without needing a pilot. The Puerto Casilda is more inclosed and goes further back inland but cannot be entered without a pilot, on account of the breakers (arrecifes) and the Mulas and Mulattas. The great mole, constructed with wood, and very useful to commerce, was damaged in discharging pieces of artillery. It is entirely destroyed, and it was undecided whether it would be best to reconstruct it with masonry, according to the project of Don Luis de Bassecourt, or to open the bar of Guaurabo by dredging it. The great disadvantage of Puerto de Casilda is the want of fresh water, which vessels have to procure at the distance of a league.

We passed a very agreeable evening in the house of one of the richest inhabitants, Don Antonio Padron, where we found assembled at a tertulia all the good company of Trinidad. We were again struck with the gaiety and vivacity that distinguish the women of Cuba. These are happy gifts of nature to which the refinements of European civilization might lend additional charms but which, nevertheless, please in their primitive simplicity. We quitted Trinidad on the night of the 15th March. The municipality caused us to be conducted to the mouth of the Rio Guaurabo in a fine carriage lined with old crimson damask; and, to add to our confusion, an ecclesiastic, the poet of the place, habited in a suit of velvet notwithstanding the heat of the climate, celebrated, in a sonnet, our voyage to the Orinoco.

On the road leading to the port we were forcibly struck by a spectacle which our stay of two years in the hottest part of the tropics might have rendered familiar to us; but previously I had nowhere seen such an innumerable quantity of phosphorescent insects.* (* Cocuyo, Elater noctilucus.) The grass that overspread the ground, the branches and foliage of the trees, all shone with that reddish and moveable light which varies in its intensity at the will of the animal by which it is produced. It seemed as though the starry firmament reposed on the savannah. In the hut of the poorest inhabitants of the country, fifteen cocuyos, placed in a calabash pierced with holes, afford sufficient light to search for anything during the night. To shake the calabash forcibly is all that is necessary to excite the animal to increase the intensity of the luminous discs situated on each side of its body. The people of the country remark, with a simple truth of expression, that calabashes filled with cocuyos are lanterns always ready lighted. They are, in fact, only extinguished by the sickness or death of the insects, which are easily fed with a little sugar-cane. A young woman at Trinidad de Cuba told us that during a long and difficult passage from the main land, she always made use of the phosphorescence of the cocuyos, when she gave suck to her child at night; the captain of the ship would allow no other light on board, from the fear of corsairs.

As the breeze freshened in the direction of north-east we sought to avoid the group of the Caymans but the current drove us towards those islands. Sailing to south 1/4 south-east, we gradually lost sight of the palm-covered shore, the hills rising above the town of Trinidad and the lofty mountains of the island of Cuba. There is something solemn in the aspect of land from which the voyager is departing and which he sees sinking by degrees below the horizon of the sea. The interest of this impression was heightened at the period to which I here advert; when Saint Domingo was the centre of great political agitations, and threatened to involve the other islands in one of those sanguinary struggles which reveal to man the ferocity of his nature. These threatened dangers were happily averted; the storm was appeased on the spot which gave it birth; and a free black population, far from troubling the peace of the neighbouring islands, has made some steps in the progress of civilization and has promoted the establishment of good institutions. Porto Rico, Cuba and Jamaica, with 370,000 whites and 885,000 men of colour, surround Hayti, where a population of 900,000 negros and mulattos have been emancipated by their own efforts. The negros, more inclined to cultivate alimentary plants than colonial productions, augment with a rapidity only surpassed by the increase of the population of the United States.

CHAPTER 3.30.
PASSAGE FROM TRINIDAD DE CUBA TO RIO SINU. CARTHAGENA. AIR VOLCANOES OF TURBACO. CANAL OF MAHATES.

On the morning of the 17th of March, we came within sight of the most eastern island of the group of the Lesser Caymans. Comparing the reckoning with the chronometric longitude, I ascertained that the currents had borne us in seventeen hours twenty miles westward. The island is called by the English pilots Cayman-brack, and by the Spanish pilots, Cayman chico oriental. It forms a rocky wall, bare and steep towards the south and south-east. The north and north-west part is low, sandy, and scantily covered with vegetation. The rock is broken into narrow horizontal ledges. From its whiteness and its proximity to the island of Cuba, I supposed it to be of Jura limestone. We approached the eastern extremity of Cayman-brack within the distance of 400 toises. The neighbouring coast is not entirely free from danger and breakers; yet the temperature of the sea had not sensibly diminished at its surface. The chronometer of Louis Berthoud gave me 82 degrees 7 minutes 37 seconds for the longitude of the eastern cape of Cayman-brack. The latitude reduced by the reckoning on the rhumbs of wind at the meridian observation, appeared to me to be 19 degrees 40 minutes 50 seconds.

As long as we were within sight of the rock of Cayman-brack sea-turtles of extraordinary dimensions swam round our vessel. The abundance of these animals led Columbus to give the whole group of the Caymans the name of Penascales de las Tortugas (rocks of the turtles.) Our sailors would have thrown themselves into the water to catch some of these animals; but the numerous sharks that accompany them rendered the attempt too perilous. The sharks fixed their jaws on great iron hooks which were flung to them; these hooks were very sharp and (for want of anzuelos encandenados* (* Fish-hooks with chains.)) they were tied to cords: the sharks were in this manner drawn up half the length of their bodies; and we were surprised to see that those which had their mouths wounded and bleeding continued to seize the bait over and over again during several hours.* (* Vidimus quoque squales, quotiescunque, hamo icti, dimidia parte corporis e fluctibus extrahebantur, cito alvo stercus emittere haud absimile excrementis caninis. Commovebat intestina (ut arbitramur) subitus pavor. Although the form and number of teeth change with age, and the teeth appear successively in the shark genus, I doubt whether Don Antonio Ulloa be correct in stating that the young sharks have two, and the old ones four rows of grinders. These, like many other sea-fish, are easily accustomed to live in fresh water, or in water slightly briny. It is observed that sharks (tiburones) abound of late in the Laguna of Maracaybo, whither they have been attracted by the dead bodies thrown into the water after the frequent battles between the Spanish royalists and the Columbian republicans.) At the sight of these voracious fish the sailors in a Spanish vessel always recollect the local fable of the coast of Venezuela, which describes the benediction of a bishop as having softened the habits of the sharks, which are everywhere else the dread of mariners. Do these wild sharks of the port of La Guayra specifically differ from those which are so formidable in the port of the Havannah? And do the former belong to the group of Emissoles with small sharp teeth, which Cuvier distinguishes from the Melandres, by the name of Musteli?

The wind freshened more and more from the south-east, as we advanced in the direction of Cape Negril and the western extremity of the great bank of La Vibora. We were often forced to diverge from our course; and, on account of the extreme smallness of our vessel, we were almost constantly under water. On the 18th of March at noon we found ourselves in latitude 18 degrees 17 minutes 40 seconds, and in 81 degrees 50 minutes longitude. The horizon, to the height of 50 degrees, was covered with those reddish vapours so common within the tropics, and which never seem to affect the hygrometer at the surface of the globe. We passed fifty miles west of Cape Negril on the south, nearly at the point where several charts indicate an insulated flat of which the position is similar to that of Sancho Pardo, opposite to Cape San Antonio de Cuba. We saw no change in the bottom. It appears that the rocky shoal at a depth of four fathoms, near Cape Negril, has no more existence than the rock (cascabel) itself, long believed to mark the western extremity of La Vibora (Pedro Bank, Portland Rock or la Sola), marking the eastern extremity. On the 19th of March, at four in the afternoon, the muddy colour of the sea denoted that we had reached that part of the bank of La Vibora where we no longer find fifteen, and indeed scarcely nine or ten, fathoms of water. Our chronometric longitude was 81 degrees 3 minutes; and our latitude probably below 17 degrees. I was surprised that, at the noon observation, at 17 degrees 7 minutes of latitude, we yet perceived no change in the colour of the water. Spanish vessels going from Batabano or Trinidad de Cuba to Carthagena, usually pass over the bank of La Vibora, on its western side, at between fifteen and sixteen fathoms water. The dangers of the breakers begin only beyond the meridian 80 degrees 45 minutes west longitude. In passing along the bank on its southern limit, as pilots often do in proceeding from Cumana or other parts of the mainland, to the Great Caymnan or Cape San Antonio, they need not ascend along the rocks, above 16 degrees 47 minutes latitude. Fortunately the currents run on the whole bank to south-west.

Considering La Vibora not as a submerged land, but as a heaved-up part of the surface of the globe, which has not reached the level of the sea, we are struck at finding on this great submarine island, as on the neighbouring land of Jamaica and Cuba, the loftiest heights towards its eastern boundary. In that direction are situated Portland Rock, Pedro Keys and South Key, all surrounded by dangerous breakers. The depth is six or eight fathoms; but, in advancing to the middle of the bank, along the line of the summit, first towards the west and then towards the north-west, the depth becomes successively ten, twelve, sixteen and nineteen fathoms. When we survey on the map the proximity of the high lands of San Domingo, Cuba and Jamaica, in the neighbourhood of the Windward Channel, the position of the island of Navaza and the bank of Hormigas, between Capes Tiburon and Morant; when we trace that chain of successive breakers, from the Vibora, by Baxo Nuevo,

Serranilla, and Quita Sueno, as far as the Mosquito Sound, we cannot but recognize in this system of islands and shoals the almost-continued line of a heaved-up ridge running from north-east to south-west. This ridge, and the old dyke, which link, by the rock of Sancho Pardo, Cape San Antonio to the peninsula of Yucatan, divide the great sea of the West Indies into three partial basins, similar to those observed in the Mediterranean.

The colour of the troubled waters on the shoal of La Vibora has not a milky appearance like the waters in the Jardinillos and on the bank of Bahama; but it is of a dirty grey colour. The striking differences of tint on the bank of Newfoundland, in the archipelago of the Bahama Islands and on La Vibora, the variable quantities of earthy matter suspended in the more or less troubled waters of the soundings, may all be the effects of the variable absorption of the rays of light, contributing to modify to a certain point the temperature of the sea. Where the shoals are 8 to 10 degrees colder at their surface than the surrounding sea, it cannot be surprising that they should produce a local change of climate. A great mass of very cold water, as on the bank of Newfoundland, in the current of the Peruvian shore (between the port of Callao and Punta Parina* (* I found the surface of the Pacific ocean, in the month of October 1802 on the coast of Truxillo, 15.8 degrees centigrade; in the port of Callao, in November, 15.5; between the parallel of Callao and Punta Parina, in December, 19 degrees; and progressively, when the current advanced towards the equator and receded towards the west-north-west, 20.5 and 22.3 degrees)), or in the African current near Cape Verd, have necessarily an influence on the atmosphere that covers the sea, and on the climate of the neighbouring land; but it is less easy to conceive that those slight changes of temperature (for instance, a centesimal degree on the bank of La Vibora) can impart a peculiar character to the atmosphere of the shoals. May not these submarine islands act upon the formation and accumulation of the vesicular vapours in some other way than by cooling the waters of the surface?

Quitting the bank of La Vibora, we passed between the Baxo Nuevo and the light-house of Camboy; and on the 22nd March we passed more than thirty leagues to westward of El Roncador (The Snorer), a name which this shoal has received from the pilots who assert, on the authority of ancient traditions, that a sound like snoring is heard from afar. If such a sound be really heard, it arises, no doubt, from a periodical issuing of air compressed by the waters in a rocky cavern. I have observed the same phenomenon on several coasts, for instance, on the promontories of Teneriffe, in the limestones of the Havannah,* (* Called by the Spanish sailors El Cordonazo de San Francisco.) and in the granite of Lower Peru between Truxillo and Lima. A project was formed at the Canary Islands for placing a machine at the issue of the compressed air and allowing the sea to act as an impelling force. While the autumnal equinox is everywhere dreaded in the sea of the West Indies (except on the coast of Cumana and Caracas), the spring equinox produces no effect on the tranquillity of those tropical regions: a phenomenon almost the inverse of that observable in high latitudes. Since we had quitted La Vibora the weather had been remarkably fine; the colour of the sea was indigo-blue and sometimes violet, owing to the quantity of medusae and eggs of fish (purga de mar) which covered it. Its surface was gently agitated. The thermometer kept up, in the shade, from 26 to 27 degrees; not a cloud arose on the horizon although the wind was constantly north, or north-north-west. I know not whether to attribute to this wind, which cools the higher layers of the atmosphere, and there produces icy crystals, the halos which were formed round the moon two nights successively. The halos were of small dimensions, 45 degrees diameter. I never had an opportunity of seeing and measuring any* of which the diameter had attained 90 degrees. (* In Captain Parry's first voyage halos were measured round the sun and moon, of which the rays were 22 1/2 degrees; 22 degrees 52 minutes; 38 degrees; 46 degrees. North-west Passage, 1821.) The disappearance of one of those lunar halos was followed by the formation of a great black cloud, from which fell some drops of rain; but the sky soon resumed its fixed serenity, and we saw a long series of falling-stars and bolides which moved in one direction and contrary to that of the wind of the lower strata.

On the 23rd March, a comparison of the reckoning with the chronometric longitude, indicated the force of a current bearing towards west-south-west. Its swiftness, in the parallel of 17 degrees, was twenty to twenty-two miles in twenty-four hours. I found the temperature of the sea somewhat diminished; in latitude 12 degrees 35 minutes it was only 25.9 degrees (air 27.0 degrees). During the whole day the firmament exhibited a spectacle which was thought remarkable even by the sailors and which I had observed on a previous occasion (June 13th, 1799). There was a total absence of clouds, even of those light vapours called dry; yet the sun coloured, with a fine rosy tint, the air and the horizon of the sea. Towards night the sea was covered with great bluish clouds; and when they disappeared we saw, at an immense height, fleecy clouds in regular spaces, and ranged in convergent bands. Their direction was from north-north-west to south-south-east, or more exactly, north 20 degrees west, consequently contrary to the direction of the magnetic meridian.

On the 24th March we entered the gulf which is bounded on the east by the coast of Santa Marta, and on the west by Costa Rica; for the mouth of the Magdalena and that of the Rio San Juan de Nicaragua are on the same parallel, nearly 11 degrees latitude. The proximity of the Pacific Ocean, the configuration of the neighbouring lands, the smallness of the isthmus of Panama, the lowering of the soil between the gulf of Papagayo and the port of San Juan de Nicaragua, the vicinity of the snowy mountains of Santa Marta, and many other circumstances too numerous to mention, combine to create a peculiar climate in this gulf. The atmosphere is agitated by violent gales known in winter by the name of the brizotes de Santa Marta. When the wind abates, the currents bear to north-east, and the conflict between the slight breezes (from east and north-east) and the current renders the sea rough and agitated. In calm weather, the vessels going from Carthagena to Rio Sinu, at the mouth of the Atrato and at Portobello, are impeded in their course by the currents of the coast. The heavy or brizote winds, on the contrary, govern the movement of the waters, which they impel in an opposite direction, towards west-south-west. It is the latter movement which Major Rennell, in his great hydrographic work, calls drift; and he distinguishes it from real currents, which are not owing to the local action of the wind, but to differences of level in the surface of the ocean; to the rising and accumulation of waters in very distant latitudes. The observations which I have collected on the force and direction of the winds, on the temperature and rapidity of the currents, on the influence of the seasons, or the variable declination of the sun, have thrown some light on the complicated system of those pelagic floods that furrow the surface of the ocean: but it is less easy to conceive the causes of the change in the movement of the waters at the same season and with the same wind. Why is the Gulf-stream sometimes borne on the coast of Florida, sometimes on the border of the shoal of Bahama? Why do the waters flow, for the space of whole weeks, from the Havannah to Matanzas, and (to cite an example of the corriente por arriba, which is sometimes observed in the most eastern part of the main land during the prevalence of gentle winds) from La Guayra to Cape Codera and Cumana?

As we advanced, on the 25th of March, towards the coast of Darien, the north-east wind increased with violence. We might have imagined ourselves transported to another climate. The sea became very rough during the night yet the temperature of the water kept up (from latitude 10 degrees 30 minutes, to 9 degrees 47 minutes) at 25.8 degrees. We perceived at sunrise a part of the archipelago* of Saint Bernard, which closes the gulf of Morrosquillo on the north. (* It is composed of the islands Mucara, Ceycen, Maravilla, Tintipan, Panda, Palma, Mangles, and Salamanquilla, which rise little above the sea. Several of them have the form of a bastion. There are two passages in the middle of this archipelago, from seventeen to twenty fathoms. Large vessels can pass between the Isla Panda and Tintipan, and between the Isla de Mangles and Palma.) A clear spot between the clouds enabled me to take the horary angles. The chronometer, at the little island of Mucara, gave longitude 78 degrees 13 minutes 54 seconds. We passed on the southern extremity of the Placer de San Bernardo. The waters were milky, although a sounding of twenty-five fathoms did not indicate the bottom; the cooling of the water was not felt, doubtless owing to the rapidity of the current. Above the archipelago of Saint Bernard and Cape Boqueron we saw in the distance the mountains of Tigua. The stormy weather and the difficulty of going up against the wind induced the captain of our frail vessel to seek shelter in the Rio Sinu, or rather, near the Punta del Zapote, situated on the eastern bank of the Ensenada de Cispata, into which flows the river Sinu or the Zenu of the early Conquistadores. It rained with violence, and I availed myself of that occasion to measure the temperature of the rain-water:

100

it was 26.3 degrees, while the thermometer in the air kept up, in a place where the bulb was not wet, at 24.8 degrees. This result differed much from that we had obtained at Cumana, where the rain-water was often a degree colder than the air.* (* As, within the tropics, it takes but little time to collect some inches of water in a vase having a wide opening, and narrowing towards the bottom, I do not think there can be any error in the observation, when the heat of the rain-water differs from that of the air. If the heat of the rain-water be less than that of the air it may be presumed that only a part of the total effect is observed. I often found at Mexico at the end of June, the rain at 19.2 or 19.4 degrees, when the air was at 17.8 and 18 degrees. In general it appeared to me that, within the torrid zone, either at the level of the sea, or on table-lands from 1200 to 1500 toises high, there is no rain but that during storms, which falls in large drops very distant from each other, and is sensibly colder than the air. These drops bring with them, no doubt, the low temperature of the high regions. In the rain which I found hotter than the air, two causes may act simultaneously. Great clouds heat by the absorption of the rays of the sun which strike their surface; and the drops of water in falling cause an evaporation and produce cold in the air. The temperature of rain-water, to which I devoted much attention during my travels, has become a more important problem since M. Boisgiraud, Professor of Experimental Philosophy at Poitiers, has proved that in Europe rain is generally sufficiently cold, relatively to the air, to cause precipitation of vapour at the surface of every drop. From this fact he traces the cause of the unequal quantity of rain collected at different heights. When we recollect that one degree only of cooling precipitates more water in the hot climate of the tropics, than by a temperature of 10 to 13 degrees, we may cease to be surprised at the enormous size of the drops of rain that fall at Cumana, Carthagena and Guayaquil.)

Our passage from the island of Cuba to the coast of South America terminated at the mouth of the Rio Sinu, and it occupied sixteen days. The roadstead near the Punta del Zapote afforded very bad anchorage; and in a rough sea, and with a violent wind, we found some difficulty in reaching the coast in our canoe. Everything denoted that we had entered a wild region rarely visited by strangers. A few scattered houses form the village of Zapote: we found a great number of mariners assembled under a sort of shed, all men of colour, who had descended the Rio Sinu in their barks, to carry maize, bananas, poultry and other provisions to the port of Carthagena. These barks, which are from fifty to eighty feet long, belong for the most part to the planters (haciendados) of Lorica. The value of their largest freight amounts to about 2000 piastres. These boats are flat-bottomed, and cannot keep at sea when it is very rough. The breezes from the north-east had, during ten days, blown with violence on the coast, while, in the open sea, as far as 10 degrees latitude, we had only had slight gales, and a constantly calm sea. In the aerial, as in the pelagic currents, some layers of fluids move with extreme swiftness, while others near them remain almost motionless. The zambos of the Rio Sinu wearied us with idle questions respecting the purpose of our voyage, our books, and the use of our instruments: they regarded us with mistrust; and to escape from their importunate curiosity we went to herborize in the forest, although it rained. They had endeavoured, as usual, to alarm us by stories of boas (traga-venado), vipers and the attacks of jaguars; but during a long residence among the Chayma Indians of the Orinoco we were habituated to these exaggerations, which arise less from the credulity of the natives, than from the pleasure they take in tormenting the whites. Quitting the coast of Zapote, covered with mangroves,* (* Rhizophora mangle.) we entered a forest remarkable for a great variety of palm-trees. We saw the trunks of the Corozo del Sinu* pressed against each other, which formed heretofore our species Alfonsia, yielding oil in abundance (* In Spanish America palm-trees with leaves the most different in kind and species are called Corozo: the Corozo del Sinu, with a short, thick, glossy trunk, is the Elaeis melanococca of Martius, Palm. page 64 tab. 33, 55. I cannot believe it to be identical with the Elaeis guineensis (Herbal of Congo River page 37) since it vegetates spontaneously in the forests of the Rio Sinu. The Corozo of Caripe is slender, small and covered with thorns; it approaches the Cocos aculeata of Jacquin. The Corozo de los Marinos of the valley of Cauca, one of the tallest palm-trees, is the Cocus butyracea of Linnaeus.); the Cocos butyracea, called here palma dolce or palma real, and very different from the palma real of the island of Cuba; the palma amarga, with fan-leaves that serve to cover the roofs of houses, and the latta,* (*

Perhaps of the species of Aiphanes.) resembling the small piritu palm-tree of the Orinoco. This variety of palm-trees was remarked by the first Conquistadores.* (* Pedro de Cieca de Leon, a native of Seville, who travelled in 1531, at the age of thirteen years, in the countries I have described, observes that Las tierras comarcanas del Rio Cenu y del Golfo de Uraba estan llena de unos palmares muy grandes y espessos, que son unos arboles gruessos, y llevan unas ramas como palma de datiles. [The lands adjacent to the Rio Cenu and the Gulf of Uraba are full of very tall, spreading palm-trees. They are of vast size and are branched like the date-palm.] See La Cronica del Peru nuevamenta escrita, Antwerp 1554 pages 21 and 204.) The Alfonsia, or rather the species of Elais, which we had nowhere else seen, is only six feet high, with a very large trunk; and the fecundity of its spathes is such that they contain more than 200,000 flowers. Although a great number of those flowers (one tree bearing 600,000 at the same time) never come to maturity,* the soil remains covered with a thick layer of fruits. (* I have carefully counted how many flowers are contained in a square inch on each amentum, from 100 to 120 of which are found united in one spathe.) We often made a similar observation under the shade of the mauritia palm-tree, the Cocos butyracea, the Seje and the Pihiguao of the Atabapo. No other family of arborescent plants is so prolific in the development of the organs of flowering. The almond of the Corozo del Sinu is peeled in the water. The thick layer of oil that swims in the water is purified by boiling, and yields the butter of Corozo (manteca de Corozo) which is thicker than the oil of the cocoa-tree, and serves to light churches and houses. The palm-trees of the section of Cocoinies of Mr. Brown are the olive-trees of the tropical regions. As we advanced in the forest, we began to find little pathways, looking as though they had been recently cleared out by the hatchet. Their windings displayed a great number of new plants: Mougeotia mollis, Nelsonia albicans, Melampodium paludosum, Jonidium anomalum, Teucrium palustre, Gomphia lucens, and a new kind of Composees, the Spiracantha cornifolia. A fine Pancratium embalmed the air in the humid spots, and almost made us forget that those gloomy and marshy forests are highly dangerous to health.

After an hour's walk we found, in a cleared spot, several inhabitants employed in collecting palm-tree wine. The dark tint of the zambos formed a strong contrast with the appearance of a little man with light hair and a pale complexion who seemed to take no share in the labour. I thought at first that he was a sailor who had escaped from some North American vessel; but I was soon undeceived. This fair-complexioned man was my countryman, born on the coast of the Baltic; he had served in the Danish navy and had lived for several years in the upper part of the Rio Sinu, near Santa Cruz de Lorica. He had come, to use the words of the loungers of the country para ver tierras, y pasear, no mas (to see other lands, and to roam about, nothing else.) The sight of a man who could speak to him of his country seemed to have no attraction for him; and, as he had almost forgotten German without being able to express himself clearly in Spanish, our conversation was not very animated. During the five years of my travels in Spanish America I found only two opportunities of speaking my native language. The first Prussian I met with was a sailor from Memel who served on board a ship from Halifax, and who refused to make himself known till after he had fired some musket-shot at our boat. The second, the man we met at the Rio Sinu, was very amicably disposed. Without answering my questions he continued repeating, with a smile, that the country was hot and humid; that the houses in the town of Pomerania were finer than those of Santa Cruz de Lorica; and that, if we remained in the forest, we should have the tertian fever (calentura) from which he had long suffered. We had some difficulty in testifying our gratitude to this good man for his kind advice; for according to his somewhat aristocratic principles, a white man, were he bare-footed, should never accept money "in the presence of those vile coloured people!" (gente parda). Less disdainful than our European countryman, we saluted politely the group of men of colour who were employed in drawing off into large calabashes, or fruits of the Crescentia cujete, the palm-tree wine from the trunks of felled trees. We asked them to explain to us this operation, which we had already seen practised in the missions of the Cataracts. The vine of the country is the palma dolce, the Cocos butyracea, which, near Malgar, in the valley of the Magdalena, is called the wine palm-tree, and here, on account of its majestic height, the royal palm-tree. After having thrown down the trunk, which diminishes but little towards the top, they make just below the point whence the leaves (fronds) and spathes issue, an excavation in the ligneous part, eighteen inches long, eight broad, and six in depth. They work in the hollow of the tree, as though they were making a canoe; and three days afterwards this cavity is found filled with a yellowish-white juice, very limpid, with a sweet and vinous flavour. The fermentation appears to commence as soon as the trunk falls, but the vessels preserve their vitality; for we saw that the sap flowed even when the summit of the palm-tree (that part whence the leaves sprout out) is a foot higher than the lower end, near the roots. The sap continues to mount as in the arborescent Euphorbia recently cut. During eighteen to twenty days, the palm-tree wine is daily collected; the last is less sweet but more alcoholic and more highly esteemed. One tree yields as much as eighteen bottles of sap, each bottle containing forty-two cubic inches. The natives affirm that the flowing is more abundant when the petioles of the leaves, which remain fixed to the trunk, are burnt.

The great humidity and thickness of the forest forced us to retrace our steps and to gain the shore before sunset. In several places the compact limestone rock, probably of tertiary formation, is visible. A thick layer of clay and mould rendered observation difficult; but a shelf of carburetted and shining slate seemed to me to indicate the presence of more ancient formations. It has been affirmed that coal is to be found on the banks of the Sinu. We met with Zambos carrying on their shoulders the cylinders of palmetto, improperly called the cabbage palm, three feet long and five to six feet thick. The stem of the palm-tree has been for ages an esteemed article of food in those countries. I believe it to be wholesome although historians relate that, when Alonso Lopez de Ayala was governor of Uraba, several Spaniards died after having eaten immoderately of the palmetto, and at the same time drinking a great quantity of water. In comparing the herbaceous and nourishing fibres of the young undeveloped leaves of the palm-trees with the sago of the Mauritia, of which the Indians make bread similar to that of the root of the Jatropha manihot, we involuntarily recollect the striking analogy which modern chemistry has proved to exist between ligneous matter and the amylaceous fecula. We stopped on the shore to collect lichens, opegraphas and a great number of mosses (Boletus, Hydnum, Helvela, Thelephora) that were attached to the mangroves, and there, to my great surprise, vegetating, although moistened by the sea-water.

Before I quit this coast, so seldom visited by travellers and described by no modern voyager, I may here offer some information which I acquired during my stay at Carthagena. The Rio Sinu in its upper course approaches the tributary streams of the Atrato which, to the auriferous and platiniferous province of Choco, is of the same importance as the Magdalena to Cundinamarca, or the Rio Cauca to the provinces of Antioquia and Popayan. The three great rivers here mentioned have heretofore been the only commercial routes, I might almost add, the only channels of communication for the inhabitants. The Rio Atrato receives, at twelve leagues distance from its mouth, the Rio Sucio on the east; the Indian village of San Antonio is situated on its banks. Proceeding upward beyond the Rio Pabarando, you arrive in the valley of Sinu. After several fruitless attempts on the part of the Archbishop Gongora to establish colonies in Darien del Norte and on the eastern coast of the gulf of Uraba, the Viceroy Espeleta recommended the Spanish Government to fix its whole attention on the Rio Sinu; to destroy the colony of Cayman; to fix the planters in the Spanish village of San Bernardo del Viento in the jurisdiction of Lorica; and from that post, which is the most westerly, to push forward the peaceful conquests of agriculture and civilization towards the banks of the Pabarando, the Rio Sucio and the Atrato.* (* I will here state some facts which I obtained from official documents during my stay at Carthagena, and which have not yet been published. In the sixteenth and seventeenth centuries the name of Darien was given vaguely to the whole coast extending from the Rio Damaquiel to the Punta de San Blas, on 2 1/4 degrees of longitude. The cruelties exercised by Pedrarias Davila rendered almost inaccessible to the Spaniards a country which was one of the first they had colonized. The Indians (Dariens and Cunas-Cunas) remained masters of the coast, as they still are at Poyais, in the land of the Mosquitos. Some Scotchmen formed in 1698 the settlements of New Caledonia, New Edinburgh and Scotch Port, in the most eastern part of the isthmus, a little west of Punta Carreto. They were soon driven away by the Spaniards but, as the latter occupied no part of the coast, the Indians continued their attacks against

Choco's boats, which from time to time descended the Rio Atrato, The sanguinary expedition of Don Manuel de Aldarete in 1729 served only to augment the resentment of the natives. A settlement for the cultivation of the cocoa-tree, attempted in the territory of Urabia in 1740 by some French planters under the protection of the Spanish Government, had no durable success; and the court, excited by the reports of the archbishop-viceroy, Gongora, ordered, by the cedule of the 15th August, 1783, either the conversion and conquest, or the destruction (reduccion o extincion) of the Indians of Darien. This order, worthy of another age, was executed by Don Antonio de Arebalo: he experienced little resistance and formed, in 1785, the four settlements and forts of Cayman on the eastern coast of the Gulf of Urabia, Concepcion, Carolina and Mandinga. The Lele, or high-priest of Mandinga, took an oath of fidelity to the King of Spain; but in 1786 the war with the Darien Indians recommenced and was terminated by a treaty concluded July 27th, 1787, between the archbishop-viceroy and the cacique Bernardo. The forts and new colonies, which figured only on the maps sent to Madrid, augmented the debt of the treasury of Santa Fe de Bogota, in 1789, to the sum of 1,200,000 piastres. The viceroy, Gil Lemos, wiser than his predecessor, obtained permission from the Court to abandon Carolina, Concepcion and Mandinga. The settlement of Cayman only was preserved, on account of the navigation of the Atrato, and it was declared free, under the government of the archbishop-viceroy: it was proposed to transfer this settlement to a more healthy spot, that of Uraba; but lieutenant-general Don Antonio Arebalo, having proved that the expense of this removal would amount to the sum of 40,000 piastres, the fort of Cayman was also destroyed, by order of the viceroy Espeleta in 1791, and the planters were compelled to join those of the village of San Bernardo.) The number of independent Indians who inhabit the lands between Uraba, Rio Atrato, Rio Sucio and Rio Sinu was, according to a census made in 1760, at least 1800. They were distributed in three small villages, Suraba, Toanequi and Jaraguia. This population was computed, at the period when I travelled there, to be 3000. The natives, comprehended in the general name of Caymans, live at peace with the inhabitants of San Bernardo del Viento (pueblo de Espanoles), situated on the western bank of the Rio Sinu, lower than San Nicolas de Zispata, and near the mouth of the river. These people have not the ferocity of the Darien and Cunas Indians, on the left bank of the Atrato; who often attack the boats trading with the town of Quidbo in the Choco; they also make incursions on the territory of Uraba, in the months of June and November, to collect the fruit of the cacao-trees. The cacao of Uraba is of excellent quality; and the Darien Indians sometimes come to sell it, with other productions, to the inhabitants of Rio Sinu, entering the valley of that river by one of its tributary streams, the Jaraguai.

It cannot be doubted that the Gulf of Darien was considered, at the beginning of the sixteenth century, as a nook in the country of the Caribs. The word Caribana is still preserved in the name of the eastern cape of that gulf. We know nothing of the languages of the Darien, Cunas and Cayman Indians: and we know not whether Carib or Arowak words are found in their idioms; but it is certain, notwithstanding the testimony of Anghiera on the identity of the race of the Caribs of the Lesser Antilles and the Indians of Uraba, that Pedro de Cieca, who lived so long among the latter, never calls them Caribs nor cannibals. He describes the race of that tribe as being naked with long hair, and going to the neighbouring countries to trade; and says the women are cleanly, well dressed and extremely engaging (amorosas y galanas). "I have not seen," adds the Conquistador, "any women more beautiful* in all the Indian lands I have visited: they have one fault, however, that of having too frequent intercourse with the devil." (* Cronica del Peru pages 21 and 22. The Indians of Darien, Uraba, Zenu (Sinu), Tatabe, the valleys of Nore and of Guaca, the mountains of Abibe and Antioquia, are accused, by the same author, of the most ferocious cannibalism; and perhaps that circumstance alone gives rise to the idea that they were of the same race as the Caribs of the West Indies. In the celebrated Provision Real of the 30th of October, 1503, by which the Spaniards are permitted to make slaves of the anthropophagic Indians of the archipelago of San Bernardo, opposite the mouth of the Rio Sinu, the Isla Fuerte, Isla Bura (Baru) and Carthagena, there is more of a question of morals than of race, and the denomination of Caribs is altogether avoided. Cieca asserts that the natives of the valley of Nore seized the women of neighbouring tribes, in order first to devour the children who were born of the union with foreign wives, and then the women themselves. Foreseeing that this horrible depravity would not be believed, although it had been observed by Columbus in the West Indies, he cites the testimony of Juan de Vadillo, who had observed the same facts and who was still living in 1554 when the Cronica del Peru appeared in Dutch. With respect to the etymology of the word cannibal, it seems to me entirely cleared up by the discovery of the journal kept by Columbus during his first voyage of discovery, and of which Bartholomew de las Casas has left us an abridged copy. Dice mas el Almirante que en las islas passadas estaban con gran temor de carib: y en algunas los llamaban caniba; pero en la Espanola carib y son gente arriscada, pues andan por todas estas islas y comen la gente que pueden haber. [And the Admiral moreover says that in the islands they passed, great apprehension was entertained on account of the caribs. Some call them canibas; but in Spanish they are called caribs. They are a very bold people, and they travel about these islands, and devour all the persons whom they capture.] Navarete tome 1 page 135. In this primitive form of words it is easy to perceive that the permutation of the letters r and n, resulting from the imperfection of the organs in some nations, might change carib into canib, or caniba. Geraldini who, according to the tendency of that age, sought, like Cardinal Bembo, to latinize all barbarous denominations, recognizes in the Cannibals the manners of dogs (canes) just as St. Louis desired to send the Tartars ad suas tartareas sedes unde exierint.)

The Rio Sinu, owing to its position and its fertility, is of the highest importance for provisioning Carthagena. In time of war the enemy usually stationed their ships between the Morro de Tigua and the Boca de Matunilla, to intercept barques laden with provisions. In that station they were, however, sometimes exposed to the attack of the gun-boats of Carthagena: these gun-boats can pass through the channel of Pasacaballos which, near Saint Anne, separates the isle of Baru from the continent. Lorica has, since the sixteenth century, been the principal town of Rio Sinu; but its population which, in 1778, under the government of Don Juan Diaz Pimienta, amounted to 4000 souls, has considerably diminished, because nothing has been done to secure the town from inundations and the deleterious miasmata they produce.

The gold-washings of the Rio Sinu, heretofore so important above all, between its source and the village of San Geronimo, have almost entirely ceased, as well as those of Cienega de Tolu, Uraba and all the rivers descending from the mountains of Abibe. "The Darien and the Zenu," says the bachelor Enciso in his geographical work published at the beginning of the sixteenth century, "is a country so rich in gold pepites that, in the running waters, that metal can be fished with nets." Excited by these narratives, the governor Pedrarias sent his lieutenant, Francisco Becerra, in 1515, to the Rio Sinu. This expedition was most unfortunate for Becerra and his troop were massacred by the natives, of whom the Spaniards, according to the custom of the time, had carried away great numbers to be sold as slaves in the West Indies. The province of Antioquia now furnishes, in its auriferous veins, a vast field for mining speculations; but it might be well worth while to relinquish gold-washings for the cultivation of colonial productions in the fertile lands of Sinu, the Rio Damaquiel, the Uraba and the Darien del Norte; above all, that of cacao, which is of a superior quality. The proximity of the port of Carthagena would also render the neglected cultivation of cinchona an object of great importance to European trade. That precious tree vegetates at the source of the Rio Sinu, as in the mountains of Abibe and Maria. The real febrifuge cinchona, with a hairy corolla, is nowhere else found so near the coast, if we except the Sierra Nevada of Santa Marta.

The Rio Sinu and the Gulf of Darien were not visited by Columbus. The most eastern point at which that great man touched land, on the 26th November, 1503, is the Puerto do Retreto, now called Punta de Escribanos, near the Punta of San Blas, in the isthmus of Panama. Two years previously, Rodrigo de Bastidas and Alanso do Ojeda, accompanied by Amerigo Vespucci, had discovered the whole coast of the main land, from the Gulf of Maracaybo as far as the Puerto do Retreto. Having often had occasion in the preceding volumes to speak of New Andalusia, I may here mention that I found that denomination, for the first time, in the convention made by Alonso de Ojeda with the Conquistador Diego de Sicuessa, a powerful man, say the historians of his time, because he was a flattering courtier and a wit. In 1508 all the country from the Cabo de la Vela to the Gulf of Uraba, where the Castillo del Oro begins, was called New Andalusia, a name since restricted to the province of Cumana.

A fortunate chance led me to see, during the course of my travels, the two extremities of the main land, the mountainous and verdant coast of Paria, which Columbus supposes to have been the cradle of the human race, and the low and humid coast extending from the mouth of the Sinu towards the Gulf of Darien. The comparison of these scenes, which have again relapsed into a savage state, confirms what I have elsewhere advanced relative to the strange and sometimes retrograde nature of civilization in America. On one side, the coast of Paria, the islands of Cubagua and Marguerita; on the other, the Gulf of Uraba and Darien, received the first Spanish colonists. Gold and pearls, which were there found in abundance, because from time immemorial they had been accumulated in the hands of the natives, gave those countries a popular celebrity from the beginning of the sixteenth century. At Seville, Toledo, Pisa, Genoa and Antwerp those countries were viewed like the realms of Ormuz and of Ind. The pontiffs of Rome mentioned them in their bulls; and Bembo has celebrated them in those historical pages which add lustre to the glory of Venice.

At the close of the fifteenth, and the beginning of the sixteenth century, Europe saw, in those parts of the New World discovered by Columbus, Ojeda, Vespucci and Rodrigo de Bastidas, only the advanced capes of the vast territories of India and eastern Asia. The immense wealth of those territories in gold, diamonds, pearls and spices had been vaunted in the narratives of Benjamin de Tudela, Rubruquis, Marco Polo and Mandeville. Columbus, whose imagination was excited by these narrations, caused a deposition to be made before a notary, on the 12th of June, 1494, in which sixty of his companions, pilots, sailors and passengers certified upon oath that the southern coast of Cuba was a part of the continent of India. The description of the treasures of Cathay and Cipango, of the celestial town of Quinsay and the province of Mango, which had fired the admiral's ambition in early life, pursued him like phantoms in his declining days. In his fourth and last voyage, on approaching the coast of Cariay (Poyais or Mosquito Coast), Veragua and the Isthmus, he believed himself to be near the mouth of the Ganges.* (* Tambien dicen que la mar baxa a Ciguare, y de alli a diez jornadas es el Rio de Guangues: para que estas tierras estan con Veragua como Tortosa con Fuenterabia o Pisa con Venecia." [Also it is said that the sea lowers at Ciguara, and from thence it is a ten days' journey to the river Ganges; for these lands are, with reference to Veragua, like Tortosa with respect to Fuenterabia, or Pisa, with respect to Venice.] These words are taken from the Lettera Rarissima of Columbus, of which the original Spanish was lately found, and published by the learned M. Navarrete, in his Coleccion de Viages volume 1 page 299.) These geographical illusions, this mysterious veil, which enveloped the first discoveries, contributed to magnify every object, and to fix the attention of Europe on regions, the very names of which are, to us, scarcely known. New Cadiz, the principal seat of the pearl-fishery, was on an island which has again become uninhabited. The extremity of the rocky coast of Paria is also a desert. Several towns were founded at the mouth of the Rio Atrato, by the names of Antigua del Darien, Uraba or San Sebastian de Buenavista. In these spots, so celebrated at the beginning of the sixteenth century, the historians of the conquest tell us that the flower of the Castilian heroes were found assembled: thence Balboa set out to discover the South Sea; Pizarro marched from thence to conquer and ravage Peru; and Pedro de Cieca constantly followed the chain of the Andes, by Autioquia, Popayan and Cuzco, as far as La Plata, after having gone 900 leagues by land. These towns of Darien are destroyed; some ruins scattered on the hills of Uraba, the fruit-trees of Europe mixed with native trees, are all that mark to the traveller the spots on which those towns once stood. In almost all Spanish America the first lands peopled by the Conquistadores, have retrograded into barbarism.* (* In carefully collating the testimonies of the historians of the Conquest, some contradictions are observed in the periods assigned to the foundation of the towns of Darien. Pedro de Cieca, who had been on the spot, affirms that, under the government of Alonzo de Ojeda and Nicuessa, the town of Nuestra Senora Santa Maria el Antigua del Darien was founded on the western coast of the Gulf or Culata de Uraba, in 1509; and that later (despues desto passado) Ojeda passed to the eastern coast of the Culata to construct the town of San Sebastian de Uraba. The former, called by abbreviation Ciudad del Antigua, had soon a population of 2000 Spaniards; while the latter, the Ciudad del Uraba, remained uninhabited, because Francisco Pizarro, since known as the conqueror of Peru, was forced to abandon it, having vainly demanded succour from St. Domingo. The historian Herrera, after having said that the foundation of Antigua had preceded by one year that of Uraba or San Sebastian, affirms the contrary in the following chapter and in the Chronicle itself. It was, according to the Chronicle, in 1501 that Ojeda, accompanied by Vespucci, and penetrating for the first time the Gulf of Uraba or Darien, resolved to construct, with wood and unbaked bricks, a fort at the entrance of Culata. It appears, however, that this enterprise was not executed; for, in 1508, in the convention made by Ojeda and Nicuessa, they each promised to build two fortresses on the limits of New Andalusia and of Castillo del Oro. Herrera, in the 7th and 8th books of the first Decade, fixes the foundation of San Sebastian de Uraba at the beginning of 1510, and mentions it as the most ancient town of the continent of America, after that of Ceragua, founded by Columbus in 1503, on the Rio Belen. He relates how Francisco Pizarro abandoned that town, and how the foundation of the Ciudad del Antigua by Entiso, towards the end of the year 1510, was the consequence of that event. Leo X made Antigua a bishopric in 1514; and this was the first episcopal church of the continent. In 1519 Pedrarius Davila persuaded the court of Madrid, by false reports, that the site of the new town of Panama was more healthful than that of Antigua, the inhabitants were compelled to abandon the latter town, and the bishopric was transferred to Panama. The Gulf of Uraba was deserted during thirteen years, till the founder of the town of Carthagena, Pedro de Heredia, after having dug up the graves, or huacas, of the Rio Sinu, to collect gold, sent his brother Alonzo, in 1532, to repeople Uraba, and reconstruct on that spot a town under the name of San Sebastian de Buenavista.) Other countries, discovered later, attract the attention of the colonists: such is the natural progress of things in peopling a vast continent. It may be hoped that on several points the people will return to the places that were first chosen. It is difficult to conceive why the mouth of a great river, descending from a country rich in gold and platina, should have remained uninhabited. The Atrato, heretofore called Rio del Darien, de San Juan or Dabayba, has had the same fate as the Orinoco. The Indians who wander around the delta of those rivers continue in a savage state.

We weighed anchor in the road of Zapote, on the 27th March, at sunrise. The sea was less stormy, and the weather rather warmer, although the fury of the wind was undiminished. We saw on the north a succession of small cones of extraordinary form, as far as the Morro de Tigua; they are known by the name of the Paps (tetas) of Santero, Tolu, Rincon and Chichimar. The two latter are nearest the coast. The Tetas de Tolu rise in the middle of the savannahs. There, from the trunks of the Toluifera balsamum, is collected the precious balsam of Tolu, heretofore so celebrated in the pharmacopoeias of Europe, and in which is a profitable article of trade at Corozal, Caimito and the town of Tocasuan. In the savannahs (altas del Tolu) oxen and mules wander half wild. Several of those hills between Cienega de Pesquero and the Punta del Comissario are linked two-and-two together, like basaltic columns; it is, however, very probable that they are calcareous, like the Tetas de Managua, south of the Havannah. In the archipelago of San Bernardo we passed between the island of Salamanquilla and Cape Boqueron. We had scarcely quitted the gulf of Morosquillo when the sea became so rough that the waves frequently washed over the deck of our little vessel. It was a fine moonlight night. Our captain sought in vain a sheltering-place on the coast to the north of the village of Rincon. We cast anchor at four fathoms but, having discovered that we were lying over a reef of coral, we preferred the open sea.

The coast has a singular configuration beyond the Morro de Tigua, the terminatory point of the group of little mountains which rise like islands from the plain. We found at first a marshy soil extending over a square of eight leagues between the Bocas de Matuna and Matunilla. These marshes are connected by the Cienega de la Cruz, with the Dique of Mahates and the Rio Magdalena. The island of Baru which, with the island of Tierra Bomba, forms the vast port of Carthagena, is, properly speaking, but a peninsula fourteen miles long, separated from the continent by the narrow channel of Pasacaballos. The archipelago of San Bernardo is situated opposite Cape Boqueron. Another archipelago, called Rosario, lies off the southern point of the peninsula of Baru. These rents in the coast are repeated at the 10

3/4 and 11 degrees of latitude. The peninsulas near the Ensenada of Galera de Zamba and near the port of Savanilla have the same aspect as the peninsula Baru. Similar causes have produced similar effects; and the geologist must not neglect those analogies, in the configuration of a coast which, from Punta Caribana in the mouth of the Atrato, beyond the cape of La Vela, along an extent of 120 leagues, has a general direction from south-west to north-east.

The wind having dropped during the night we could only advance to the island of Arenas where we anchored. I found it was 78 degrees 2 minutes 10 seconds of longitude. The weather became stormy during the night. We again set sail on the morning of the 29th of March, hoping to be able to reach Boca Chica that day. The gale blew with extreme violence, and we were unable to proceed with our frail bark against the wind and the current, when, by a false manoeuvre in setting the sails (we had but four sailors), we were during some minutes in imminent danger. The captain, who was not a very bold mariner, declined to proceed further up the coast and we took refuge, sheltered from the wind, in a nook of the island of Baru south of Punta Gigantes. It was Palm Sunday and the Zambo, who had accompanied us to the Orinoco and did not leave us till we returned to France, reminded us that on the same Sunday in the preceding year, we had nearly been lost on the north of the mission of Uruana.

There was to be an eclipse of the moon during the night, and the next day an occultation of alpha Virginis. The observation of the latter phenomenon might have been very important in determining the longitude of Carthagena. In vain I urged the captain to allow one of his sailors to accompany me by land to the foot of Boca Chica, a distance of five miles. He objected on account of the wild state of the country in which there is neither habitation nor path. A little incident which might have rendered Palm-Sunday more fatal justified the prudence of the captain. We went by moonlight to collect plants on the shore; as we approached the land, we saw a young negro issue from the thicket. He was quite naked, loaded with chains, and armed with a machete. He invited us to land on a part of the beach covered with large mangroves, as being a spot where the surf did not break, and offered to conduct us to the interior of the island of Baru if we would promise to give him some clothes. His cunning and wild appearance, the often-repeated question whether we were Spaniards, and certain unintelligible words which he addressed to some of his companions who were concealed amidst the trees, inspired us with some mistrust. These blacks were no doubt maroon negroes: slaves escaped from prison. This unfortunate class are much to be feared: they have the courage of despair, and a desire of vengeance excited by the severity of the whites. We were without arms; the negroes appeared to be more numerous than we were and, thinking that possibly they invited us to land with the desire of taking possession of our canoe, we thought it most prudent to return on board. The aspect of a naked man wandering on an uninhabited beach, unable to free himself from the chains fastened round his neck and the upper part of his arm, was an object calculated to excite the most painful impressions. Our sailors wished to return to the shore for the purpose of seizing the fugitives, to sell them secretly at Carthagena. In countries where slavery exists the mind is familiarized with suffering and that instinct of pity which characterizes and enobles our nature is blunted.

Whilst we lay at anchor near the island of Baru in the meridian of Punta Gigantes I observed the eclipse of the moon of the 29th of March, 1801. The total immersion took place at 11 hours 30 minutes 12.6 seconds mean time. Some groups of vapours, scattered over the azure vault of the sky, rendered the observation of the immersion uncertain.

During the total eclipse the lunar disc displayed, as almost always happens, a reddish tint, without disappearing; the edges, examined with a sextant, were strongly undulating, notwithstanding the considerable altitude of the orb. It appeared to me that the moon was more luminous than I had ever seen it in the temperate zone. The vividness of the light, it may be conceived, does not depend solely on the state of the atmosphere, which reflects, more or less feebly, the solar rays, by inflecting them in the cone of the shade. The light is also modified by the variable transparency of that part of the atmosphere across which we perceived the moon eclipsed. Within the tropics great serenity of the sky and a perfect dissolution of the vapours diminish the extinction of the light sent back to us by the lunar disc. I was singularly struck during the eclipse by the want of uniformity in the distribution of the refracted light by the terrestrial atmosphere. In the central region of the disc there was a shadow like a round cloud, the movement of which was from east to west. The part where the immersion was to take place was consequently a few minutes prior to the immersion much more brightly illumined than the western edges. Is this phenomenon to be attributed to an inequality of our atmosphere; to a partial accumulation of vapour which, by absorbing a considerable part of the solar light, inflects less on one side the cone of the shadow of the earth? If a similar cause, in the perigee of central eclipses, sometimes renders the disc invisible, may it not happen also that only a small portion of the moon is seen; a disc, irregularly formed, and of which different parts were successively enlightened?

On the morning of the 30th of March we doubled Punta Gigantes, and made for the Boca Chica, the present entrance of the port of Carthagena. From thence the distance is seven or eight miles to the anchorage near the town; and although we took a practico to pilot us, we repeatedly touched on the sandbanks. On landing I learned, with great satisfaction, that the expedition appointed to take the survey of the coast under the direction of M. Fidalgo, had not yet put to sea. This circumstance not only enabled me to ascertain the astronomical position of several towns on the shore which had served me as points of departure in fixing chronometrically the longitude of the Llanos and the Orinoco, but also served to guide me with respect to the future direction of my journey to Peru. The passage from Carthagena to Porto Bello and that of the isthmus by the Rio Chagres and Cruces, are alike short and easy; but it was to be feared that we might stay long at Panama before we found an opportunity of proceeding to Guayaquil, and in that case the voyage on the Pacific would be extremely lingering, as we should have to sail against contrary winds and currents. I relinquished with regret the hope of levelling by the barometer the mountains of the isthmus, though it would then have been difficult to foresee that at the present time (1827), while measurements have been effected on so many other points of Mexico and Columbia, we should remain in ignorance of the height of the ridge which divides the waters in the isthmus. The persons we consulted all agreed that the journey by land along the Cordilleras by Santa Fe de Bogota, Popayan, Quito and Caxamarca would be preferable to the sea-voyage, and would furnish an immense field for exploration. The predilection of Europeans for the tierras frias, that is to say, the cold and temperate climate that prevails on the back of the Andes, gave further weight to these counsels. The distances were known, but we were deceived with respect to the time it would take to traverse them on mules' backs. We did not imagine that it would require more than eighteen months to go from Carthagena to Lima. Notwithstanding this delay, or rather owing to the slowness with which we passed through Cundinamarca, the provinces of Popayan and Quito, I did not regret having sacrificed the passage of the isthmus to the route of Bogota, for every step of the journey was full of interest both geographically and botanically. This change of direction gave me occasion to trace the map of the Rio Magdalena, to determine astronomically the position of eighty points situated in the inland country between Carthagena, Popayan, and the upper course of the river Amazon and Lima, to discover the error in the longitude of Quito, to collect several thousand new plants, and to observe on a vast scale the relations between the rocks of syenitic porphyry and trachyte with the fire of volcanoes.

The result of those labours of which it is not for me to appreciate the importance have long since been published. My map of the Rio Magdalena, multiplied by the copies of the year 1802 in America and Spain, and comprehending the country between Almaguer and Santa Marta, from 1 degree 54 minutes to 11 degrees 15 minutes latitude, appeared in 1816. Till that period no traveller had undertaken to describe New Grenada; and the public, except in Spain, knew the navigation of the Magdalena only by some lines traced by Bouguer. That learned traveller had descended the river from Honda; but, being in want of astronomical instruments, he had ascertained but four or five latitudes, by means of small dials hastily constructed. The narratives of travels in America are now singularly multiplied. Political events have led numbers of persons to those countries: and travellers have perhaps too hastily published their journals on returning to Europe. They have described the towns where they resided, and landscape scenery remarkable for beauty; they have furnished information respecting the inhabitants and the different modes of travelling in barks, on mules or on men's backs. These works, several of which are agreeable and instructive, have familiarized the nations of the Old World with those of Spanish America, from Buenos Ayres and Chili as far as Zacatecas and New Mexico. But unfortunately, in many instances, the want of a thorough knowledge of the Spanish language and the little care taken to acquire the names of places, rivers and tribes, have occasioned extraordinary mistakes.

During the six days of our stay at Carthagena our most interesting excursions were to the Boca Grande and the hill of Popa; the latter commands the town and a very extensive view. The port, or rather the bahia, is nearly nine miles and a half long, if we compute the length from the town (near the suburb of Jehemani or Xezemani) to the Cienega of Cacao. The Cienega is one of the nooks of the isle of Baru, south-west of the Estero de Pasacaballos, by which we reach the opening of the Dique de Mahates. Two extremities of the small island of Tierra Bomba form, on the north, with a neck of land of the continent, and on the south, with a cape of the island of Baru, the only entrances to the Bay of Carthagena; the former is called Boca Grande, the second Boca Chica. This extraordinary conformation of the land has given birth, for the space of a century, to theories entirely contradictory respecting the defence of a place which, next to the Havannah and Porto Cabello, is the most important of the main land and the West Indies. Engineers differed respecting the choice of the opening which should be closed; and it was not, as some writers have stated, after the landing of Admiral Vernon, in 1741, that the idea was first conceived* of filling up the Boca Grande. (* Don Jorge Juan in his Secret Notices addressed to the Marques de la Ensenada says: La entrada antigua era por un angosto canal que llaman Boca Chica; de resultas de esta invasion se acordo deja cioga y impassable la Boca Grande, y volver a abrir la antigua fortificandola. [The old entrance was by a narrow channel called the Boca Chica; but after this invasion it was determined to close up the Boca Grande and to open the old passage, fortifying it.] Secr. Not. volume 1 page 4.) The English forced the small entrance when they made themselves masters of the bay; but being unable to take the town of Carthagena, which made a gallant resistance, they destroyed the Castillo Grande (called also Santa Cruz) and the two forts of San Luis and San Jose which defended the Boca Chica.

The apprehension excited by the proximity of the Boca Grande to the town determined the court of Madrid, after the English expedition, to shut up the entrance along a distance of 2640 varas. From two and a half to three fathoms of water were found; and a wall, or rather a dyke, in stone, from fifteen to twenty feet high, was raised on piles. The slope on the side of the water is unequal, and seldom 45 degrees. This immense work was completed under the Viceroy Espeleta in 1795. But art could not vanquish nature; the sea is unceasingly though gradually silting up the Boca Chica, while it labours unceasingly to open and enlarge the Boca Grande. The currents which, during a great part of the year, especially when the bendavales blow with violence, ascend from south-west to north-east, throw sand into the Boca Chica, and even into the bay itself. The passage, which is from seventeen to eighteen fathoms deep, becomes more and more narrow,* and if a regular cleansing be not established by dredging machines, vessels will not be able to enter without risk. (* At the foot of the two forts San Jose and San Fernando, constructed for the defence of the Boca Chica, it may be seen how much the land has gained upon the sea. Necks of land are formed on both sides, and also before the Castillo del Angel which, northward, commands the fort of San Fernando.) It is this small entrance which should have been closed; its opening is only 250 toises, and the passage or navigable channel is 110 toises. If it should one day be determined to abandon the Boca Chica, and re-establish the Boca Grande in the state which nature seems to prescribe, new fortifications must be constructed on the south-south-west of the town. This fortress has always required great pecuniary outlays to keep it up.

The insalubrity of Carthagena varies with the state of the great marshes that surround the town on the east and north. The Cienega de Tesca is more than fifteen miles long; it communicates with the ocean where it approaches the village of Guayeper. When, in years of drought, the heaped-up earth prevents the salt water from covering the whole plain, the emanations that rise during the heat of the day when the thermometer stands between 28 and 32 degrees are very pernicious to the health of the inhabitants. A small portion of hilly land separates the town of Carthagena and the islet of Manga from the Cienega de Tesca.

Those hills, some of which are more than 500 feet high, command the town. The Castillo de San Lazaro is seen from afar rising like a great rocky pyramid; when examined nearer its fortifications are not very formidable. Layers of clay and sand, belonging to the tertiary formation of nagelfluhe, are covered with bricks and furnish a kind of construction which has little stability. The Cerro de Santa Maria de la Popa, crowned by a convent and some batteries, rises above the fort of San Lazaro and is worthy of more solid and extensive works. The image of the Virgin, preserved in the church of the convent, has been long revered by mariners. The hill itself forms a prolonged ridge from west to east. The calcareous rock, with cardites, meandrites and petrified corals, somewhat resembles the tertiary limestone of the peninsula of Araya near Cumana. It is split and decomposed in the steep parts of the rock, and the preservation of the convent on so unsolid a foundation is considered by the people as one of the miracles of the patron of the place. Near the Cerro de la Popa there appears, on several points, breccia with a limestone cement containing angular fragments of Lydian stone. Whether this formation of nagelfluhe is superposed on tertiary limestone of coral, and whether the fragments of the Lydian stone come from secondary limestone analogous to that of Zacatecas and the Moro de Nueva Barcelona, are questions which I have not had leisure to investigate. The view from the Popa is extensive and varied, and the windings and rents of the coast give it a peculiar character. I was assured that sometimes from the windows of the convent and even in the open sea, before the fort of Boca Chica, the snowy tops of the Sierra Nevada de Santa Marta are discernible. The distance of the Horqueta to the Popa is seventy-eight nautical miles. This group of colossal mountains is most frequently wrapped in thick clouds: and it is most veiled at the season when the gales blow with violence. Although only forty-five miles distant from the coast, it is of little service as a signal to mariners who seek the port of Saint Marta. Hidalgo during the whole time of his operations near the shore could take only one observation of the Nevados.

A gloomy vegetation of cactus, Jatropha gossypifolia, croton and mimosa covers the barren declivity of Cerro de la Popa. In herbalizing in those wild spots, our guides showed us a thick bush of Acacia cornigera, which had become celebrated by a deplorable event. Of all the species of mimosa the acacia is that which is armed with the sharpest thorns; they are sometimes two inches long; and being hollow, serve for the habitation of ants of an extraordinary size. A woman, annoyed by the jealousy and well founded reproaches of her husband, conceived a project of the most barbarous vengeance. With the assistance of her lover she bound her husband with cords, and threw him, at night, into a bush of Mimosa cornigera. The more violently he struggled, the more the sharp woody thorns of the tree tore his skin. His cries were heard by persons who were passing, and he was found after several hours of suffering, covered with blood, and dreadfully stung by the ants. This crime is perhaps without example in the history of human turpitude: it indicates a violence of passion less assignable to the climate than to the barbarism of manners prevailing among the lower class of the people.

My most important occupation at Carthagena was the comparison of my observations with the astronomical positions fixed by the officers of the expedition of Fidalgo. In the year 1783 (under the ministry of M. Valdes) Don Josef Espinosa, Don Dionisio Galiano and Don Josef de Lanz proposed to the Spanish government a plan for taking a survey of the coast of America, in order to extend the atlas of Tofino to the western colonies. The plan was approved; but it was not till 1792 that an expedition was fitted out at Cadiz, and they were enabled to commence their scientific operations at the island of Trinidad.

CHAPTER 3.31. CUBA AND THE SLAVE TRADE.

I might enumerate among the causes of the lowering of the temperature at Cuba during the winter months, the great number of shoals with which the island is surrounded, and on which the heat is diminished several degrees of centesimal temperature. This diminished heat may be assigned to the molecules of water locally cooled, which go to the bottom; to the polar currents, which are borne toward the abyss of the tropical ocean, or to the mixture of the deep waters with those of the surface at the declivities of the banks. But the lowering of the temperature is partly compensated by the flood of hot water, the Gulf Stream, which runs along the north-west coast, and the swiftness of which is often diminished by the north and north-east winds. The chain of shoals which encircles the island and which appears on our maps like a penumbra, is fortunately broken on several points, and those interruptions afford free access to the shore. In the south-east part the proximity of the lofty primitive mountains renders the coast more precipitous. In that direction are situated the ports of Santiago de Cuba, Guantanamo, Baitiqueri and (in turning the Punta Maysi) Baracoa. The latter is the place most early peopled by Europeans. The entrance to the Old Channel, from Punta de Mulas, west-north-west of Baracoa, as far as the new settlement which has taken the name of Puerto de las Nuevitas del Principe, is alike free from shoals and breakers. Navigators find excellent anchorage a little to the east of Punta de Mulas, in the three rocks of Tanamo, Cabonico, and Nipe; and on the west of Punta de Mulas in the ports of Sama, Naranjo, del Padre and Nuevas Grandes. It is remarkable that near the latter port, almost in the same meridian where, on the southern side of the island, are situated the shoals of Buena Esperanza and of Las doce Leguas, stretching as far as the island of Pinos, we find the commencement of the uninterrupted series of the cayos of the Old Channel, extending to the length of ninety-four leagues, from Nuevitas to Punta Icacos. The Old Channel is narrowest opposite to Cayo Cruz and Cayo Romano; its breadth is scarcely more than five or six leagues. On this point, too, the Great Bank of Bahama takes its greatest development. The Cayos nearest the island of Cuba and those parts of the bank not covered with water (Long Island, Eleuthera) are, like Cuba, of a long and narrow shape. Were they only twenty or thirty feet higher, an island much larger than St. Domingo would appear at the surface of the ocean. The chain of breakers and cayos that bound the navigable part of the Old Channel towards the south leave between the channel and the coast of Cuba small basins without breakers, which communicate with several ports having good anchorage, such as Guanaja, Moron and Remedios.

Having passed through the Old Channel, or rather the Channel of San Nicolas, between Cruz del Padre and the bank of the Cayos de Sel, the lowest of which furnish springs of fresh water, we again find the coast, from Punta de Icacos to Cabanas, free from danger. It affords, in the interval, the anchorage of Matanzas, Puerto Escondido, the Havannah and Mariel. Further on, westward of Bahia Honda, the possession of which might well tempt a maritime enemy of Spain, the chain of shoals recommences* (* They are here called Bajos de Santa Isabel y de los Colorados.) and extends without interruption as far as Cape San Antonio. From that cape to Punta de Piedras and Bahia de Cortez, the coast is almost precipitous, and does not afford soundings at any distance; but between Punta de Piedras and Cabo Cruz almost the whole southern part of Cuba is surrounded with shoals of which the isle of Pinos is but a portion not covered with water. These shoals are distinguished on the west by the name of Gardens (Jardines y Jardinillos); and on the east, by the names Cayo Breton, Cayos de las doce Leguas, and Bancos de Buena Esperanza. On all this southern line the coast is exempt from danger with the exception of that part which lies between the strait of Cochinos and the mouth of the Rio Guaurabo. These seas are very difficult to navigate. I had the opportunity of determining the position of several points in latitude and longitude during the passage from Batabano to Trinidad of Cuba and to Carthagena. It would seem that the resistance of the currents of the highlands of the island of Pines, and the remarkable out-stretching of Cabo Cruz, have at once favoured the accumulation of sand, and the labours of the coralline polypes which inhabit calm and shallow water. Along this extent of the southern coast a length of 145 leagues, only one-seventh affords entirely free access; namely that part between Cayo de Piedras and Cayo Blanco, a little to the east of Puerto Casilda. There are found anchorages often frequented by small barks; for example, the Surgidero del Batabano, Bahia de Xagua, and Puerto Casilda, or Trinidad de Cuba. Beyond this latter port, towards the mouth of the Rio Cauto and Cabo Cruz (behind the Cayos de doce Leguas), the coast, covered with lagoons, is not very accessible, and is almost entirely desert.

At the island of Cuba, as heretofore in all the Spanish possessions in America, we must distinguish between the ecclesiastic, politico-military, and financial divisions. We will not add those of the judicial hierarchy which have created so much confusion amongst modern geographers, the island having but one Audiencia, residing since the year 1797 at Puerto Principe, whose jurisdiction extends from Baracoa to Cape San Antonio. The division into two bishoprics dates from 1788 when Pope Pius VI nominated the first bishop of the Havannah. The island of Cuba was formerly, with Louisiana and Florida, under the jurisdiction of the archbishop of San Domingo, and from the period of its discovery it had only one bishopric, founded in 1518, in the most western part at Baracoa by Pope Leo X. The translation of this bishopric to Santiago de Cuba, took place four years later; but the first bishop, Fray Juan de Ubite, arrived only in 1528. In the beginning of the nineteenth century (1804), Santiago de Cuba was made an archbishopric. The ecclesiastical limit between the diocese of the Havannah and Cuba passes in the meridian of Cayo Romano, nearly in the 80 3/4 degree of longitude west of Paris, between the Villa de Santo Espiritu and the city of Puerto Principe. The island, with relation to its political and military government, is divided into two goviernos, depending on the same capitan-general. The govierno of the Havannah comprehends, besides the capital, the district of the Quatro Villas (Trinidad, Santo Espiritu, Villa Clara and San Juan de los Remedios) and the district of Puerto Principe. The Capitan-general y Gobernador of the Havannah has the privilege of appointing a lieutenant in Puerto Principe (Teniente Gobernador), as also at Trinidad and Nueva Filipina. The territorial jurisdiction of the capitan-general extends, as the jurisdiction of a corregidor, to eight pueblos de Ayuntamiento (the ciudades of Matanzas, Jaruco, San Felipe y Santiago, Santa Maria del Rosario; the villas of Guanabacoa, Santiago de las Vegas, Guines, and San Antonio de los Banos). The govierno of Cuba comprehends Santiago de Cuba, Baracoa, Holguin and Bayamo. The present limits of the goviernos are not the same as those of the bishoprics. The district of Puerto Principe, with its seven parishes, for instance, belonged till 1814 to the govierno of the Havannah and the archbishopric of Cuba. In the enumerations of 1817 and 1820 we find Puerto Principe joined with Baracoa and Bayamo, in the jurisdiction of Cuba. It remains for me to speak of a third division altogether financial. By the cedula of the 23rd March, 1812, the island was divided into three Intendencias or Provincias; those of the Havannah, Puerto Principe and Santiago de Cuba, of which the respective length from east to west is about ninety, seventy and sixty-five sea-leagues. The intendant of the Havannah retains the prerogatives of Superintendente general subdelegado de Real Hacienda de la Isla de Cuba. According to this division, the Provincia de Cuba comprehends Santiago de Cuba, Baracoa, Holguin, Bayamo, Gibara, Manzanillo, Jiguani, Cobre, and Tiguaros; the Provincia de Puerto Principe, the town of that name, Nuevitas, Jagua, Santo Espiritu, San Juan de los Remedios, Villa de Santa Clara and Trinidad. The most westerly intendencia, or Provincia de la Havannah, occupies all that part situated west of the Quatro Villas, of which the intendant of the capital has lost the financial administration. When the cultivation of the land shall be more uniformly advanced, the division of the island into five departments, namely: the vuelta de abaxo (from Cape San Antonio to the fine village of Guanajay and Mariel), the Havannah (from Mariel to Alvarez), the Quintas Villas (from Alvarez to Moron), Puerto Principe (from Moron to Rio Cauto), and Cuba (from Rio Cauto to Punta Maysi), will perhaps appear the most fit, and most consistent with the historical remembrances of the early times of the Conquest.

My map of the island of Cuba, however imperfect it may be for the interior, is yet the only one on which are marked the thirteen ciudades; and also seven villas, which are included in the divisions I have just enumerated. The boundary between the two bishoprics (linea divisoria de los dos obispados de la Havana y de Santiago de Cuba) extends from the mouth of the small river of Santa Maria (longitude 80 degrees 49 minutes), on the southern coast, by the parish of San Eugenio de la Palma, and by the haciendas of Santa Anna, Dos Hermanos, Copey, and Cienega, to La Punta de Judas (longitude 80 degrees 46 minutes) on the northern coast opposite Cayo Romano. During the regime of the Spanish Cortes it was agreed that this ecclesiastical limit should be also that of the two Deputaciones provinciales of the Havannah and of Santiago. (Guia Constitucional de la isla de Cuba, 1822 page 79). The diocese of the Havannah comprehends forty, and that of Cuba twenty-two, parishes. Having been established at a time when the greater part of the island was occupied by farms of cattle (haciendas de ganado), these parishes are of too great extent, and little adapted to the requirements of present civilization. The bishopric of Santiago de Cuba contains the five cities of Baracoa, Cuba, Holguin, Guiza, Puerto Principe and the Villa de Bayamo. In the bishopric of San Cristoval de la Havannah are included the eight cities of the Havannah, namely: Santa Maria del Rosario, San Antonio Abad or de los Banos, San Felipe y Santiago del Bejucal, Matanzas, Jaruco, La Paz and Trinidad, and the six villas of Guanabacoa, namely: Santiago de las Vegas or Compostela, Santa Clara, San Juan de los Remedios, Santo Espiritu and S. Julian de los Guines. The territorial division most in favour among the inhabitants of the Havannah, is that of vuelta de arriba and de abaxo, east and west of the meridian of the Havannah. The first governor of the island who took the title of Captain-general (1601) was Don Pedro Valdes. Before him there were sixteen other governors, of whom the series begins with the famous Poblador and Conquistador, Diego Velasquez, native of Cuellar, who was appointed by Columbus in 1511.

In the island of Cuba free men compose 0.64 of the whole population; and in the English islands, scarcely 0.19. In the whole archipelago of the West Indies the copper-coloured men (blacks and mulattos, free and slaves) form a mass of 2,360,000, or 0.83 of the total population. If the legislation of the West Indies and the state of the men of colour do not shortly undergo a salutary change; if the legislation continue to employ itself in discussion instead of action, the political preponderance will pass into the hands of those who have strength to labour, will to be free, and courage to endure long privations. This catastrophe will ensue as a necessary consequence of circumstances, without the intervention of the free blacks of Hayti, and without their abandoning the system of insulation which they have hitherto followed. Who can venture to predict the influence which may be exercised on the politics of the New World by an African Confederation of the free states of the West Indies, situated between Columbia, North America, and Guatimala? The fear of this event may act more powerfully on the minds of many, than the principles of humanity and justice; but in every island the whites believe that their power is not to be shaken. All simultaneous action on the part of the blacks appears to them impossible; and every change, every concession granted to the slave population, is regarded as a sign of weakness. The horrible catastrophe of San Domingo is declared to have been only the effect of the incapacity of its government. Such are the illusions which prevail amidst the great mass of the planters of the West Indies, and which are alike opposed to an amelioration of the condition of the blacks in Georgia and in the Carolinas. The island of Cuba, more than any other of the West India Islands, might escape the common wreck. That island contains 455,000 free men and 160,000 slaves: and there, by prudent and humane measures, the gradual abolition of slavery might be brought about. Let us not forget that since San Domingo has become free there are in the whole archipelago of the West Indies more free negroes and mulattos than slaves. The whites, and above all, the free men, whose cause it would be easy to link with that of the whites, take a very rapid numerical increase at Cuba. The slaves would have diminished, since 1820, with great rapidity, but for the fraudulent continuation of the slave-trade. If, by the progress of human civilization, and the firm resolution of the new states of free America, this infamous traffic should cease altogether, the diminution of the slave population would become more considerable for some time, on account of the disproportion existing between the two sexes, and the continuance of emancipation. It would cease only when the relation between the deaths and births of slaves should be such that even the effects of enfranchisement would be counterbalanced. The whites and free men now form two-thirds of the whole population of the island, and this increase marks in some degree the diminution of the slaves. Among the latter, the women are to the men (exclusive of the mulatto slaves), scarcely in the proportion of 1 : 4, in the sugar-cane plantations; in the whole island, as 1 : 1.7; and in the towns and farina where the negro slaves serve as domestics, or work by the day on their own account as well as that of their masters, the proportion is as 1 : 1.4; even (for instance at the Havannah),* as 1 : 1.2. (* It appears probable that at the end of 1825, of the total population of men of colour (mulattos and negroes, free and slaves), there were nearly 160,000 in the towns, and 230,000 in the fields. In 1811 the Consulado, in a statement presented to the Cortes of Spain, computed at 141,000, the number of men of colour in the towns, and 185,000 in the fields. Documentes sobre los Negros page 121.) This great accumulation of mulattos, free negros and slaves in the towns is a characteristic feature in the island of Cuba.) The developments that follow will show that these proportions are founded on numerical statements which may be regarded as the limit-numbers of the maximum.

The prognostics which are hazarded respecting the diminution of the total population of the island, at the period when the slave-trade shall be really abolished, and not merely according to the laws, as since 1820, respecting the impossibility of continuing the cultivation of sugar on a large scale, and respecting the approaching time when the agricultural industry of Cuba shall be restrained to plantations of coffee and tobacco, and the breeding of cattle, are founded on arguments which do not appear to me to be perfectly just. Instead of indulging in gloomy presages the planters would do well to wait till the government shall have procured positive statistical statements. The spirit in which even very old enumerations were made, for instance that of 1775, by the distinction of age, sex, race, and state of civil liberty, deserves high commendation. Nothing but the means of execution were wanting. It was felt that the inhabitants were powerfully interested in knowing partially the occupations of the blacks, and their numerical distribution in the sugar-settlements, farms and towns. To remedy evil, to avoid public danger, to console the misfortunes of a suffering race, who are feared more than is acknowledged, the wound must be probed; for in the social body, when governed by intelligence, there is found, as in organic bodies, a repairing force, which may be opposed to the most inveterate evils.

In the year 1811 the municipality and the Tribunal of Commerce of the Havannah computed the total population of the island of Cuba to be 600,000, including 326,000 people of colour, free or slaves, mulattos or blacks. At that time, nearly three-fifths of the people of colour resided in the jurisdiction of the Havannah, from Cape Saint Antonio to Alvarez. In this part it appears that the towns contained as many mulattos and free negroes as slaves, but that the coloured population of the towns was to that of the fields as two to three. In the eastern part of the island, on the contrary, from Alvarez to Santiago de Cuba and Cape Maysi, the men of colour inhabiting the towns nearly equalled in number those scattered in the farms. From 1811 till the end of 1825, the island of Cuba has received along the whole extent of its coast, by lawful and unlawful means, 185,000 African blacks, of whom the custom-house of the Havannah, only, registered from 1811 to 1820, about 116,000. This newly introduced mass has no doubt been spread more in the country than in the towns; it must have changed the relations which persons well informed of the localities had established in 1811, between the eastern and western parts of the island, between the towns and the fields. The negro slaves have much augmented in the eastern plantations; but the fact that, notwithstanding the importation of 185,000 bozal negroes, the mass of men of colour, free and slaves, has not augmented, from 1811 to 1825, more than 64,000, or one-fifth, shows that the changes in the relation of partial distribution are restrained within narrower limits than one would at first be inclined to admit.

The proportions of the castes with respect to each other will remain a political problem of high importance till such time as a wise legislation shall have succeeded in calming inveterate animosities and in granting equality of rights to the oppressed classes. In 1811, the number of whites in the island of Cuba exceeded that of the slaves by 62,000, whilst it nearly equalled the number of the people of colour, both free and slaves. The whites, who in the French and English islands formed at the same period nine-hundredths of the total population, amounted in the island of Cuba to forty-five hundredths. The free men of colour amounted to nineteen hundredths, that is, double the number of those in Jamaica and Martinique. The numbers given in the enumeration of 1817, modified by the Deputacion Provincial, being only 115,700 freedmen and 225,300 slaves, the comparison proves, first, that the freedmen have been estimated with little precision either in 1811 or in 1817; and, secondly, that the mortality of the negroes is so great, that notwithstanding the introduction of more than 67,700 African negroes registered at the custom-house, there were only 13,300 more slaves in 1817 than in 1811.

In 1817 a new enumeration was substituted for the approximative estimates attempted in 1811. From the census of 1817 it appears that the total population of the island of Cuba amounted to 572,363. The number of whites was 257,380; of free men of colour, 115,691, and of slaves 199,292.

In no part of the world where slavery prevails is emancipation so frequent as in the island of Cuba. The Spanish legislature favours liberty, instead of opposing it, like the English and French legislatures. The right of every slave to choose his own master, or set himself free, if he can pay the purchase-money, the religious feeling which disposes many masters in easy circumstances to liberate some of their slaves, the habit of keeping a multitude of blacks for domestic service, the attachments which arise from this intercourse with the whites, the facility with which slaves who are mechanics accumulate money, and pay their masters a certain sum daily, in order to work on their own account—such are the principal causes which in the towns convert so many slaves into free men of colour. I might add the chances of the lottery, and games of hazard, but that too much confidence in those means often produces the most fatal effects.

The primitive population of the West India Islands having entirely disappeared (the Zambo Caribs, a mixture of natives and negroes, having been transported in 1796, from St. Vincent to the island of Ratan), the present population of the islands (2,850,000) must be considered as composed of European and African blood. The negroes of pure race form nearly two-thirds; the whites one-fifth; and the mixed race one-seventh. In the Spanish colonies of the continent, we find the descendants of the Indians who disappear among the mestizos and zambos, a mixture of Indians with whites and negroes. The archipelago of the West Indies suggests no such consolatory idea. The state of society was there such, at the beginning of the sixteenth century, that, with some rare exceptions, the new planters paid as little attention to the natives as the English now do in Canada. The Indians of Cuba have disappeared like the Guanches of the Canaries, although at Guanabacoa and Teneriffe false pretensions were renewed forty years ago, by several families, who obtained small pensions from the government on pretext of having in their veins some drops of Indian or Guanche blood. It is impossible now to form an accurate judgment of the population of Cuba or Hayti in the time of Columbus. How can we admit, with some, that the island of Cuba, at its conquest in 1511, had a million of inhabitants, and that there remained of that million, in 1517, only 14,000! The statistic statements in the writings of the bishop of Chiapa are full of contradictions. It is related that the Dominican monk, Fray Luys Bertram, who was persecuted* by the encomenderos, as the Methodists now are by some English planters, predicted that the 200,000 Indians which Cuba contained, would perish the victims of the cruelty of Europeans. (* See the curious revelations in Juan de Marieta, Hist. de todos los Santos de Espana libro 7 page 174.) If this be true, we may at least conclude that the native race was far from being extinct between the years 1555 and 1569; but according to Gomara (such is the confusion among the historians of those times) there were no longer any Indians on the island of Cuba in 1553. To form an idea of the vagueness of the estimates made by the first Spanish travellers, at a period when the population of no province of the peninsula was ascertained, we have but to recollect that the number of inhabitants which Captain Cook and other navigators assigned to Otaheite and the Sandwich Islands, at a time when statistics furnished the most exact comparisons, varied from one to five. We may conceive that the island of Cuba, surrounded with coasts adapted for fishing, might, from the great fertility of its soil, afford sustenance for several millions of those Indians who have no desire for animal food, and who cultivate maize, manioc, and other nourishing roots; but had there been that amount of population, would it not have been manifest by a more advanced degree of civilization than the narrative of Columbus describes? Would the people of Cuba have remained more backward in civilization than the inhabitants of the Lucayes Islands? Whatever activity may be attributed to causes of destruction, such as the tyranny of the conquistadores, the faults of governors, the too severe labours of the gold-washings, the small-pox and the frequency of suicides,* it would be difficult to conceive how in thirty or forty years three or four hundred thousand Indians could entirely disappear. (* The rage of hanging themselves by whole families, in huts and caverns, as related by Garcilasso, was no doubt the effect of despair; yet instead of lamenting the barbarism of the sixteenth century, it was attempted to exculpate the conquistadores, by attributing the disappearance of the natives to their taste for suicide. See Patriota tome 2 page 50. Numerous sophisms of this kind are found in a work published by M. Nuix on the humanity of the Spaniards in the conquest of America. This work is entitled Reflexiones imparciales sobre la humanidad de los Epanoles contra los pretendidos filosofos y politicos, para illustrar las historias de Raynal y Robertson; escrito en Italiano por el Abate Don Juan Nuix, y traducido al castellano par Don Pedro Varela y Ulloa, del Consejo de S.M. 1752. [Impartial reflections on the humanity of the Spaniards, intended to controvert pretended philosophers and politicians, and to illustrate the histories of Raynal and Robertson; written in Italian by the Abate Don Juan Nuix and translated into Castilian by Don Pedro Varela y Ulloa, member of His Majesty's Council.] The author, who calls the expulsion of the Moors under Philip III a meritorious and religious act, terminates his work by congratulating the Indians of America "on having fallen into the hands of the Spaniards, whose conduct has been at all times the most humane, and their government the wisest." Several pages of this book recall the salutary rigour of the Dragonades; and that odious passage, in which a man distinguished for his talents and his private virtues, the Count de Maistre (Soirees de St. Petersbourg tome 2 page 121) justifies the Inquisition of Portugal "which he observes has only caused some drops of guilty blood to flow." To what sophisms must they have recourse, who would defend religion, national honour or the stability of governments, by exculpating all that is offensive to humanity in the actions of the clergy, the people, or kings! It is vain to seek to destroy the power most firmly established on earth, namely, the testimony of history.) The war with the Cacique Hatuey was short and was confined to the most eastern part of the island. Few complaints arose against the administration of the two first Spanish governors, Diego Velasquez and Pedro de Barba. The oppression of the natives dates from the arrival of the cruel Hernando de Soto about the year 1539. Supposing, with Gomara, that fifteen years later, under the government of Diego de Majariegos (1554 to 1564), there were no longer any Indians in Cuba, we must necessarily admit that considerable remains of that people saved themselves by means of canoes in Florida, believing, according to ancient traditions, that they were returning to the country of their ancestors. The mortality of the negro slaves, observed in our days in the West Indies, can alone throw some light on these numerous contradictions. To Columbus and Velasquez the island of Cuba must have appeared well peopled,* if, for instance, it contained as many inhabitants as were found there by the English in 1762. (* Columbus relates that the island of Hayti was sometimes attacked by a race of black men (gente negra), who lived more to the south or south-west. He hoped to visit them in his third voyage because those black men possessed a metal of which the admiral had procured some pieces in his second voyage. These pieces were sent to Spain and found to be composed of 0.63 of gold, 0.14 of silver and 0.19 of copper. In fact, Balboa discovered this black tribe in the Isthmus of Darien. "That conquistador," says Gomara, "entered the province of Quareca: he found no gold, but some blacks, who were slaves of the lord of the place. He asked this lord whence he had received them; who replied, that men of that colour lived near the place, with whom they were constantly at war...These negroes," adds Gomara, "exactly resemble those of Guinea; and no others have since been seen in America (en las Indios yo pienso que no se han visto negros despues.") The passage is very remarkable. Hypotheses were formed in the sixteenth century, as now; and Petrus Martyr imagined that these men seen by Balboa (the Quarecas), were Ethiopian blacks who, as pirates, infested the seas, and had been shipwrecked on the coast of America. But the negroes of Soudan are not pirates; and it is easier to conceive that Esquimaux, in their boats of skins, may have gone to Europe, than the Africans to Darien. Those learned speculators who believe in a mixture of the Polynesians with the Americans rather consider the Quarecas as of the race of Papuans, similar to the negritos of the Philippines. Tropical migrations from west to east, from the most western part of Polynesia to the Isthmus of Darien, present great difficulties, although the winds blow during whole weeks from the west. Above all, it is essential to know whether the Quarecas were really like the negroes of Soudan, as Gomara asserts, or whether they were only a race of very dark Indians (with smooth and glossy hair), who from time to time, before 1492, infested the coasts of the island of Hayti which has become in our days the domain of Ethiopians.) The first travellers were easily deceived by the crowds which the appearance of European vessels brought together on some points of the coast. Now, the island of Cuba, with the same ciudades and villas which it possesses at present, had not in 1762 more than 200,000 inhabitants; and yet, among a people treated like slaves, exposed to the violence and brutality of their masters, to excess of labour, want of nourishment, and the ravages of the small-pox—forty-two years would not suffice to obliterate all but the remembrance of their misfortunes on the earth. In several of the Lesser Antilles the population diminishes under English domination five and

117

six per cent annually; at Cuba, more than eight per cent; but the annihilation of 200,000 in forty-two years supposes an annual loss of twenty-six per cent, a loss scarcely credible, although we may suppose that the mortality of the natives of Cuba was much greater than that of negroes bought at a very high price.

In studying the history of the island we observe that the movement of colonization has been from east to west; and that here, as everywhere in the Spanish colonies, the places first peopled are now the most desert. The first establishment of the whites was in 1511 when, according to the orders of Don Diego Columbus, together with the conquistador and poblador Velasquez, he landed at Puerto de Palmas, near Cape Maysi, then called Alfa y Omega, and subdued the cacique Hatuey who, an emigrant and fugitive from Hayti, had withdrawn to the eastern part of the island of Cuba, and had become the chief of a confederation of petty native princes. The building of the town of Baracoa was begun in 1512; and later, Puerto Principe, Trinidad, the Villa de Santo Espiritu, Santiago de Cuba (1514), San Salvador de Bayamo, and San Cristoval de la Havana. This last town was originally founded in 1515 on the southern coast of the island, in the Partido of Guines, and transferred, four years later, to Puerto de Carenas, the position of which at the entrance of the two channels of Bahama (el Viejo y de Nuevo) appears to be much more favourable to commerce than the coast on the south-west of Batabano.* (* A tree is still shown at the Havannah (at Puerto de Carenas) under the shade of which the Spaniards celebrated their first mass. The island, now called officially The ever-faithful island of Cuba, was after its discovery named successively Juana Fernandina, Isla de Santiago, and Isla del Ave Maria. Its arms date from the year 1516.) The progress of civilization since the sixteenth century has had a powerful influence on the relations of the castes with each other; these relations vary in the districts which contain only farms for cattle, and in those where the soil has been long cleared; in the sea-ports and inland towns, in the spots where colonial produce is cultivated, and in such as produce maize, vegetables and forage.

Until the latter part of the eighteenth century the number of female slaves in the sugar plantations of Cuba was extremely limited; and what may appear surprising is that a prejudice, founded on religious scruples, opposed the introduction of women, whose price at the Havannah was generally one-third less than that of men. The slaves were forced to celibacy on the pretext of avoiding moral disorder. The Jesuits and the Bethlemite monks alone renounced that fatal prejudice, and encouraged negresses in their plantations. If the census, no doubt imperfect, of 1775, yielded 15,562 female, and 29,366 male slaves, we must not forget that that enumeration comprehended the totality of the island, and that the sugar plantations occupy even now but a quarter of the slave population. After the year 1795, the Consulado of the Havannah began to be seriously occupied with the project of rendering the increase of the slave population more independent of the variations of the slave-trade. Don Francisco Arango, whose views were ever characterized by wisdom, proposed a tax on the plantations in which the number of slaves was not comprised of one-third females. He also proposed a tax of six piastres on every negro brought into the island, and from which the women (negras bozales) should be exempt. These measures were not adopted because the colonial assembly refused to employ coercive means; but a desire to promote marriages and to improve the condition of the children of slaves has existed since that period, when a cedula real (of the 22nd April, 1804) recommended those objects "to the conscience and humanity of the planters."

The first introduction of negroes into the eastern part of the island of Cuba took place in 1521 and their number did not exceed 300. The Spaniards were then much less eager for slaves than the Portuguese; for, in 1539, there was a sale of 12,000 negroes at Lisbon, as in our days (to the eternal shame of Christian Europe) the trade in Greek slaves is carried on at Constantinople and Smyrna. In the sixteenth century the slave-trade was not free in Spain; the privilege of trading, which was granted by the Court, was purchased in 1586, for all Spanish America, by Gaspar de Peralta; in 1595, by Gomez Reynel; and in 1615, by Antonio Rodriguez de Elvas. The total importation then amounted to only 3500 negroes annually; and the inhabitants of Cuba, who were wholly engaged in rearing cattle, scarcely received any. During the war of succession, French ships were accustomed to stop at the Havannah and to exchange slaves for tobacco. The Asiento treaty with the English in some degree augmented the introduction of negroes; yet in 1763, although the taking of the Havannah and the sojourn of strangers gave rise to new wants, the number of slaves in the jurisdiction of the Havannah did not amount to 25,000; and in the whole island, not to 32,000. The total number of African negroes imported from 1521 to 1763 was probably 60,000; their descendants survive among the free mulattos, who inhabit for the most part the eastern side of the island. From the year 1763 to 1790, when the negro-trade was declared free, the Havannah received 24,875 (by the Compania de Tobacos 4957, from 1763 to 1766; by the contract of the Marquess de Casa Enrile, 14,132, from 1773 to 1779; by the contract of Baker and Dawson, 5786, from 1786 to 1789). If we estimate the introduction of slaves in the eastern part of the island during those twenty-seven years (1763 to 1790) at 6000, we find from the discovery of the island of Cuba, or rather from 1521 to 1790, a total of 90,875. We shall soon see that by the ever-increasing activity of the slave-trade the fifteen years that followed 1790 furnished more slaves than the two centuries and a half which preceded the period of the free trade. That activity was redoubled when it was stipulated between England and Spain that the slave-trade should be prohibited north of the equator, from November 22nd, 1817, and entirely abolished on the 30th May, 1820. The King of Spain accepted from England (which posterity will one day scarcely believe) a sum of 400,000 pounds sterling, as a compensation for the loss which might result from the cessation of that barbarous commerce.

Jamaica received from Africa in the space of three hundred years 850,000 blacks; or, to fix on a more certain estimate, in one hundred and eight years (from 1700 to 1808) nearly 677,000; and yet that island does not now possess 380,000 blacks, free mulattos and slaves. The island of Cuba furnishes a more consoling result; it has 130,000 free men of colour, whilst Jamaica, on a total population half as great, contains only 35,000.

On comparing the island of Cuba with Jamaica, the result of the comparison seems to be in favour of the Spanish legislation, and the morals of the inhabitants of Cuba. These comparisons demonstrate a state of things in the latter island more favorable to the physical preservation, and to the liberation of the blacks; but what a melancholy spectacle is that of Christian and civilized nations, discussing which of them has caused the fewest Africans to perish during the interval of three centuries, by reducing them to slavery! Much cannot be said in commendation of the treatment of the blacks in the southern parts of the United States; but there are degrees in the sufferings of the human species. The slave who has a hut and a family is less miserable than he who is purchased as if he formed part of a flock. The greater the number of slaves established with their families in dwellings which they believe to be their own property, the more rapidly will their numbers increase.

The annual increase of the last ten years in the United States (without counting the manumission of 100,000), was twenty-six on a thousand, which produces a doubling in twenty-seven years. Now, if the slaves at Jamaica and Cuba had multiplied in the same proportion, those two islands (the former since 1795, and the latter since 1800) would possess almost their present population, without 400,000 blacks having been dragged from the coast of Africa, to Port-Royal and the Havannah.

The mortality of the negroes is very different in the island of Cuba, as in all the West Indies, according to the nature of their treatment, the humanity of masters and overseers, and the number of negresses who can attend to the sick. There are plantations in which fifteen to eighteen per cent perish annually. I have heard it coolly discussed whether it were better for the proprietor not to subject the slaves to excessive labour and consequently to replace them less frequently, or to draw all the advantage possible from them in a few years, and replace them oftener by the acquisition of bozal negroes. Such are the reasonings of cupidity when man employs man as a beast of burden! It would be unjust to entertain a doubt that within fifteen years negro mortality has greatly diminished in the island of Cuba. Several proprietors have made laudable efforts to improve the plantation system.

It has been remarked how much the population of the island of Cuba is susceptible of being augmented in the lapse of ages. As the native of a northern country, little favoured by nature, I may observe that the Mark of Brandebourg, for the most part sandy, contains, under an administration favourable to the progress of agricultural industry, on a surface only one-third of that of Cuba, a population nearly double. The extreme inequality in the distribution of the population, the want of inhabitants on a great part of the coast, and its immense development, render the military defence of the whole island impossible: neither the landing of an enemy nor illicit trade can be prevented. The Havannah is well defended, and its works rival those of the most important fortified towns of Europe; the Torreones, and the fortifications of Cogimar, Jaruco, Matanzas, Mariel, Bahia Honda, Batabano, Xagua and Trinidad might resist for a considerable time the assaults of an enemy; but on the other hand two-thirds of the island are almost without defence, and could scarcely be protected by the best gun-boats.

Intellectual cultivation is almost entirely limited to the whites, and is as unequally distributed as the population. The best society of the Havannah may be compared for easy and polished manners with the society of Cadiz and with that of the richest commercial towns of Europe; but on quitting the capital, or the neighbouring plantations, which are inhabited by rich proprietors, a striking contrast to this state of partial and local civilization is manifest, in the simplicity of manners prevailing in the insulated farms and small towns. The Havaneros or natives of the Havannah were the first among the rich inhabitants of the Spanish colonies who visited Spain, France and Italy; and at the Havannah the people were always well informed of the politics of Europe. This knowledge of events, this prescience of future chances, have powerfully aided the inhabitants of Cuba to free themselves from some of the burthens which check the development of colonial prosperity. In the interval between the peace of Versailles and the beginning of the revolution of San Domingo, the Havannah appeared to be ten times nearer to Spain than to Mexico, Caracas and New Grenada. Fifteen years later, at the period of my visit to the colonies, this apparent inequality of distance had considerably diminished; now, when the independence of the continental colonies, the importation of foreign manufactures and the financial wants of the new states have multiplied the intercourse between Europe and America; when the passage is shortened by improvements in navigation; when the Columbians, the Mexicans and the inhabitants of Guatimala rival each other in visiting Europe; the ancient Spanish colonies—those at least that are bathed by the Atlantic—seem alike to have drawn nearer to the continent. Such are the changes which a few years have produced, and which are proceeding with increasing rapidity. They are the effects of knowledge and of long-restrained activity; and they render less striking the contrast in manners and civilization which I observed at the beginning of the century, at Caracas, Bogota, Quito, Lima, Mexico and the Havannah. The influences of the Basque, Catalanian, Galician and Andalusian origin become every day more imperceptible.

The island of Cuba does not possess those great and magnificent establishments the foundation of which is of very remote date in Mexico; but the Havannah can boast of institutions which the patriotism of the inhabitants, animated by a happy rivalry between the different centres of American civilization, will know how to extend and improve whenever political circumstances and confidence in the preservation of internal tranquillity may permit. The Patriotic Society of the Havannah (established in 1793); those of Santo Espiritu, Puerto Principe, and Trinidad, which depend on it; the university, with its chairs of theology, jurisprudence, medicine and mathematics, established since 1728, in the convent of the Padres Predicedores;* (* The clergy of the island of Cuba is neither numerous nor rich, if we except the Bishop of the Havannah and the Archbishop of Cuba, the former of whom has 110,000 piastres, and the latter 40,000 piastres per annum. The canons have 3000 piastres. The number of ecclesiastics does not exceed 1100, according to the official enumeration in my possession.) the chair of political economy, founded in 1818; that of agricultural botany; the museum and the school of descriptive anatomy, due to the enlightened zeal of Don Alexander Ramirez; the public library, the free school of drawing and painting; the national school; the Lancastrian schools, and the botanic garden, are institutions partly new, and partly old. Some stand in need of progressive amelioration, others require a total reform to place them in harmony with the spirit of the age and the wants of society.

AGRICULTURE.

When the Spaniards began their settlements in the islands and on the continent of America those productions of the soil chiefly cultivated were, as in Europe, the plants that serve to nourish man. This primitive stage of the agricultural life of nations has been preserved till the present time in Mexico, in Peru, in the cold and temperate regions of

Cundinamarca, in short, wherever the domination of the whites comprehends a vast extent of territory. The alimentary plants, bananas, manioc, maize, the cereals of Europe, potatoes and quinoa, have continued to be, at different heights above the level of the sea, the basis of continental agriculture within the tropics. Indigo, cotton, coffee and sugar-cane appear in those regions only in intercalated groups. Cuba and the other islands of the archipelago of the Antilles presented during the space of two centuries and a half a uniform aspect: the same plants were cultivated which had nourished the half-wild natives and the vast savannahs of the great islands were peopled with numerous herds of cattle. Piedro de Atienza planted the first sugar-canes in Saint Domingo about the year 1520; and cylindrical presses, moved by water-wheels, were constructed.* (* On the trapiches or molinos de agua of the sixteenth century see Oviedo, Hist. nat. des Ind. lib. 4 cap. 8.) But the island of Cuba participated little in these efforts of rising industry; and what is very remarkable, in 1553, the historians of the Conquest* mention no exportation of sugar except that of Mexican sugar for Spain and Peru. (* Lopez de Gomara, Conquista de Mexico (Medina del Campo 1353) fol. 129.) Far from throwing into commerce what we now call colonial produce, the Havannah, till the eighteenth century, exported only skins and leather. The rearing of cattle was succeeded by the cultivation of tobacco and the rearing of bees, of which the first hives (colmenares) were brought from the Floridas. Wax and tobacco soon became more important objects of commerce than leather, but were shortly superseded in their turn by the sugar-cane and coffee. The cultivation of these productions did not exclude more ancient cultivation; and, in the different phases of agricultural industry, notwithstanding the general tendency to make the coffee plantations predominate, the sugar-houses furnish the greatest amount in the annual profits. The exportation of tobacco, coffee, sugar and wax, by lawful and illicit means, amounts to fourteen millions of piastres, according to the actual price of those articles.

Three qualities of sugar are distinguished in the island of Cuba, according to the degree of purity attained by refining (grados de purga). In every loaf or reversed cone the upper part yields the white sugar; the middle part the yellow sugar, or quebrado; and the lower part, or point of the cone, the cucurucho. All the sugar of Cuba is consequently refined; a very small quantity is introduced of coarse or muscovado sugar (by corruption, azucar mascabado). The forms being of a different size, the loaves (panes) differ also in weight. They generally weigh an arroba after refining. The refiners (maestros de azucar) endeavour to make every loaf of sugar yield five-ninths of white, three-ninths of quebrado, and one-ninth of cucurucho. The price of white sugar is higher when sold alone than in the sale called surtido, in which three-fifths of white sugar and two-fifths of quebrado are combined in the same lot. In the latter case the difference of the price is generally four reals (reales de plata); in the former, it rises to six or seven reals. The revolution of Saint Domingo, the prohibitions dictated by the Continental System of Napoleon, the enormous consumption of sugar in England and the United States, the progress of cultivation in Cuba, Brazil, Demerara, the Mauritius and Java, have occasioned great fluctuations of price. In an interval of twelve years it was from three to seven reals in 1807, and from twenty-four to twenty-eight reals in 1818, which proves fluctuations in the relation of one to five.

During my stay in the plains of Guines, in 1804, I endeavoured to obtain some accurate information respecting the statistics of the making of cane-sugar. A great yngenio producing from 32,000 to 40,000 arrobas of sugar is generally fifty caballerias,* or 650 hectares in extent, of which the half (less than one-tenth of a square sea league) is allotted to sugar-making properly so called (canaveral) and the other half for alimentary plants and pasturage (potrero). (* The agrarian measure, called caballeria, is eighteen cordels, (each cordel includes twenty-four varas) or 432 square varas; consequently, as 1 vara = 0.835m., according to Rodriguez, a caballeria is 186,624 square varas, or 130,118 square metres, or thirty-two and two-tenths English acres.) The price of land varies, naturally, according to the quality of the soil and the proximity of the ports of the Havannah, Matanzas and Mariel. In a circuit of twenty-five leagues round the Havannah the caballeria may be estimated at two or three thousand piastres. For a produce* of 32,000 arrobas (or 2000 cases of sugar) the yngenio must have at least three hundred negroes. (* There are very few plantations in the whole island of Cuba capable of furnishing 40,000 arrobas; among these few are the yngenio of Rio Blanco, or of the Marquess del Arca, and those belonging to Don Rafael Ofarrel and Dona Felicia Jaurregui. Sugar-houses are thought to be very considerable that yield 2000 cases annually, or 32,000 arrobas (nearly 368,000 kilogrammes.) In the French colonies it is generally computed that the third or fourth part only of the land is allotted for the plantation of food (bananas, ignames and batates); in the Spanish colonies a greater surface is lost in pasturage; this is the natural consequence of the old habits of the haciendas de ganado.) An adult and acclimated slave is worth from four hundred and fifty to five hundred piastres; a bozal negro, adult, not acclimated, three hundred and seventy to four hundred piastres. It is probable that a negro costs annually, in nourishment, clothing and medicine, forty-five to fifty piastres; consequently, with the interest of the capital, and deducting the holidays, more than twenty-two sous per day. The slaves are fed with tasajo (meat dried in the sun) of Buenos Ayres and Caracas; salt-fish (bacalao) when the tasajo is too dear; and vegetables (viandas) such as pumpkins, munatos, batatas, and maize. An arroba of tasajo was worth ten to twelve reals at Guines in 1804; and from fourteen to sixteen in 1825. An yngenio, such as we here suppose (with a produce of 32,000 to 40,000 arrobas), requires, first, three machines with cylinders put in motion by oxen (trapiches) or two water-wheels; second, according to the old Spanish method, which, by a slow fire causes a great consumption of wood, eighteen cauldrons (piezas); according to the first method of reverberation (introduced since the year 1801 by Mr. Bailli of Saint Domingo under the auspices of Don Nicolas Calvo) three clarificadoras, three peilas and two traines de tachos (each train has three piezas), in all twelve fondos. It is commonly asserted that three arrobas of refined sugar yield one barrel of miel, and that the molasses are sufficient for the expenses of the plantation: this is especially the case where they produce brandy in abundance. Thirty-two thousand arrobas of sugar yield 15,000 bariles de miel (at two arrobas) of which five hundred pipas de aguardiente de cana are made, at twenty-five piastres.

In establishing an yngenio capable of furnishing two thousand caxas yearly, a capitalist would draw, according to the old Spanish method, and at the present price of sugar, an interest of six and one-sixth per cent; an interest no way considerable for an establishment not merely agricultural, and of which the expense remains the same, although the produce sometimes diminishes more than a third. It is very rarely that one of those great yngenios can make 32,000 cases of sugar during several successive years. It cannot therefore be matter of surprise that when the price of sugar in the island of Cuba has been very low (four or five piastres the quintal), the cultivation of rice has been preferred to that of the sugar-cane. The profit of the old landowners (haciendados) consists, first, in the circumstance that the expenses of the settlement were much less twenty or thirty years ago, when a caballeria of good land cost only 1200 or 1600 piastres, instead of 2500 to 3000; and the adult negro 300 piastres, instead of 450 to 500; second, in the balance of the very low and the very high prices of sugar. These prices are so different in a period of ten years that the interest of the capital varies from five to fifteen per cent. In the year 1804, for instance, if the capital employed had been only 100,000 piastres, the raw produce, according to the value of sugar and rum, would have amounted to 94,000 piastres. Now, from 1797 to 1800, the price of a case of sugar was sometimes, mean value, forty piastres instead of twenty-four, which I was obliged to suppose in the calculation for the year 1825. When a sugar-house, a great manufacture or a mine is found in the hands of the person who first formed the establishment, the estimate of the rate of interest which the capital employed yields to the proprietor, can be no guide to those who, purchasing afterwards, balance the advantages of different kinds of industry.

In soils that can be watered, or where plants with tuberose roots have preceded the cultivation of the sugar-cane, a caballeria of fertile land yields, instead of 1500 arrobas, 3000 or 4000, making 2660 or 3340 kilogrammes of sugar (blanco and quebrado) per hectare. In fixing on 1500 arrobas and estimating the case of sugar at 24 piastres, according to the price of the Havannah, we find that the hectare produces the value of 870 francs in sugar; and that of 288 francs in wheat, in the supposition of an octuple harvest, and the price of 100 kilogrammes of wheat being 18 francs. I have observed elsewhere that in this comparison of the two branches of cultivation it must not be forgotten that the cultivation of sugar requires great capital; for instance, at present 400,000 piastres for an annual production of 32,000 arrobas, or 368,000 kilogrammes, if this quantity be made in one single settlement. At Bengal, in watered lands, an acre (4044 square metres) renders 2300 kilogrammes of coarse sugar, making 5700 kilogrammes per hectare. If this fertility is common in lands of great extent we must not be surprised at the low price of sugar in the East Indies. The produce of a hectare is double that of the best soil in the West Indies and the price of a free Indian day-labourer is not one-third the price of the day-labour of a negro slave in the island of Cuba.

In Jamaica in 1825 a plantation of five hundred acres (or fifteen and a half caballerias), of which two hundred acres are cultivated in sugar-cane, yields, by the labour of two hundred slaves, one hundred oxen and fifty mules 2800 hundredweight, or 142,200 kilogrammes of sugar, and is computed to be worth, with its slaves, 43,000 pounds sterling. According to this estimate of Mr. Stewart, one hectare would yield 1760 kilogrammes of coarse sugar; for such is the quality of the sugar furnished for commerce at Jamaica. Reckoning in a great sugar-fabric of the Havannah 25 caballerias or 325 hectares for a produce of from 32,000 to 40,000 cases, we find 1130 or 1420 kilogrammes of refined sugar (blanco and quebrado) per hectare. This result agrees sufficiently with that of Jamaica, if we consider the loss sustained in the weight of sugar by refining, in converting the coarse sugar into azucar blanco y quebrado) or refined sugar. At San Domingo a square (3403 square toises = 1.29 hectare) is estimated at forty, and sometimes at sixty quintals: if we fix on 5000 pounds, we still find 1900 kilogrammes of coarse sugar per hectare. Supposing, as we ought to do when speaking of the produce of the whole island of Cuba, that, in soils of average fertility, the caballeria (at 13 hectares) yields 1500 arrobas of refined sugar (mixed with blanco and quebrado), or 1330 kilogrammes per hectare, it follows that 60,872 hectares, or nineteen five-fourths square sea leagues, (nearly a ninth of the extent of a department of France of middling size), suffice to produce the 440,000 cases of refined sugar furnished by the island of Cuba for its own consumption and for lawful and illicit exportation. It seems surprising that less than twenty square sea leagues should yield an annual produce of more than the value of fifty-two millions of francs (counting one case, at the Havannah, at the rate of twenty-four piastres). To furnish coarse sugar for the consumption of thirty millions of French (which is actually from fifty-six to sixty millions of kilogrammes) it requires within the tropics but nine and five-sixths square sea leagues cultivated with sugar-cane; and in temperate climates but thirty-seven and a half square sea leagues cultivated with beet-root. A hectare of good soil, sown or planted with beet-root, produces in France from ten to thirty thousand kilogrammes of beet-root. The mean fertility is 20,000 kilogrammes, which furnish 2 1/2 per cent, or five hundred kilogrammes of coarse sugar. Now, one hundred kilogrammes of that sugar yield fifty kilogrammes of refined sugar, thirty of sugar vergeoise, and twenty of muscovade; consequently, a hectare of beet-root produces 250 kilogrammes of refined sugar.

A short time before my arrival at the Havannah there had been sent from Germany some specimens of beet-root sugar which were said to menace the existence of the Sugar Islands in America. The planters had learned with alarm that it was a substance entirely similar to sugar-cane, but they flattered themselves that the high price of labour in Europe and the difficulty of separating the sugar fit for crystallization from so great a mass of vegetable pulp would render the operation on a grand scale little profitable. Chemistry has, since that period, succeeded in overcoming those difficulties; and, in the year 1812, France alone had more than two hundred beet-root sugar factories working with very unequal success and producing a million of kilogrammes of coarse sugar, that is, a fifty-eighth part of the actual consumption of sugar in France. Those two hundred factories are now reduced to fifteen or twenty, which yield a produce of 300,000 kilogrammes.* (* Although the actual price of cane-sugar not refined is 1 franc 50 cents the kilogramme, in the ports, the production of beetroot-sugar offers a still greater advantage in certain localities, for instance, in the vicinity of Arras. These establishments would be introduced in many other parts of France if the price of the sugar of the West Indies rose to 2 francs, or 2 francs 25 cents the kilogramme, and if the government laid no tax on the beetroot-sugar, to compensate the loss on the consumption of colonial sugar. The making of beetroot-sugar is especially profitable when combined with a general system of rural economy, with the improvement of the soil and the nourishment of cattle: it is not a cultivation independent of local circumstances, like that of the sugar-cane in the tropics.) The inhabitants of the West Indies, well informed of the affairs of Europe, no longer fear beet-root, grapes, chesnuts, and mushrooms, the coffee of Naples nor the indigo of the south of France. Fortunately the improvement of the condition of the West India slaves does not depend on the success of these branches of European cultivation.

Previously to the year 1762 the island of Cuba did not furnish more commercial produce than the three least industrious and most neglected provinces with respect to cultivation, Veragua, the isthmus of Panama and Darien, do at present. A political event which appeared extremely unfortunate, the taking of the Havannah by the English, roused the public mind. The town was evacuated in 1784 and its subsequent efforts of industry date from that memorable period. The construction of new fortifications on a gigantic plan* threw a great deal of money suddenly into circulation (* It is affirmed that the construction of the fort of Cabana alone cost fourteen millions of piastres.); later the slave-trade became free and furnished hands for the sugar factories. Free trade with all the ports of Spain and occasionally with neutral states, the able administration of Don Luis de Las Casas, the establishment of the Consulado and the Patriotic Society, the destruction of the French colony of Saint Domingo,* (* In three successive attempts, in August 1791, June 1793, and October 1803. Above all the unfortunate and sanguinary expedition of Generals Leclerc and Rochambeau completed the destruction of the sugar factories of Saint Domingo.) the rise in the price of sugar which was the natural consequence, the improvement in machines and ovens, due in great part to the refugees of Cape Francois, the more intimate connection formed between the proprietors of the sugar factories and the merchants of the Havannah, the great capital employed by the latter in agricultural establishments (sugar and coffee plantations), such have been successively the causes of the increasing prosperity of the island of Cuba, notwithstanding the conflict of the authorities, which serves to embarrass the progress of affairs.

The greatest changes in the plantations of sugar-cane and in the sugar factories, took place from 1796 to 1800. First, mules were substituted (trapiches de mulas) for oxen (trapiches de bueyes); and afterwards hydraulic wheels were introduced (trapiches de agua), which the first conquistadores had employed at Saint Domingo; finally the action of steam-engines was tried at Ceibabo, at the expense of Count Jaruco y Mopex. There are now twenty-five of those machines in the different sugar mills of the island of Cuba. The culture of the sugar-cane of Otaheite in the meantime increased. Boilers of preparation (clarificadoras) were introduced and the reverberating furnaces better arranged. It must be said, to the honour of wealthy proprietors, that in a great number of plantations, a kind solicitude is manifested for sick slaves, for the introduction of negresses, and for the education of children.

The number of sugar factories (yngenios) in 1775 was 473 in the whole island; and in 1817 more than 780. Among the former, none produced the fourth part of the sugar now made in the yngenios of second rank; it is consequently not the number of factories that can afford an accurate idea of the progress of that branch of agricultural industry.

The first sugar-canes carefully planted on virgin soil yield a harvest during twenty to twenty-five years, after which they must be replanted every three years. There existed in 1804, at the Hacienda de Matamoros, a square (canaveral) worked during forty-five years. The most fertile soil for the production of sugar is now in the vicinity of Mariel and Guanajay. That variety of sugar-cane known by the name of Cana de Otahiti, recognised at a distance by a fresher green, has the advantage of furnishing, on the same extent of soil, one-fourth more juice, and a stem more woody, thicker, and consequently richer in combustible matter. The refiners (maestros de azucar), pretend that the vezou (guarapo) of the Cana de Otahiti is more easily worked, and yields more crystallized sugar by adding less lime or potass to the vezou. The South Sea sugar-cane furnishes, no doubt, after five or six years' cultivation, the thinnest stubble, but the knots remain more distant from each other than in the Cana creolia or de la tierra. The apprehension at first entertained of the former degenerating by degrees into ordinary sugar-cane is happily not realized. The sugar-cane is planted in the island of Cuba in the rainy season, from July to October; and the harvest is gathered from February to May.

In proportion as by too rapid clearing the island has become unwooded, the sugar-houses have begun to want fuel. A little stalk (sugar-cane destitute of its juice) used to be employed to quicken the fire beneath the old cauldrons (tachos); but it is only since the introduction of reverberating furnaces by the emigrants of Saint Domingo that the attempt has been made to dispense altogether with wood and burn only refuse sugar-cane. In the old construction of furnaces and cauldrons, a tarea of wood, of one hundred and sixty cubic feet, is burnt to produce five arrobas of sugar, or, for a hundred kilogrammes of raw sugar, 278 cubic feet of the wood of the lemon and orange trees are required. In the reverberating furnaces of Saint Domingo a cart of refuse-cane of 495 cubic feet produced 640 pounds of coarse sugar, which make 158 cubic feet of refuse-cane for 100 kilogrammes of sugar. I attempted, during my stay at Guines, and especially at Rio Blanco, with the Count de Mopex, several new constructions, with the view of diminishing the expense of fuel, surrounding the focus with substances which do not powerfully conduct the heat, and thus diminish the sufferings of the slaves who keep up the fire. A long residence in the salt-producing districts of Europe, and the labours of practical halurgy, to which I have been devoted since my early youth, suggested to me the idea of those constructions, which have been imitated with some success. Cuvercles of wood, placed on clarificadoras, accelerated the evaporation, and led me to believe that a system of cuvercles and moveable frames, furnished with counter-weights, might extend to other cauldrons. This object merits further examination; but the quantity of vezou (guarapo) of the crystallized sugar extracted, and that which is destroyed, the fuel, the time and the pecuniary expense, must be carefully estimated.

An error, very general through Europe and one which influences opinion respecting the effects of the abolition of the slave-trade, is that in those West India islands called sugar colonies, the majority of the slaves are supposed to be employed in the production of sugar. The cultivation of the sugar-cane is no doubt a powerful incentive to the activity of the slave trade; but a very simple calculation suffices to prove that the total mass of slaves contained in the West Indies is nearly three times greater than the number employed in the production of sugar. I showed seven years ago that, if the 200,000 cases of sugar exported from the island of Cuba in 1812 were produced in the great establishments, less than 30,000 slaves would have sufficed for that kind of labour. It ought to be borne in mind for the interests of humanity that the evils of slavery weigh on a much greater number of individuals than agricultural labours require, even admitting, which I am very far from doing, that sugar, coffee, indigo and cotton can be cultivated only by slaves. At the island of Cuba it is generally supposed that one hundred and fifty negroes are required to produce 1000 cases (184,000 kilogrammes) of refined sugar; or, in round numbers, a little more than 1200 kilogrammes, by the labour of each adult slave. The production of 440,000 cases would consequently require only 66,000 slaves. If we add 36,000 to that number for the cultivation of coffee and tobacco in the island of Cuba, we find that about 100,000 of the 260,000 slaves now there would suffice for the three great branches of colonial industry on which the activity of commerce depends.

COFFEE.

The cultivation of coffee takes its date, like the improved construction of cauldrons in the sugar houses, from the arrival of the emigrants of San Domingo, especially after the years 1796 and 1798. A hectare yields 860 kilogrammes, the produce of 3500 plants. The province of the Havannah reckoned:

In 1800 60 cafetales.

In 1817 779 cafetales.

The coffee tree being a shrub that yields a good harvest only in the fourth year, the exportation of coffee from the port of the Havannah was, in 1804, only 50,000 arrobas. It rose:

In 1809 to 320,000 arrobas.

In 1815 to 918,263 arrobas.

In 1815, when the price of coffee was fifteen piastres the quintal, the value of the exportation from the Havannah exceeded the sum of 3,443,000 piastres. In 1823, the exportation from the port of Matanzas was 84,440 arrobas; so that it seems not doubtful that, in years of medium fertility, the total exportation of the island, lawful and contraband, is more than fourteen millions of kilogrammes.

From this calculation it results that the exportation of coffee from the island of Cuba is greater than that from Java, estimated by Mr. Crawfurd, in 1820, at 190,000 piculs, 11 4/5 millions of kilogrammes. It likewise exceeds the exportation from Jamaica, which amounted, in 1823, according to the registers of the custom-house, only to 169,734 hundredweight, or 8,622,478 kilogrammes. In the same year Great Britain received, from all the English islands, 194,820 hundredweight; or 9,896,856 kilogrammes; which proves that Jamaica only produced six-sevenths. Guadaloupe sent, in 1810, to the mother country, 1,017,190 kilogrammes; Martinico, 671,336 kilogrammes. At Hayti, where the production of coffee before the French revolution was 37,240,000 kilogrammes, Port-au-Prince exported, in 1824, only 91,544,000 kilogrammes. It appears that the total exportation of coffee from the archipelago of the West Indies, by lawful means only, now amounts to more than thirty-eight millions of kilogrammes; nearly five times the consumption of France, which, from 1820 to 1823, was, on the yearly average, 8,198,000 kilogrammes. The consumption of Great Britain is yet* only 3 1/2 millions of kilogrammes. (* Before the year 1807, when the tax on coffee was reduced, the consumption of Great Britain was not 8000 hundredweight (less than 1/2 million of kilogrammes); in 1809, it rose to 45,071 hundredweight; in 1810, to 49,147 hundredweight; in 1823, to 71,000 hundredweight, in 1824, to 66,000 hundredweight (or 3,552,800 kilogrammes.)

The exportation of 1814 was 60 1/2 millions of kilogrammes, which we may suppose was at that period nearly the consumption of the whole of Europe. Great Britain (taking that denomination in its true sense, as denoting only England and Scotland) now consumes nearly two-thirds less coffee and three times more sugar than France.

The price of sugar at the Havannah is always by the arroba of 25 Spanish pounds (or 11.49 kilogrammes), and the price of coffee by the quintal (or 45.97 kilogrammes). The latter has been known to vary from 4 to 30 piastres; it even fell, in 1808, below 24 reals. The price of 1815 and 1819 was between 13 and 17 piastres the quintal; coffee is now at 12 piastres. It is probable that the cultivation of coffee scarcely employs in the whole island of Cuba 28,000 slaves, who produce, on the yearly average, 305,000 Spanish quintals (14 millions of kilogrammes), or, according to the present value, 3,660,000 piastres; while 66,000 negroes produce 440,000 cases (81 millions of kilogrammes) of sugar, which, at the price of 24 piastres, is worth 10,560,000 piastres. It results from this calculation that a slave now produces the value of 130 piastres of coffee, and 160 piastres of sugar. It is almost useless to observe that these relations vary with the price of the two articles, of which the variations are often opposite and that, in calculations which may throw some light on agriculture in the tropical region, I comprehend in the same point of view interior consumption, exportation lawful and contraband.

TOBACCO.

The tobacco of the island of Cuba is celebrated throughout Europe. The custom of smoking, borrowed from the natives of Hayti, was introduced into Europe about the end of the sixteenth and beginning of the seventeenth century. It was generally hoped that the cultivation of tobacco, freed from an oppressive monopoly, would be to the Havannah a very profitable object of commerce. The good intentions displayed by the government in abolishing, within six years, the Factoria de tabacos, have not been attended by the improvement which was expected in that branch of industry. The cultivators want capital, the farms have become extremely dear, and the predilection for the cultivation of coffee is prejudicial to that of tobacco.

The oldest information we possess respecting the quantity of tobacco which the island of Cuba has thrown into the magazines of the mother country go back to 1748. According to the Abbe Raynal, a much more exact writer than is generally believed, that quantity, from 1748 to 1753 (average year) was 75,000 arrobas. From 1789 to 1794 the produce of the island amounted annually to 250,000 arrobas; but from that period to 1803 the increased price of land, the attention given exclusively to the coffee plantations and the sugar factories, little vexations in the exercise of the royal monopoly (estanco), and impediments in the way of export trade, have progressively diminished the produce by more than one-half. The total produce of tobacco in the island is, however, believed to have been, from 1822 to 1825, again from 300,000 to 400,000 arrobas.

In good years, when the harvest rose to 350,000 arrobas of leaves, 128,000 arrobas were prepared for the Peninsula, 80,000 for the Havannah, 9200 for Peru, 6000 for Panama, 3000 for Buenos Ayres, 2240 for Mexico, and 1000 for Caracas and Campeachy. To complete the sum of 315,000,000 (for the harvest loses 10 per cent of its weight in merma y aberias, during the preparation and the transport) we must suppose that 80,000 arrobas were consumed in the interior of the island (en los campos), whither the monopoly and the taxes did not extend. The maintenance of 120 slaves and the expense of the manufacture amounted only to 12,000 piastres annually; the persons employed in the factoria cost 54,100 piastres. The value of 128,000 arrobas, which in good years was sent to Spain, either in cigars or in snuff (rama y polvos), often exceeded 5,000,000 piastres, according to the common price of Spain. It seems surprising to see that the statements of exportation from the Havannah (documents published by the Consulado) mark the exportations for 1816, at only 3400 arrobas; for 1823, only 13,900 arrobas of tabaco en rama, and 71,000 pounds of tabaco torcida, estimated together, at the custom-house, at 281,000 piastres; for 1825, only 70,302 pounds of cigars, and 167,100 pounds of tobacco in leaves; but it must be remembered that no branch of contraband is more active than that of cigars. Although the tobacco of the Vuelta de abaxo is the most famous, a considerable exportation takes place in the eastern part of the island. I rather doubt the total exportation of 200,000 boxes of cigars (value 2,000,000 piastres) as stated by several travellers during latter years. If the harvests were thus abundant, why should the island of Cuba receive tobacco from the United States for the consumption of the lower class of people?

I shall say nothing of the cotton, the indigo, or the wheat of the island of Cuba. These branches of colonial industry are of comparatively little importance; and the proximity of the United States and Guatimala renders competition almost impossible. The state of Salvador, belonging to the Confederation of Central America, now throws 12,000 tercios annually, or 1,800,000 pounds of indigo into trade; an exportation which amounts to more than 2,000,000 piastres. The cultivation of wheat succeeds (to the great astonishment of travellers who have passed through Mexico), near the Quatro Villas, at small heights above the level of the ocean, though in general it is very limited. The flour is fine; but colonial productions are more tempting, and the plains of the United States—that Crimea of the New World—yield harvests too abundant for the commerce of native cereals to be efficaciously protected by the prohibitive system of the custom-house, in an island near the mouth of the Mississippi and the Delaware. Analogous difficulties oppose the cultivation of flax, hemp, and the vine. Possibly the inhabitants of Cuba are themselves ignorant of the fact that, in the first years of the conquest by the Spaniards, wine was made in their island of wild grapes.* (* De muchas parras monteses con ubas se ha cogido vino, aunque algo agrio. [From several grape-bearing vines which grow in the mountains, they extract a kind of wine; but it is very acid.] Herera Dec. 1 page 233. Gabriel de Cabrera found a tradition at Cuba similar to that which the people of Semitic race have of Noah experiencing for the first time the effect of a fermented liquor. He adds that the idea of two races of men, one naked, another clothed, is linked to the American tradition. Has Cabrera, preoccupied by the rites of the Hebrews, imperfectly interpreted the words of the natives, or, as seems more probable, has he added something to the analogies of the woman-serpent, the conflict of two brothers, the cataclysm of water, the raft of Coxcox, the exploring bird, and many other things that teach us incontestably that there existed a community of antique traditions between the nations of the two worlds? Views of the Cordilleras and Monuments of America.) This kind of vine, peculiar to America, has given rise to the general error that the true Vitis vinifera is common to the two continents. The Parras monteses which yields the somewhat sour wine of the island of Cuba, was probably gathered on the Vitis tiliaefolia which Mr. Willdenouw has described from our herbals. In no part of the northern hemisphere has the vine hitherto been cultivated with the view of producing wine south of the 27 degrees 48 minutes, or the latitude of the island of Ferro, one of the Canaries, and of 29 degrees 2 minutes, or the latitude of Bushire in Persia.

WAX.

This is not the produce of native bees (the Melipones of Latreille), but of bees brought from Europe by way of Florida. The trade in wax has only become important since 1772. The exportation of the whole island, which from 1774 to 1779 was only 2700 arrobas (average year), was estimated in 1803, including contraband, at 42,700 arrobas, of which 25,000 were destined for Vera Cruz. In the churches of Mexico there is a great consumption of Cuban wax. The price varies from sixteen to twenty piastres the arroba.

Trinidad and the small port of Baracoa also carry on a considerable trade in wax, furnished by the almost uncultivated regions on the east of the island. In the proximity of the sugar-factories many bees perish of inebriety from the molasses, of which they are extremely fond. In general the production of wax diminishes in proportion as the cultivation of the land augments. The exportation of wax, according to the present price, amounts to about 500,000 of piastres.

COMMERCE.

It has already been observed that the importance of the commerce of the island of Cuba depends not solely on the riches of its productions, the wants of the population in the articles and merchandize of Europe, but also in great part on the favourable position of the port of the Havannah. This port is situated at the entrance of the Gulf of Mexico, where the high roads of the commercial nations of the old and the new worlds cross each other. It was remarked by the Abbe Raynal, at a period when agriculture and industry were in their infancy, and scarcely threw into commerce the value of 2,000,000 piastres in sugar and tobacco, that the island of Cuba alone might be worth a kingdom to Spain. There seems to have been something prophetic in those memorable words; and since the parent state has lost Mexico, Peru and so many other colonies declared independent, they demand the serious consideration of statesmen who are called upon to discuss the political interests of the Peninsula.

The island of Cuba, to which for a long time the court of Madrid wisely granted great freedom of trade, exports, lawfully and by contraband, of its own native productions, in sugar, coffee, tobacco, wax and skins, to the value of more than 14,000,000 piastres; which is about one-third less than the value of the precious metals furnished by Mexico at the period of the greatest prosperity of its mines.* (* In 1805 gold and silver specie was struck at Mexico to the value of 27,165,888 piastres; but, taking an average of ten years of political tranquillity, we find from 1800 to 1810 scarcely 24 1/2 million of piastres.) It may be said that the Havannah and Vera Cruz are to the rest of America what New York is to the United States. The tonnage of 1000 to 1200 merchant ships which annually enter the port of the Havannah, amounts (excluding the small coasting-vessels), to 150,000 or 170,000 tons.* (* In 1816 the tonnage of the commerce of New York was 299,617 tons; that of Boston, 143,420 tons. The amount of tonnage is not always an exact measure of the wealth of commerce. The countries which export rice, flour, hewn wood and cotton require more capaciousness than the tropical regions of which the productions (cochineal, indigo, sugar and coffee) are of little bulk, although of considerable value.) In time of peace from 120 to 150 ships of war are frequently seen at anchor at the Havannah. From 1815 to 1819 the productions registered at the custom-house of that port only (sugar, rum, molasses, coffee, wax and butter) amounted, on the average, to the value of 11,245,000 piastres per annum. In 1823 the exportation registered two-thirds less than their actual price, amounted (deducting 1,179,000 piastres in specie) to more than 12,500,000 piastres. It is probable that the importations of the whole island (lawful and contraband), estimated at the real price of the articles, the merchandize and the slaves, amount at present to 15,000,000 or 16,000,000 piastres, of which scarcely 3,000,000 or 4,000,000 are re-exported. The Havannah purchases from abroad far beyond its own wants, and exchanges its colonial articles for the productions of the manufactures of Europe, to sell a part of them at Vera Cruz, Truxillo, Guayra, and Carthagena.

On comparing, in the commercial tables of the Havannah, the great value of merchandise imported, with the little value of merchandise re-exported, one is surprised at the vast internal consumption of a country containing only 325,000 whites and 130,000 free men of colour. We find, in estimating the different articles, according to the real current prices: in cotton and linen (bretanas, platillas, lienzos y hilo), two and a half to three millions of piastres; in tissues of cotton (zarazas musulinas), one million of piastres; in silk (rasos y generos de seda), 400,000 piastres; and in linen and woollen tissues, 220,000 piastres. The wants of the island, in European tissues, registered as exported to the port of the Havannah only, consequently exceeded, in these latter years, from four millions to four and a half millions of piastres. To these importations of the Havannah we must add: hardware and furniture, more than half a million of piastres; iron and steel, 380,000 piastres; planks and great timber, 400,000 piastres; Castile soap, 300,000 piastres. With respect to the importation of provisions and drinks to the Havannah, it appears to me to be well worthy the attention of those who would know the real state of those societies which are called sugar or slave colonies. Such is the composition of those societies established on the most fruitful soil which nature can furnish for the nourishment of man, such the direction of agricultural labours and industry in the West Indies, that, in the best climate of the equinoctial region, the population would want subsistence but for the freedom and activity of external commerce. I do not speak of the introduction of wines at the port of the Havannah, which amounted (according to the registers of the custom-house), in 1803, to 40,000 barrels; in 1823, to 15,000 pipas and 17,000 barrels, to the value of 1,200,000 piastres; nor of the introduction of 6000 barrels of brandy from Spain and Holland, and 113,000 barrels (1,864,000 piastres) of flour. These wines, liquors and flour are consumed by the opulent part of the nation. The cereals of the United States have become articles of absolute necessity in a zone where maize, manioc and bananas were long preferred to every other amylaceous food. The development of a luxury altogether European, cannot be complained of amidst the prosperity and increasing civilization of the Havannah; but, along with the introduction of the flour, wine, and spirituous liquors of Europe, we find, in the year 1816, 1 1/2millions of piastres; and, in the year 1823, 3 1/2 millions for salt meat, rice and dried vegetables. In the last mentioned year, the importation of rice was 323,000 arrobas; and the importation of dried and salt meat (tasajo), for the slaves, 465,000 arrobas.

The scarcity of necessary articles of subsistence characterizes a part of the tropical climates where the imprudent activity of Europeans has inverted the order of nature: it will diminish in proportion as the inhabitants, more enlightened respecting their true interests, and discouraged by the low price of colonial produce, will vary the cultivation, and give free scope to all the branches of rural economy. The principles of that narrow policy which guides the government of very small islands, inhabited by men who desert the soil whenever they are sufficiently enriched, cannot be applicable to a country of an extent nearly equal to that of England, covered with populous cities, and where the inhabitants, established from father to son during ages, far from regarding themselves as strangers to the American soil, cherish it as their own country. The population of the island of Cuba, which in fifty years will perhaps exceed a million, may open by its own consumption an immense field to native industry. If the slave-trade should cease altogether, the slaves will pass by degrees into the class of free men; and society, being reconstructed, without suffering any of the violent convulsions of civil dissension, will follow the path which nature has traced for all societies that become numerous and enlightened. The cultivation of the sugar-cane and of coffee will not be abandoned; but it will no longer remain the principal basis of national existence than the cultivation of cochineal in Mexico, of indigo in Guatimala, and of cacao in Venezuela. A free, intelligent and agricultural population will progressively succeed a slave population, destitute of foresight and industry. Already the capital which the commerce of the Havannah has placed within the last twenty-five years in the hands of cultivators, has begun to change the face of the country; and to that power, of which the action is constantly increasing, another will be necessarily joined, inseparable from the progress of industry and national wealth—the development of human intelligence. On these united powers depend the future destinies of the metropolis of the West Indies.

In reference to what has been said respecting external commerce, I may quote the author of a memoir which I have often mentioned, and who describes the real situation of the island. "At the Havannah, the effects of accumulated wealth begin to be felt; the price of provisions has been doubled in a small number of years. Labour is so dear that a bozal negro, recently brought from the coast of Africa, gains by the labour of his hands (without having learned any trade) from four to five reals (two francs thirteen sous to three francs five sous) a day. The negroes who follow mechanical trades, however common, gain from five to six francs. The patrician families remain fixed to the soil: a man who has enriched himself does not return to Europe taking with him his capital. Some families are so opulent that Don Matheo de Pedroso, who died lately, left in landed property above two millions of piastres. Several commercial houses of the Havannah purchase, annually, from ten to twelve thousand cases of sugar, for which they pay at the rate of from 350,000 to 420,000 piastres." (De la situacion presente de Cuba in manuscript.) Such was the state of public wealth at the end of 1800. Twenty-five years of increasing prosperity have elapsed since that period, and the population of the island is nearly doubled. The exportation of registered sugar had not, in any year before 1800, attained the extent of 170,000 cases (31,280,000 kilogrammes); in these latter times it has constantly surpassed 200,000 cases, and even attained 250,000 and 300,000 cases (forty-six to fifty-five millions of kilogrammes). A new branch of industry has sprung up (that of plantations of the coffee tree) which furnishes an exportation of the value of three millions and a half of piastres. Industry, guided by a greater mass of knowledge, has been better directed. The system of taxation that weighed on national industry and exterior commerce has been made lighter since 1791, and been improved by successive changes. Whenever the mother-country, mistaking her own interests, has attempted to make a retrograde step, courageous voices have arisen not only among the Havaneros, but often among the Spanish rulers, in defence of the freedom of American commerce. A new channel has recently been opened for capital by the enlightened zeal and patriotic views of the intendant Don Claudio Martinez de Pinillos, and the commerce of entrepot has been granted to the Havannah on the most advantageous conditions.

The difficult and expensive interior communications of the island render its own productions dearer at the ports, notwithstanding the short distance between the northern and southern coasts. A project of canalization which unites the double advantage of connecting the Havannah and Batabano by a navigable line, and diminishing the high price of the transport of native produce, merits here a special mention. The idea of the Canal of Guines had been conceived for more than half a century with the view of furnishing timber at a more moderate price for ship-building in the arsenal of the Havannah. In 1796 the Count de Jaruco y Mopox, an enterprising man, who had acquired great influence by his connection with the Prince of the Peace, undertook to revive this project. The survey was made in 1798 by two very able engineers, Don Francisco and Don Felix Lemaur. These officers ascertained that the canal in its whole development would be nineteen leagues long (5000 varas or 4150 metres), that the point of partition would be at the Taverna del Rey, and that it would require nineteen locks on the north, and twenty-one on the south. The distance from the Havannah to Batabano is only eight and a half sea-leagues. The canal of Guines would be very useful for the transport of agricultural productions by steam-boats,* because its course would be in proximity with the best cultivated lands. (* Steam-boats are established from the Havannah to Matanzas, and from the Havannah to Mariel. The government granted to Don Juan O'Farrill (March 24th, 1819) a privilege on the barcos de vapor.) The roads are nowhere worse in the rainy season than in this part of the island, where the soil is of friable limestone, little fitted for the construction of solid roads. The transport of sugar from Guines to the Havannah, a distance of twelve leagues, now costs one piastre per quintal. Besides the advantage of facilitating internal communications, the canal would also give great importance to the surgidero of Batabano, into which small vessels laden with salt provisions (tasajo) from Venezuela, would enter without being obliged to double Cape Saint Antonio. In the bad season and in time of war, when corsairs are cruising between Cape Catoche, Tortugas and Mariel, the passage from the Spanish main to the island of Cuba would be shortened by entering, not at the Havannah, but at some port of the southern coast. The cost of constructing the canal de Guines was estimated in 1796 at one million, or 1,200,000 piastres: it is now thought that the expense would amount to more than one million and a half. The productions which might annually pass the canal have been estimated at 75,000 cases of sugar, 25,000 arrobas of coffee, and 8000 bocoyes of molasses and rum. According to the first project, that of 1796, it was intended to link the canal with the small river of Guines, to be brought from the Ingenio de la Holanda to Quibican, three leagues south of Bejucal and Santa Rosa. This idea is now relinquished, the Rio de los

Guines losing its waters towards the east in the irrigation of the savannahs of Hato de Guanamon. Instead of carrying the canal east of the Barrio del Cerro and south of the fort of Atares, in the bay of the Havannah, it was proposed at first to make use of the bed of the Chorrera or Rio Armendaris, from Calabazal to the Husillo, and then of the Zanja Real, not only for conveying the boats to the centre of the arrabales and of the city of the Havannah, but also for furnishing water to the fountains which require to be supplied during three months of the year. I visited several times, with MM. Lemaur, the plains through which this line of navigation is intended to pass. The utility of the project is incontestable if in times of great drought a sufficient quantity of water can be brought to the point of partition.

At the Havannah, as in every place where commerce and the wealth it produces increase rapidly, complaints are heard of the prejudicial influence exercised by them on ancient manners. We cannot here stop to compare the first state of the island of Cuba, when covered with pasturage, before the taking of the capital by the English, and its present condition, since it has become the metropolis of the West Indies; nor to throw into the balance the candour and simplicity of manners of an infant society, against the manners that belong to the development of an advanced civilization. The spirit of commerce, leading to the love of wealth, no doubt brings nations to depreciate what money cannot obtain. But the state of human things is happily such that what is most desirable, most noble, most free in man, is owing only to the inspirations of the soul, to the extent and amelioration of its intellectual faculties. Were the thirst of riches to take absolute possession of every class of society, it would infallibly produce the evil complained of by those who see with regret what they call the preponderance of the industrious system; but the increase of commerce, by multiplying the connections between nations, by opening an immense sphere to the activity of the mind, by pouring capital into agriculture, and creating new wants by the refinement of luxury, furnishes a remedy against the supposed dangers.

FINANCE.

The increase of the agricultural prosperity of the island of Cuba and the influence of the accumulation of wealth on the value of importations, have raised the public revenue in these latter years to four millions and a half, perhaps five millions of piastres. The custom-house of the Havannah, which before 1794 yielded less than 600,000 piastres, and from 1797 to 1800, 1,900,000 piastres, pours into the treasury, since the declaration of free trade, a revenue (importe liquido) of more than 3,100,000 piastres.* (* The custom-house of Port-au-Prince, at Hayti, produced in 1825, the sum of 1,655,764 piastres; that of Buenos Ayres, from 1819 to 1821, average year, 1,655,000 piastres. See Centinela de La Plata, September 1822 Number 8; Argos de Buenos Ayres Number 85.)

The island of Cuba as yet contains only one forty-second part of the population of France; and one half of its inhabitants, being in the most abject indigence, consume but little. Its revenue is nearly equal to that of the Republic of Columbia, and it exceeds the revenue of all the custom houses of the United States* before the year 1795, when that confederation had 4,500,000 inhabitants, while the island of Cuba contained only 715,000. (* The custom-houses of the United States, which yielded in 1801 to 1808 sixteen millions of dollars, produced in 1815 but 7,282,000.) The principal source of the public revenue of this fine colony is the custom-house, which alone produces above three-fifths, and amply suffices for all the wants of the internal administration and military defence. If in these latter years, the expense of the general treasury of the Havannah amounted to more than four millions of piastres, this increase of expense is solely owing to the obstinate struggle maintained between the mother country and her freed colonies. Two millions of piastres were employed to pay the land and sea forces which poured back from the American continent, by the Havannah, on their way to the Peninsula. As long as Spain, unmindful of her real interests, refuses to recognize the independence of the New Republics, the island of Cuba, menaced by Columbia and the Mexican Confederation, must support a military force for its external defence, which ruins the colonial finances. The Spanish naval force stationed in the port of the Havannah generally costs above 650,000 piastres. The land forces require nearly one million and a half of piastres. Such a state of things cannot last indefinitely if the Peninsula do not relieve the burden that presses upon the colony.

From 1789 to 1797 the produce of the custom-house at the Havannah never rose to more than 700,000 piastres. In 1814 it was 1,855,117. From 1815 to 1819 the royal taxes in the port of the Havannah amounted to 11,575,460 piastres; total 18,284,807 piastres; or, average year, 3,657,000 piastres, of which the municipal taxes formed 0.36.

The public revenue of the Administracion general de Rentas of the jurisdiction of Havannah amounted:

in 1820 to 3,631,273 piastres. in 1821 to 3,277,639 piastres. in 1822 to 3,378,228 piastres.

The royal and municipal taxes of importation at the custom-house of the Havannah in 1823 were 2,734,563 piastres.

The total amount of the revenue of the Havannah in 1824 was 3,025,300 piastres.

In 1825 the revenue of the town and jurisdiction of the Havannah was 3,350,300 piastres.

These partial statements show that from 1789 to 1824 the public revenue of Cuba has been increased sevenfold.

According to the estimates of the Cajas matrices, the public revenue in 1822 was, in the province of the Havannah alone, 4,311,862 piastres; which arose from the custom-house (3,127,918 piastres), from the ramos de directa entrada, as lottery, tithes, etc. (601,808 piastres), and anticipations on the charges of the Consulado and the Deposito (581,978 piastres). The expenditure in the same year, for the island of Cuba, was 2,732,738 piastres, and for the succour destined to maintain the struggle with the continental colonies declared independent, 1,362,029 piastres. In the first class of expenditure we find 1,355,798 piastres for the subsistence of the military forces kept up for the defence of the Havannah and the neighbouring places; and 648,908 piastres for the royal navy stationed in the port of the Havannah. In the second class of expense foreign to the local administration we find 1,115,672 piastres for the pay of 4234 soldiers who, after having evacuated Mexico, Columbia and other parts of the Continent formerly Spanish possessions, passed by the Havannah to return to Spain; 164,000 piastres is the cost of the defence of the castle of San Juan de Ulloa.

I here terminate the Political Essay on the island of Cuba, in which I have traced the state of that important Spanish possession as it now is. My object has been to throw light on facts and give precision to ideas by the aid of comparisons and statistical tables. That minute investigation of facts is desirable at a moment when, on the one hand enthusiasm exciting to benevolent credulity, and on the other animosities menacing the security of the new republics, have given rise to the most vague and erroneous statements. I have as far as possible abstained from all reasoning on future chances, and on the probability of the changes which external politics may produce in the situation of the West Indies. I have merely examined what regards the organization of human society; the unequal partition of rights and of the enjoyments of life; the threatening dangers which the wisdom of the legislator and the moderation of free men may ward off, whatever be the form of the government. It is for the traveller who has been an eyewitness of the suffering and the degradation of human nature to make the complaints of the unfortunate reach the ear of those by whom they can be relieved. I observed the condition of the blacks in countries where the laws, the religion and the national habits tend to mitigate their fate; yet I retained, on quitting America, the same horror of slavery which I had felt in Europe. In vain have writers of ability, seeking to veil barbarous institutions by ingenious turns of language, invented the expressions negro peasants of the West Indies, black vassalage, and patriarchal protection: that is profaning the noble qualities of the mind and the imagination, for the purpose of exculpating by illusory comparisons or captious sophisms excesses which afflict humanity, and which prepare the way for violent convulsions. Do they think that they have acquired the right of putting down commiseration, by comparing* the condition of the negroes with that of the serfs of the middle ages, and with the state of oppression to which some classes are still subjected in the north and east of Europe? (* Such comparisons do not satisfy those secret partisans of the slave trade who try to make light of the miseries of the black race, and to resist every emotion those miseries awaken. The permanent condition of a caste founded on barbarous laws and institutions is often confounded with the excesses of a power temporarily exercised on individuals. Thus Mr. Bolingbroke, who lived seven years at Demerara and who visited the West India Islands, observes that "on board an English ship of war, flogging is more frequent than in the plantations of the English colonies." He adds "that in general the negroes are but little flogged, but that very reasonable means of correction have been imagined, such as making them take boiling soup strongly peppered, or obliging them to drink, with a very small spoon, a solution of Glauber-salts." Mr. Bolingbroke regards the slave-trade as a universal benefit; and he is persuaded that if negroes who have enjoyed, during twenty years, all the comforts of slave life at Demerara, were permitted to return to the coast of Africa, they would effect recruiting on a large scale, and bring whole nations to the English possessions. Voyage to Demerara, 1807. Such is the firm and frank profession of faith of a planter; yet Mr. Bolingbroke, as several passages of his book prove, is a moderate man, full of benevolent intentions towards the slaves.) These comparisons, these artifices of language, this disdainful impatience with which even a hope of the gradual abolition of slavery is repulsed as chimerical, are useless arms in the times in which we live. The great revolutions which the continent of America and the Archipelago of the West Indies have undergone since the commencement of the nineteenth century, have had their influence on public feeling and public reason, even in countries where slavery exists and is beginning to be modified. Many sensible men, deeply interested in the tranquillity of the sugar and slave islands, feel that by a liberal understanding among the proprietors, and by judicious measures adopted by those who know the localities, they might emerge from a state of danger and uneasiness which indolence and obstinacy serve only to increase.

Slavery is no doubt the greatest evil that afflicts human nature, whether we consider the slave torn from his family in his native country and thrown into the hold of a slave ship,* or as making part of a flock of black men, parked on the soil of the West Indies; but for individuals there are degrees of suffering and privation. (* "If the slaves are whipped," said one of the witnesses before the Parliamentary Committee of 1789, "to make them dance on the deck of a slave ship—if they are forced to sing in chorus; 'Messe, messe, mackerida,' [how gaily we live among the whites], this only proves the care we take of the health of those men." This delicate attention reminds me of the description of an auto-da-fe in my possession. In that curious document a boast is made of the prodigality with which refreshments are distributed to the condemned, and of the staircase which the inquisitors have had erected in the interior of the pile for the accommodation of the relazados (the relapsed culprits.)) How great is the difference in the condition of the slave who serves in the house of a rich family at the Havannah or at Kingston, or one who works for himself, giving his master but a daily retribution, and that of the slave attached to a sugar estate! The threats employed to correct an obstinate negro mark this scale of human privations. The coachman is menaced with the coffee plantation; and the slave working on the latter is menaced with the sugar house. The negro, who with his wife inhabits a separate hut, whose heart is warmed by those feelings of affection which for the most part characterize the African race, finds that after his labour some care is taken of him amidst his indigent family, is in a position not to be compared with that of the insulated slave lost in the mass. This diversity of condition escapes the notice of those who have not had the spectacle of the West Indies before their eyes. Owing to the progressive amelioration of the state even of the captive caste in the island of Cuba, the luxury of the masters and the possibility of gain by their work, have drawn more than eighty thousand slaves to the towns; and the manumission of them, favoured by the wisdom of the laws, is become so active as to have produced, at the present period, more than 130,000 free men of colour. By considering the individual position of each class, by recompensing, by the decreasing scale of privations, intelligence, love of labour and the domestic virtues, the colonial administration will find the best means of improving the condition of the blacks. Philanthropy does not consist in giving a little more salt-fish, and some fewer lashes: the real amelioration of the captive caste ought to extend over the whole moral and physical position of man.

The impulse may be given by those European governments which have a right comprehension of human dignity, and who know that whatever is unjust bears with it a germ of destruction; but this impulse, it is melancholy to add, will be powerless if the union of the planters, if the colonial assemblies or legislatures, fail to adopt the same views and to act by a well-concerted plan, having for its ultimate aim the cessation of slavery in the West Indies. Till then it will be in vain to register the strokes of the whip, to diminish the number that may be given at one time, to require the presence of witnesses and to appoint protectors of slaves; all these regulations, dictated by the most benevolent intentions, are easily eluded: the isolated position of the plantations renders their execution impossible. They pre-suppose a system of domestic inquisition incompatible with what is understood in the colonies by the phrase established rights. The state of slavery cannot be altogether peaceably ameliorated except by the simultaneous action of the free men (white men and coloured) residing in the West Indies; by colonial assemblies and legislatures; by the influence of those who, enjoying great moral consideration among their countrymen and acquainted with the localities, know how to vary the means of improvement conformably with the manners, habits, and the position of every island. In preparing the way for the accomplishment of this task, which ought to embrace a great part of the archipelago of the West Indies, it may be useful to cast a retrospective glance on the events by which the freedom of a considerable part of the human race was obtained in Europe in the middle ages. In order to ameliorate without commotion new institutions must be made, as it were, to rise out of those which the barbarism of centuries has consecrated. It will one day seem incredible that until the year 1826 there existed no law in the Great Antilles to prevent the sale of young infants and their separation from their parents, or to prohibit the degrading custom of marking the negroes with a hot iron, merely to enable these human cattle to be more easily recognized. Enact laws to obviate the possibility of a barbarous outrage; fix, in every sugar estate, the proportion between the least number of negresses and that of the labouring negroes; grant liberty to every slave who has served fifteen years, to every negress who has reared four or five children; set them free on the condition of working a certain number of days for the profit of the plantation; give the slaves a part of the net produce, to interest them in the increase of agricultural riches;* fix a sum on the budget of the public funds, destined for the ransom of slaves, and the amelioration of their condition—such are the most urgent objects for colonial legislation. (* General Lafayette, whose name is linked with all that promises to contribute to the liberty of man and the happiness of mankind, conceived, in the year 1785, the project of purchasing a settlement at Cayenne, and to divide it among the blacks by whom it was cultivated and in whose favour the proprietor renounced for himself and his descendants all benefit whatever. He had interested in this noble enterprise the priests of the Mission of the Holy Ghost, who themselves possessed lands in French Guiana. A letter from Marshal de Castries, dated 6th June, 1785, proves that the unfortunate Louis XVI, extending his beneficent intentions to the blacks and free men of colour, had ordered similar experiments to be made at the expense of Government. M. de Richeprey, who was appointed by M. de Lafayette to superintend the partition of the lands among the blacks, died from the effects of the climate at Cayenne.)

The Conquest on the continent of Spanish America and the slave-trade in the West Indies, in Brazil, and in the southern parts of the United States, have brought together the most heterogeneous elements of population. This strange mixture of Indians, whites, negroes, mestizos, mulattoes and zambos is accompanied by all the perils which violent and disorderly passion can engender, at those critical periods when society, shaken to its very foundations, begins a new era. At those junctures, the odious principle of the Colonial System, that of security, founded on the hostility of castes, and prepared during ages, has burst forth with violence. Fortunately the number of blacks has been so inconsiderable in the new states of the Spanish continent that, with the exception of the cruelties exercised in Venezuela, where the royalist party armed their slaves, the struggle between the independents and the soldiers of the mother country was not stained by the vengeance of the captive population. The free men of colour (blacks, mulattoes and mestizoes) have warmly espoused the national cause; and the copper-coloured race, in its timid distrust and passiveness, has taken no part in movements from which it must profit in spite of itself. The Indians, long before the revolution, were poor and free agriculturists; isolated by their language and manners they lived apart from the whites. If, in contempt of Spanish laws, the cupidity of the corregidores and the tormenting system of the missionaries often restricted their liberty, that state of vexatious oppression was far different from personal slavery like that of the slavery of the blacks, or of the vassalage of the peasantry in the Sclavonian part of Europe. It is the small number of blacks, it is the liberty of the aboriginal race, of which America has preserved more than eight millions and a half without mixture of foreign blood, that characterizes the ancient continental possessions of Spain, and renders their moral and political situation entirely different from that of the West Indies, where, by the disproportion between the free men and the slaves, the principles of the Colonial System have been developed with more energy. In the West Indian archipelago as in Brazil (two portions of America which contain near 3,200,000 slaves) the fear of [?] among the blacks, and the perils that surround the whites, have been hitherto the most powerful causes of the security of the mother countries and of the maintenance of the Portuguese dynasty. Can this security, from its nature, be of long duration? Does it justify the inertness of governments who neglect to remedy the evil while it is yet time? I doubt this. When, under the influence of extraordinary circumstances, alarm is mitigated, when countries in which the accumulation of slaves has produced in society the fatal mixture of heterogeneous elements may be led, perhaps unwillingly, into an exterior struggle, civil dissensions will break forth in all their violence and European families, innocent of an order of things which they have had no share in creating, will be exposed to the most imminent dangers.

We can never sufficiently praise the legislative wisdom of the new republics of Spanish America which, since their birth, have been seriously intent on the total extinction of slavery. That vast portion of the earth has, in this respect, an immense advantage over the southern part of the United States, where the whites, during the struggle with England, established liberty for their own profit, and where the slave population, to the number of 1,600,000, augments still more rapidly than the whites.* (* In 1769, forty-six years before the declaration of the Congress at Vienna, and thirty-eight years before the abolition of the slave-trade, decreed in London and at Washington, the Chamber of Representatives of Massachusetts had declared itself against "the unnatural and unwarrantable custom of enslaving mankind." See Walsh's Appeal to the United States, 1819 page 312. The Spanish writer, Avendano, was perhaps the first who declaimed forcibly not only against the slave-trade, abhorred even by the Afghans (Elphinstone's Journey to Cabul page 245), but against slavery in general, and "all the iniquitous sources of colonial wealth." Thesaurus Ind. tom. 1 tit. 9 cap. 2.) If civilization, instead of extending, were to change its place; if, after great and deplorable convulsions in Europe, America, between Cape Hatteras and the Missouri, were to become the principal seat of the light of Christianity, what a spectacle would be presented by that centre of civilization, where, in the sanctuary of liberty, we could attend a sale of negroes after the death of a master, and hear the sobbings of parents who are separated from their children! Let us hope that the generous principles which have so long animated the legislatures of the northern parts of the United States will extend by degrees southward and towards those western regions where, by the effect of an imprudent and fatal law, slavery and its iniquities have passed the chain of the Alleghenies and the banks of the Mississippi: let us hope that the force of public opinion, the progress of knowledge, the softening of manners, the legislation of the new continental republics and the great and happy event of the recognition of Hayti by the French government, will, either from motives of prudence and fear, or from more noble and disinterested sentiments, exercise a happy influence on the amelioration of the state of the blacks in the rest of the West Indies, in the Carolinas, Guiana, and Brazil.

In order to slacken gradually the bonds of slavery the laws against the slave-trade must be most strictly enforced, and punishments inflicted for their infringement; mixed tribunals must be formed, and the right of search exercised with equitable reciprocity. It is melancholy to learn that, owing to the culpable indifference of some of the governments of Europe, the slave-trade (more cruel from having become more secret) has dragged from Africa, within ten years, almost the same number of negroes as before 1807; but we must not from this fact infer the inutility, or, as the secret partisans of slavery assert, the practical impossibility of the beneficent measures adopted first by Denmark, the United States and Great Britain, and successively by all the rest of Europe. What passed from 1807 till the time when France recovered possession of her ancient colonies, and what passes in our days in nations whose governments sincerely desire the abolition of the slave-trade and its abominable practices, proves the fallacy of this conclusion. Besides, is it reasonable to compare numerically the importation of slaves in 1825 and in 1806? With the activity prevailing in every enterprise of industry, what an increase would the importation of negroes have taken in the English West Indies and the southern provinces of the United States if the slave-trade, entirely free, had continued to supply new slaves, and had rendered the care of their preservation and the increase of the old population, superfluous? Can we believe that the English trade would have been limited, as in 1806, to the sale of 53,000 slaves; and that of the United States, to the sale of 15,000? It is pretty well ascertained that the English islands received in the 106 years preceding 1786 more than 2,130,000 negroes, forcibly carried from the coast of Africa. At the period of the French revolution, the slave-trade furnished (according to Mr. Norris) 74,000 slaves annually, of which the English colonies absorbed 38,000, and the French 20,000. It would be easy to prove that the whole of the West Indian archipelago, which now comprises scarcely 2,400,000 negroes and mulattoes (free and slaves), received, from 1670 to 1825, nearly 5,000,000 of Africans. These revolting calculations respecting the consumption of the human species do not include the number of unfortunate slaves who have perished in the passage or have been thrown into the sea as damaged merchandize.* (* Volume 7 page 151. See also the eloquent speech of the Duke de Broglie, March 28th, 1822 pages 40, 43 and 96.) By how many thousands must we have augmented the loss, if the two nations most distinguished for ardour and intelligence in the development of commerce and industry, the English and the inhabitants of the United States, had continued, from 1807, to carry on the trade as freely as some other nations of Europe? Sad experience has proved how much the treaties of the 15th July, 1814, and of the 22nd January, 1815, by which Spain and Portugal reserved to themselves the trade in blacks during a certain number of years, have been fatal to humanity.

The local authorities, or rather the rich proprietors, forming the Ayuntamiento of the Havannah, the Consulado and the Patriotic Society, have on several occasions shown a disposition favourable to the amelioration of the condition of the slaves.* (* Dicen nuestros Indios del Rio Caura cuando se confiesan que ya entienden que es pecado comer carne humana; pero piden qua se les permita desacostumbrarse poco a poco; quieren comer la carne humana una vez al mes, despues cada tres meses, hasta qua sin sentirlo pierdan la costumbre. Cartas de los Rev Padres Observantes Number 7 manuscript. [Our negroes of the River Caura say, when they confess, that they know it is sinful to eat human flesh; they beg to be permitted to break themselves of the custom, little by little: they wish to eat human flesh once a month, and afterwards once every three months, until they feel they have cured themselves of the practice.]) If the government of the mother-country, instead of dreading the least appearance of innovation, had taken advantage of those propitious circumstances, and of the ascendancy of some men of abilities over their countrymen, the state of society would have undergone progressive changes; and in our days, the inhabitants of the island of Cuba would have enjoyed some of the improvements which have been under discussion for the space of thirty years. The movement at Saint Domingo in 1790 and those which took place in Jamaica in 1794 caused so great an alarm among the hacendados of the island of Cuba that in a Junta economica it was warmly debated what measure could be adopted to secure the tranquillity of the country. Regulations were made respecting the pursuit of fugitive slaves,* which, till then, had given rise to the most revolting excesses (* Reglamento sobre los Negros Cimmarones de 26 de Dec. de 1796. Before the year 1788 there were great numbers of fugitive negroes (cimmarones) in the mountains of Jaruco, where they were sometimes apalancados, that is, where several of those unfortunate creatures formed small intrenchments for their common defence by heaping up trunks of trees. The maroon negroes, born in Africa (bozales), are easily taken; for the greater number, in the vain hope of finding their native land, march day and night in the direction of the east. When taken they are so exhausted by fatigue and hunger that they are only saved by giving them, during several days, very small quantities of soup. The creole maroon negroes conceal themselves by day in the woods and steal provisions during the night. Till 1790, the right of taking the fugitive negroes belonged only to the Alcalde mayor provincial, an hereditary office in the family of the Count de Bareto. At present any of the inhabitants can seize the maroons and the proprietor of the slave pays four piastres per head, besides the food. If the name of the master is not known, the Consulado employs the maroon negro in the public works. This man-hunting, which, at Hayti and Jamaica, has given so much fatal celebrity to the dogs of Cuba, was carried on in the most cruel manner before the regulation which I have mentioned above.); it was proposed to augment the number of negresses on the sugar estates, to direct more attention to the education of children, to diminish the introduction of African negroes, to bring white planters from the Canaries, and Indian planters from Mexico, to establish country schools with the view of improving the manners of the lower class, and to mitigate slavery in an indirect way. These propositions had not the desired effect. The junta opposed every system of immigration, and the majority of the proprietors, indulging their old illusions of security, would not restrain the slave-trade when the high price of the produce gave a hope of extraordinary profit. It would, however, be unjust not to acknowledge in this struggle between private interests and the views of wise policy, the desires and the principles manifested by some inhabitants of the island of Cuba, either in their own name or in the name of some rich and powerful corporations. "The humanity of our legislation," says M. d'Arango nobly,* in a memoir written in 1796 (* Informe sobre negros fugitivos (de 9 de Junio de 1769), par Don Francisco de Arango y Pareno, Oidor honorario y syndico del Consulado.), "grants the slave four rights (quatro consuelos) which somewhat assuage his sufferings and which have always been refused him by a foreign policy. These rights are, the choice of a master less severe* (* The right of buscar amo. When a slave has found a new master who will purchase him, he may quit the master of whom he has to complain; such is the sense and spirit of a law, beneficent, though often eluded, as are all the laws that protect the slaves. In the hope of enjoying the privilege of buscar amo, the blacks often address to the travellers they meet, a question, which in civilized Europe, where a vote or an opinion is sometimes sold, is more equivocally expressed; Quiere Vm comprarme? [Will you buy me, Sir?]); the privilege of marrying according to his own inclination; the possibility of purchasing his liberty* by his labour (* A slave in the Spanish colonies ought, according to law, to be estimated at the lowest price; this estimate, at the time of my journey, was, according to the locality, from 200 to 380 piastres. In 1825 the price of an adult negro at the island of Cuba, was 450 piastres. In 1788 the French trade furnished a negro for 280 to 300 piastres. A slave among the Greeks cost 300 to 600 drachmes (54 to 108 piastres), when the day-labourer was paid one-tenth of a piastre. While the Spanish laws and institutions favour manumission in every way, the master, in the other islands, pays the fiscal, for every freed slave, five to seven hundred piastres!), and of paying, with an acquired property, for the liberty of his wife and children.* (* What a contrast is observable between the humanity of the most ancient Spanish laws concerning slavery, and the traces of barbarism found in every page of the Black Code and in some of the provincial laws of the English islands! The laws of Barbadoes, made in 1686, and those of Bermuda, in 1730, decreed that the master who killed his negro in chastising him, could not even be sued, while the master who killed his slave wilfully should pay ten pounds sterling to the royal treasury. A law of saint Christopher's, of March 11th, 1784, begins with these words: "Whereas some persons have of late been guilty of cutting off and depriving slaves of their ears, we order that whoever shall extirpate an eye, tear out the tongue, or cut off the nose of a slave, shall pay five hundred pounds sterling, and be condemned to six months imprisonment." It is unnecessary to add that these English laws, which were in force thirty or forty years ago, are abolished and superseded by laws more humane. Why can I not say as much of the legislation of the French islands, where six young slaves, suspected of an intention to escape, were condemned, by a sentence pronounced in 1815, to have their hamstrings cut!) Notwithstanding the wisdom and mildness of Spanish legislation, to how many excesses the slave is exposed in the solitude of a plantation or a farm, where a rude capatez, armed with a cutlass (machete) and a whip, exercises absolute authority with impunity! The law neither limits the punishment of the slave, nor the duration of labour; nor does it prescribe the quality and quantity of his food.* (* A royal cedula of May 31st, 1789 had attempted to regulate the food and clothing; but that cedula was never executed.) It permits the slave, it is true, to have recourse to a magistrate, in order that he may enjoin the master to be more equitable; but this recourse is nearly illusory; for there exists another law according to which every slave may be arrested and sent back to his master who is found without permission at the distance of a league and a half from the plantation to which he belongs. How can a slave, whipped, exhausted by hunger, and excess of labour, find means to appear before the magistrate? and if he did reach him, how would he be defended against a powerful master who calls the hired accomplices of his cruelties as witnesses."

In conclusion I may quote a very remarkable extract from the Representacion del Ayuntamiento, Consulado, y Sociedad patriotica, dated July 20th, 1811. "In all that relates to the changes to be introduced in the captive class, there is much less question of our fears on the diminution of agricultural wealth, than of the security of the whites, so easy to be compromised by imprudent measures. Besides, those who accuse the consulate and the municipality of the Havannah of obstinate resistance forget that, in the year 1799, the same authorities proposed fruitlessly that the government would divert attention to the state of the blacks in the island of Cuba (del arreglo de este delicado asunto.) Further, we are far from adopting the maxims which the nations of Europe, who boast of their civilization, have regarded as incontrovertible; that, for instance, without slaves there could be no colonies. We declare, on the contrary, that without slaves, and even without blacks, colonies might have existed, and that the whole difference would have been comprised in more or less profit, by the more or less rapid increase of the products. But such being our firm persuasion, we ought also to remind your Majesty that a social organization into which slavery has been introduced as an element cannot be changed with inconsiderate precipitation. We are far from denying that it was an evil contrary to all moral principles to drag slaves from one continent to another; that it was a political error not to have listened to the remonstrances of Ovando, the governor of Hispaniola, who complained of the introduction and accumulation of so many slaves in proximity with a small number of free men; but, these evils being now inveterate, we ought to avoid rendering our position and that of our slaves worse, by the employment of violent means. What we ask of your Majesty is conformable to the wish proclaimed by one of the most ardent protectors of the rights of humanity, by the most determined enemy of slavery; we desire, like him, that the civil laws should deliver us at the same time from abuses and dangers."

On the solution of this problem depends, in the West India Islands only, and exclusive of the republic of Hayti, the security of 875,000 free men (whites and men of colour* (* Namely: 452,000 whites, of which 342,000 are in the two Spanish Islands (Cuba and Porto Rico), and 423,000 free men of colour, mulattoes, and blacks.)) and the mitigation of the sufferings of 1,150,000 slaves. It is evident that these objects can never be attained by peaceful means, without the concurrence of the local authorities, either colonial assemblies, or meetings of proprietors designated by less dreaded names, by the old parent state. The direct influence of the authorities is indispensable; and it is a fatal error to believe that we may leave it to time to act. Time will act simultaneously on the slaves, on the relations between the islands and the inhabitants of the continent, and on events which cannot be controlled, when they have been waited for with the inaction of apathy. Wherever slavery is long established, the increase of civilization solely has less influence on the treatment of slaves than many are disposed to admit. The civilization of a nation seldom extends to a great number of individuals; and does not reach those who in the plantations are in immediate contact with the blacks. I have known very humane proprietors shrink from the difficulties that arise in the great plantations; they hesitate to disturb established order, to make innovations, which, if not simultaneous, not supported by the legislation, or (which would be more powerful) by public feeling, would fail in their end, and perhaps aggravate the wretchedness of those whose sufferings they were meant to alleviate. These considerations retard the good that might be effected by men animated by the most benevolent intentions, and who deplore the barbarous institutions which have devolved to them by inheritance. They well know that to produce an essential change in the state of the slaves, to lead them progressively to the enjoyment of liberty, requires a firm will on the part of the local authorities, the concurrence of wealthy and enlightened citizens, and a general plan in which all chances of disorder and means of repression are wisely calculated. Without this community of action and effort slavery, with its miseries and excesses, will survive as it did in ancient Rome,* along with elegance of manners, progressive intelligence, and all the charms of the civilization which its presence accuses, and which it threatens to destroy, whenever the hour of vengeance shall arrive. (* The argument deduced from the civilization of Rome and Greece in favour of slavery is much in vogue in the West Indies, where sometimes we find it adorned with all the graces of erudition. Thus, in speeches delivered in 1795, in the Legislative Assembly of Jamaica, it was alleged that from the example of elephants having been employed in the wars of Pyrrhus and Hannibal, it could not be blamable to have brought a hundred dogs and forty hunters from the island of Cuba to hunt the maroon negroes. Bryan Edwards volume 1 page 570.) Civilization, or slow national demoralization, merely prepare the way for future events; but to produce great changes in the social state there must be a coincidence of certain events, the period of the occurrence of which cannot be calculated. Such is the complication of human destiny, that the same cruelties which tarnished the conquest of America have been re-enacted before our own eyes in times which we suppose to be characterized by vast progress, information and general refinement of manners. Within the interval embraced by the span of one life we have seen the reign of terror in France, the expedition to St. Domingo,* (* The North American Review for 1821 Number 30 contains the following passage: Conflicts with slaves fighting for their freedom are not only dreadful on account of the atrocities to which they give rise on both sides; but even after freedom has been gained they help to confound every sentiment of justice and injustice. Some planters are condemning to death all the male negro population above six years of age. They affirm that those who have not borne arms will be contaminated by the example of those who have been fighting. This merciless act is the consequence of the result of the continued misfortunes of the colonies. Charault, Reflexions sur Saint Domingue.), the political re-action in Naples and Spain, I may also add, the massacres of Chio, Ipsara and Missolonghi, the work of the barbarians of Eastern Europe, which the civilized nations of the north and west did not deem it their duty to prevent. In slave countries, where the effect of long habit tends to legitimize institutions the most adverse to justice, it is vain to count on the influence of information, of intellectual culture, or refinement of manners, except in as much as all those benefits accelerate the impulse given by governments and facilitate the execution of measures once adopted. Without the directive action of governments and legislatures a peaceful revolution is a thing not to be hoped for. The danger becomes the more imminent when a general inquietude pervades the public mind; when amidst the political dissensions of neighbouring countries the faults and the duties of governments have been revealed: in such cases tranquillity can be restored only by a ruling authority which, in the noble consciousness of its power and right, sways events by entering itself on the career of improvement.

CHAPTER 3.32.
GEOGNOSTIC DESCRIPTION OF SOUTH AMERICA, NORTH OF THE RIVER AMAZON, AND EAST OF THE MERIDIAN OF THE SIERRA NEVADA DE MERIDA.

The object of this memoir is to concentrate the geological observations which I collected during my journeys among the mountains of New Andalusia and Venezuela, on the banks of the Orinoco and in the Llanos of Barcelona, Calabozo and the Apure; consequently, from the coast of the Caribbean Sea to the valley of the Amazon, between 2 and 10 1/2 degrees north latitude.

The extent of country which I traversed in different directions was more than 15,400 square leagues. It has already formed the subject of a geological sketch, traced hastily on the spot, after my return from the Orinoco, and published in 1801. At that period the direction of the Cordillera on the coast of Venezuela and the existence of the Cordillera of Parime were unknown in Europe. No measure of altitude had been attempted beyond the province of Quito; no rock of South America had been named; there existed no description of the superposition of rocks in any region of the tropics. Under these circumstances an essay tending to prove the identity of the formations of the two hemispheres could not fail to excite interest. The study of the collections which I brought back with me, and four years of journeying in the Andes, have enabled me to rectify my first views, and to extend an investigation which, by reason of its novelty, had been favourably received. That the most remarkable geological relations may be the more easily seized, I shall treat aphoristically, in different sections, the configuration of the soil, the general division of the land, the direction and inclination of the beds and the nature of the primitive, intermediary, secondary and tertiary rocks.

SECTION 1.

Configuration of the Country.

Inequalities of the Soil.

Chains and Groups of Mountains.

Divisionary Ridges.

Plains or Llanos.

South America is one of those great triangular masses which form the three continental parts of the southern hemisphere of the globe. In its exterior configuration it resembles Africa more than Australia. The southern extremities of the three continents are so placed that, in sailing from the Cape of Good Hope (latitude 33 degrees 55 minutes) to Cape Horn (latitude 55 degrees 58 minutes), and doubling the southern point of Van Diemen's Land (latitude 43 degrees 38 minutes), we see those lands stretching out towards the south pole in proportion as we advance eastward. A fourth part of the 571,000 square sea leagues* (* Almost double the extent of Europe.) which South America comprises is covered with mountains distributed in chains or gathered together in groups. The other parts are plains forming long uninterrupted bands covered with forests or gramina, flatter than in Europe, and rising progressively, at the distance of 300 leagues from the coast, between 30 and 170 toises above the level of the sea. The most considerable mountainous chain in South America extends from south to north according to the greatest dimension of the continent; it is not central like the European chains, nor far removed from the sea-shore, like the Himalaya and the Hindoo-Koosh; but it is thrown towards the western extremity of the continent, almost on the coast of the Pacific Ocean. Referring to the profile which I have given* of the configuration of South America (* Map of Columbia according to the astronomical observations of Humboldt by A.H. Brue 1823.), in the latitude of Chimborazo and Grand Para, across the plains of the Amazon, we find the land low towards the east, in an inclined plane, at an angle of less than 25 seconds on a length of 600 leagues; and if, in the ancient state of our planet, the Atlantic Ocean, by some extraordinary cause, ever rose to 1100 feet above its present level (a height one-third less than the table-lands of Spain and Bavaria), the waves must, in the province of Jaen de Bracamoros, have broken upon the rocks that bound the eastern declivity of the Cordilleras of the Andes. The rising of this ridge is so inconsiderable compared to the whole continent that its breadth in the parallel of Cape Saint Roche is 1400 times greater than the average height of the Andes.

We distinguish in the mountainous part of South America a chain and three groups of mountains, namely, the Cordillera of the Andes, which the geologist may trace without interruption from Cape Pilares, in the western part of the Straits of Magellan, to the promontory of Paria opposite the island of Trinidad; the insulated group of the Sierra Nevada de Santa Marta; the group of the mountains of the Orinoco, or of La Parime; and that of the mountains of Brazil. The Sierra de Santa Marta being nearly in the meridian of the Cordilleras of Peru and New Grenada, the snowy summits descried by navigators in passing the mouth of the Rio Magdalena are commonly mistaken for the northern extremity of the Andes. I shall soon prove that the colossal group of the Sierra de Santa Marta is almost entirely separate from the mountains of Ocana and Pamplona which belong to the eastern Cordillera of New Grenada. The hot plains through which runs the Rio Cesar, and which extend towards the valley of Upar, separate the Sierra Nevada from the Paramo de Cacota, south of Pamplona. The ridge which divides the waters between the gulf of Maracaibo and the Rio Magdalena is in the plain on the east of the Laguna Zapatoza. If, on the one hand, the Sierra de Santa Marta has been erroneously considered (on account of its eternal snow, and its longitude) to be a continuation of the Cordillera of the Andes, on the other hand, the connexion of that same Cordillera with the coast mountains of the provinces of Cumana and Caracas has not been recognized. The littoral chain of Venezuela, of which the different ranges form the Montana de Paria, the isthmus of Araya, the Silla of Caracas and the gneiss-granite mountains north and south of the lake of Valencia, is joined between Porto Cabello, San Felipe and Tocuyo to the Paramos de las Rosas and Niquitao, which form the north-east extremity of the Sierra de Merida, and the eastern Cordillera of the

Andes of New Grenada. It is sufficient here to mention this connexion, so important in a geological point of view; for the denominations of Andes and Cordilleras being altogether in disuse as applied to the chains of mountains extending from the eastern gulf of Maracaibo to the promontory of Paria, we shall continue to designate those chains (stretching from west to east) by the names of littoral chain, or coast-chain of Venezuela.

Of the three insulated groups of mountains, that is to say, those which are not branches of the Cordillera of the Andes and its continuation towards the shore of Venezuela, one is on the north, and the other two on the west of the Andes: that on the north is the Sierra Nevada de Santa Marta; the two others are the Sierra de la Parime, between 4 and 8 degrees of north latitude, and the mountains of Brazil, between 15 and 28 degrees south latitude. This singular distribution of great inequalities of soil produces three plains or basins, comprising a surface of 420,600 square leagues, or four-fifths of all South America, east of the Andes. Between the coast-chain of Venezuela and the group of the Parime, the plains of the Apure and the Lower Orinoco extend; between the group of Parime and the Brazil mountains are the plains of the Amazon, of the Rio Negro and the Madeira, and between the groups of Brazil and the southern extremity of the continent are the plains of Rio de la Plata and of Patagonia. As the group of the Parime in Spanish Guiana, and of the Brazil mountains (or of Minas Geraes and Goyaz), do not join the Cordillera of the Andes of New Grenada and Upper Peru towards the west, the three plains of the Lower Orinoco, the Amazon, and the Rio de la Plata, are connected by land-straits of considerable breadth. These straits are also plains stretching from north to south, and traversed by ridges imperceptible to the eye but forming divortia aquarum. These ridges (and this remarkable phenomenon has hitherto escaped the attention of geologists) are situated between 2 and 3 degrees north latitude, and 16 and 18 degrees south latitude. The first ridge forms the partition of the waters which fall into the Lower Orinoco on the north-east, and into the Rio Negro and the Amazon on the south and south-east; the second ridge divides the tributary streams of the right bank of the Amazon and the Rio de la Plata. These ridges, of which the existence is only manifested, as in Volhynia, by the course of the waters, are parallel with the coast-chain of Venezuela; they present, as it were, two systems of counter-slopes partially developed, in the direction from west to east, between the Guaviare and the Caqueta, and between the Mamori and the Pilcomayo. It is also worthy of remark that in the southern hemisphere the Cordillera of the Andes sends an immense counterpoise eastward in the promontory of the Sierra Nevada de Cochabamba, whence begins the ridge stretching between the tributary streams of the Madeira and the Paraguay to the lofty group of the mountains of Brazil or Minas Geraes. Three transversal chains (the coast-mountains of Venezuela, of the Orinoco or Parime, and the Brazil mountains) tend to join the longitudinal chain (the Andes) either by an intermediary group (between the lake of Valencia and Tocuyo), or by ridges formed by the intersection of counter-slopes in the plains. The two extremities of the three Llanos which communicate by land-straits, the Llanos of the Lower Orinoco, the Amazon, and the Rio de la Plata or of Buenos Ayres, are steppes covered with gramina, while the intermediary Llano (that of the Amazon) is a thick forest. With respect to the two land-straits forming bands directed from north to south (from the Apure to Caqueta across the Provincia de los Llanos, and the sources of the Mamori to Rio Pilcomayo, across the province of Mocos and Chiquitos) they are bare and grassy steppes like the plains of Caracas and Buenos Ayres.

In the immense extent of land east of the Andes, comprehending more than 480,000 square sea leagues, of which 92,000 are a mountainous tract of country, no group rises to the region of perpetual snow; none even attains the height of 1400 toises. This lowering of the mountains in the eastern region of the New Continent extends as far as 60 degrees north latitude; while in the western part, on the prolongation of the Cordillera of the Andes, the highest Summits rise in Mexico (latitude 18 degrees 59 minutes) to 2770 toises, and in the Rocky Mountains (latitude 37 to 40 degrees) to 1900 toises. The insulated group of the Alleghenies, corresponding in its eastern position and direction with the Brazil group, does not exceed 1040 toises.* (* The culminant point of the Alleghenies is Mount Washington in New Hampshire, latitude 44 1/4 degrees. According to Captain Partridge its height is 6634 English feet.) The lofty summits, therefore, thrice exceeding the height of Mont Blanc, belong only to the longitudinal chain which bounds the basin of the Pacific Ocean, from 55 degrees south to 68 degrees north latitude, that is to say, the Cordillera of the Andes. The only insulated group that can be compared with the snowy summits of the equinoctial Andes, and which attains the height of nearly 3000 toises, is the Sierra de Santa Marta; it is not situated on the east of the Cordilleras, but between the prolongation of two of their branches, those of Merida and Veragua. The Cordilleras, where they bound the Caribbean Sea, in that part which we designate by the name of Coast Chain of Venezuela, do not attain the extraordinary height (2500 toises) which they reach in their prolongation towards Chita and Merida. Considering separately the groups of the east, those of the shore of Venezuela, of the Parime, and Brazil, we see their height diminish from north to south. The highest summits of each group are the Silla de Caracas (1350 toises), the peak of Duida (1300 toises), the Itacolumi and the Itambe* (900 toises). (* According to the measure of MM. Spix and Martius the Itambe de Villa de Principe is 5590 feet high.) But, as I have elsewhere observed, it would be erroneous to judge the height of a chain of mountains solely from that of the most lofty summits. The peak of the Himalayas, accurately measured, is 676 toises higher than Chimborazo (* The Peak Iewahir, latitude 30 degrees 22 minutes 19 seconds; longitude 77 degrees 35 minutes 7 seconds east of Paris, height 4026 toises, according to MM. Hodgson and Herbert.); Chimborazo is 900 toises higher than Mont Blanc; and Mont Blanc 653 toises higher than the peak of Nethou.* (* This peak, called also peak of Anethou or Malahita, or eastern peak of Maladetta, is the highest summit of the Pyrenees. It rises 1787 toises and consequently exceeds Mont Perdu by 40 toises.) These differences do not furnish the relative average heights of the Himalayas, the Andes, the Alps and the Pyrenees, that is, the height of the back of the mountains, on which arise the peaks, needles, pyramids, or rounded domes. It is that part of the back where passes are made, which furnishes a precise measure of the minimum of the height of the great chains. In comparing the whole of my measures with those of Moorcroft, Webb, Hodgson, Saussure and Ramond, I estimate the average height of the top of the Himalayas, between the meridians of 75 and 77 degrees, at 2450 toises; the Andes* (at Peru, Quito and New Grenada), at 1850 toises (* In the passage of Quindiu, between the valley of the Magdalena and that of the Rio Cauca, I found the culminant point (la Garita del Parama) to be 1798 toises; it is however, regarded as one of the least elevated. The passages of the Andes of Guanacas, Guamani and Micuipampa, are respectively 2300, 1713, and 1817 toises above sea-level. Even in 33 degrees south latitude the road across the Andes between Mendoza and Valparaiso is 1987 toises high. I do not mention the Col de l'Assuay, where I passed, near la Ladera de Cadlud, on a ridge 2428 toises high, because it is a passage on a transverse ridge joining two parallel chains.); the summit of the Alps and Pyrenees at 1150 toises. The difference of the mean height of the Cordilleras (between 5 degrees north and 2 degrees south latitude) and the Swiss Alps, is consequently 200 toises less than the difference of their loftiest summits; and in comparing the passes of the Alps, we see that their average height is nearly the same, although peak Nethou is 600 toises lower than Mont Blanc and Mont Rosa. Between the Himalaya* (* The passes of the Himalaya that lead from Chinese Tartary into Hindostan (Nitee-Ghaut, Bamsaru, etc.) are from 2400 to 2700 toises high.) and the Andes, on the contrary, (considering those chains in the limits which I have just indicated), the difference between the mean height of the ridges and that of the loftiest summits presents nearly the same proportions.

Taking an analogous view of the groups of mountains at the east of the Andes, we find the average height of the coast-chain of Venezuela to be 750 toises; of the Sierra Parime, 500 toises; of the Brazilian group, 400 toises; whence it follows that the mountains of the eastern region of South America between the tropics are, when compared to the medium elevation of the Andes, in the relation of one to three.

The following is the result of some numerical statements, the comparison of which affords more precise ideas on the structure of mountains in general.* (* The Cols or passes indicate the minimum of the height to which the ridge of the mountains lowers in a particular country. Now, looking at the principal passes of the Alps of Switzerland (Col Terret, 1191 toises, Mont Cenis, 1060 toises; Great Saint Bernard, 1246 toises; Simplon, 1029 toises; and on the neck of the Pyrenees, Benasque, 1231 toises; Pinede, 1291 toises;

Gavarnic, 1197 toises; Cavarere, 1151 toises; it would be difficult to affirm that the Pyrenees are lower than the average height of the Swiss Alps.)

TABLE OF HEIGHTS OF VARIOUS RANGES.

COLUMN 1 : NAMES OF THE CHAINS OF MOUNTAINS. COLUMN 2 : THE HIGHEST SUMMITS IN TOISES. COLUMN 3 : MEAN HEIGHT OF THE RIDGE IN TOISES. COLUMN 4 : PROPORTION OF THE MEAN HEIGHT OF THE RIDGES TO THAT A THE HIGHEST SUMMITS.

Himalayas (between north latitude : 4026 : 2450 : 1 : 1.6. 30 degrees 18 minutes and 31 degrees 53 minutes, and longitude 75 degrees 23 minutes and 77 degrees 38 minutes)

Cordillera of the Andes (between : 3350 : 1850 : 1 : 1.8. latitude 5 and 2 degrees south)

Alps of Switzerland : 2450 : 1150 : 1 : 2.1.

Pyrenees : 1787 : 1150 : 1 : 1.5.

Littoral Chain of Venezuela : 1350 : 750 : 1 : 1.8.

Group of the Mountains of the Parime : 1300 : 500 : 1 : 2.6.

Group of the Mountains of Brazil : 900 : 500 : 1 : 2.3.

If we distinguish among the mountains those which rise sporadically, and form small insulated systems,* (* As the groups of the Canaries, the Azores, the Sandwich Islands, the Monts-Dores, and the Euganean mountains.) and those that make part of a continued chain,* (* The Himalayas, the Alps, and the Andes.) we find that, notwithstanding the immense height* of the summits of some insulated systems (* Among the insulated systems, or sporadic mountains, Mowna-Roa is generally regarded as the most elevated summit of the Sandwich Islands. Its height is computed at 2500 toises, and yet at some seasons it is entirely free from snow. An exact measure of this summit, situated in very frequented latitudes, has for 25 years been desired in vain by naturalists and geologists.), the culminant points of the whole globe belong to continuous chains—to the Cordilleras of Central Asia and South America.

In that part of the Andes with which I am best acquainted, between 8 degrees south latitude and 21 degrees north latitude, all the colossal summits are of trachyte. It may almost be admitted as a general rule that whenever the mass of mountains rises in that region of the tropics much above the limit of perpetual snow (2300 to 2470 toises), the rocks commonly called primitive (for instance, gneiss-granite or mica-slate) disappear, and the summits are of trachyte or trappean-porphyry. I know only a few rare exceptions to this law, and they occur in the Cordilleras of Quito where the Nevados of Conderasto and Cuvillan, situated opposite to the trachytic Chimborazo, are composed of mica-slate and contain veins of sulphuret of silver. Thus in the groups of detached mountains which rise abruptly from the plains the loftiest summits, such as Mowna-Roa, the Peak of Teneriffe, Etna and the Peak of the Azores, present only recent volcanic rocks. It would, however, be an error to extend that law to every other continent, and to admit, as a general rule that, in every zone, the greatest elevations have produced trachytic domes: gneiss-granite and mica-slate constitute the summits of the ridge, in the almost insulated group of the Sierra Nevada of Grenada and the Peak of Malhacen,* (* This peak, according to the survey of M. Clemente Roxas, is 1826 toises above the level of the sea, consequently 39 toises higher than the loftiest summit of the Pyrenees (the granitic peak of Nethou) and 83 toises lower than the trachytic peak of Teneriffe. The Sierra Nevada of Grenada forms a system of mountains of mica-slate, passing to gneiss and clay-slate, and containing shelves of euphotide and greenstone.), as they also do in the continuous chain of the Alps, the Pyrenees and probably the Himalayas.* (* If we may judge from the specimens of rocks collected in the gorges and passes of the Himalayas or rolled down by the torrents.) These phenomena, discordant in appearance, are possibly all effects of the same cause: granite, gneiss, and all the so-styled primitive Neptunian mountains, may possibly owe their origin to volcanic forces, as well as the trachytes; but to forces of which the action resembles less the still-burning volcanoes of our days, ejecting lava, which at the moment of its eruption comes immediately into contact with the atmospheric air; but it is not here my purpose to discuss this great theoretic question.

After having examined the general structure of South America according to considerations of comparative geology, I shall proceed to notice separately the different systems of mountains and plains, the mutual connection of which has so powerful an influence on the state of industry and commerce in the nations of the New Continent. I shall give only a general view of the systems situated beyond the limits of the region which forms the special object of this memoir. Geology being essentially founded on the study of the relations of juxtaposition and place, I could not treat of the littoral chain and the chain of the Parime separately, without touching on the other systems south and west of Venezuela.

A. SYSTEMS OF MOUNTAINS.

A.1. CORDILLERAS OF THE ANDES.

This is the most continuous, the longest, the most uniform in its direction from south to north and north-north-west, of any chain of the globe. It approaches the north and south poles at unequal distances of from 22 to 33 degrees. Its development is from 2800 to 3000 leagues (20 to a degree), a length equal to the distance from Cape Finisterre in Galicia to the north-east cape (Tschuktschoi-Noss) of Asia. Somewhat less than one half of this chain belongs to South America, and runs along its western shores. North of the isthmus of Cupica and of Panama, after an immense lowering, it assumes the appearance of a nearly central ridge, forming a rocky dyke that joins the great continent of North America to the southern continent. The low lands on the east of the Andes of Guatimala and New Spain appear to have been overwhelmed by the ocean and now form the bottom of the Caribbean Sea. As the continent beyond the parallel of Florida again widens towards the east, the Cordilleras of Durango and New Mexico, as well as the Rocky Mountains, merely a continuation of those Cordilleras, appear to be thrown still further westward, that is, towards the coast of the Pacific Ocean; but they still remain eight or ten times more remote from it than in the southern hemisphere. We may consider as the two extremities of the Andes, the rock or granitic island of Diego Ramirez, south of Cape Horn, and the mountains lying at the mouth of Mackenzie River (latitude 69 degrees, longitude 130 1/2 degrees), more than twelve degrees west of the greenstone mountains, known by the name of the Copper Mountains, visited by Captain Franklin. The colossal peak of Saint Elias and that of Mount Fairweather, in New Norfolk, do not, properly speaking, belong to the northern prolongation of the Cordilleras of the Andes, but to a parallel chain (the maritime Alps of the north-west coast), stretching towards the peninsula of California, and connected by transversal ridges with a mountainous land, between 45 and 53 degrees of latitude, with the Andes of New Mexico (Rocky Mountains). In South America the mean breadth of the Cordillera of the Andes is from 18 to 22 leagues.* (* The breadth of this immense chain is a phenomenon well worthy of attention. The Swiss Alps extend, in the Grisons and in the Tyrol, to a breadth of 36 and 40 leagues, both in the meridians of the lake at Como, the canton of Appenzell, and in the meridian of Bassano and Tegernsee.) It is only in the knots of the mountains, that is where the Cordillera is swelled by side-groups or divided into several chains nearly parallel, and reuniting at intervals, for instance, on the south of the lake of Titicaca, that it is more than 100 to 120 leagues broad, in a direction perpendicular to its axis. The Andes of South America bound the plains of the Orinoco, the Amazon, and the Rio de la Plata, on the west, like a rocky wall raised across a crevice 1300 leagues long, and stretching from south to north. This upheaved part (if I may be permitted to use an expression founded on a geological hypothesis) comprises a surface of 58,900 square leagues, between the parallel of Cape Pilesar and the northern Choco. To form an idea of the variety of rocks which this space may furnish for the observation of the traveller, we must recollect that the Pyrenees, according to the observations of M. Charpentier, occupy only 768 square sea leagues.

The name of Andes in the Quichua language (which wants the consonants d, f, and g) Antis, or Ante, appears to me to be derived from the Peruvian word anta, signifying copper or metal in general. Anta chacra signifies mine of copper; antacuri, copper mixed with gold; and puca anta, copper, or red metal. As the group of the Altai mountains* takes its name from the Turkish word altor or altyn (* Klaproth. Asia polyglotta page 211. It appears to me less probable that the tribe of the Antis gave its name to the mountains of Peru.), in the same manner the Cordilleras may have been termed "Copper-country," or Anti-suyu, on account of the abundance of that metal, which the Peruvians employed for their tools. The Inca Garcilasso, who was the son of a Peruvian princess, and who wrote the history of his native country in the first years of the conquest, gives no etymology of the name of the Andes. He only opposes Anti-suyu, or the region of summits covered with eternal snow (ritiseca), to the plains or Yuncas, that is, to the lower region of Peru. The etymology of the name of the largest mountain chain of the globe cannot be devoid of interest to the mineralogic geographer.

The structure of the Cordillera of the Andes, that is, its division into several chains nearly parallel, which are again joined by knots of mountains, is very remarkable. On our maps this structure is indicated but imperfectly; and what La Condamine and Bouguer merely guessed, during their long visit to the table-land of Quito, has been generalized and ill-interpreted by those who have described the whole chain according to the type of the equatorial Andes. The following is the most accurate information I could collect by my own researches and an active correspondence of twenty years with the inhabitants of Spanish America. The group of islands called Tierra del Fuego, in which the chain of the Andes begins, is a plain extending from Cape Espiritu Santo as far as the canal of San Sebastian. The country on the west of this canal, between Cape San Valentino and Cape Pilares, is bristled with granitic mountains covered (from the Morro de San Agueda to Cabo Redondo) with calcareous shells. Navigators have greatly exaggerated the height of the mountains of Tierra del Fuego, among which there appears to be a volcano still burning. M. de Churruca found the height of the western peak of Cape Pilares (latitude 52 degrees 45 minutes south) only 218 toises; even Cape Horn is probably not more than 500 toises* high. (* It is very distinctly seen at the distance of 60 miles, which, without calculating the effects of terrestrial refraction, would give it a height of 498 toises.) The plain extends on the northern shore of the Straits of Magellan, from the Virgin's Cape to Cabo Negro; at the latter the Cordilleras rise abruptly, and fill the whole space as far as Cape Victoria (latitude 52 degrees 22 minutes). The region between Cape Horn and the southern extremity of the continent somewhat resembles the origin of the Pyrenees between Cape Creux (near the gulf of Rosas) and the Col des Perdus. The height of the Patagonian chain is not known; it appears, however, that no summit south of the parallel of 48 degrees attains the elevation of the Canigou (1430 toises) which is near the eastern extremity of the Pyrenees. In that southern country, where the summers are so cold and short, the limit of eternal snow must lower at least as much as in the northern hemisphere, in Norway, in latitude 63 and 64 degrees; consequently below 800 toises. The great breadth, therefore, of the band of snow that envelopes these Patagonian summits, does not justify the idea which travellers form of their height in 40 degrees south latitude. As we advance towards the island of Chiloe, the Cordilleras draw near the coast; and the archipelago of Chonos or Huaytecas appears like the vestiges of an immense group of mountains overwhelmed by water. Narrow estuaries fill the lower valleys of the Andes, and remind us of the fjords of Norway and Greenland. We there find, running from south to north, the Nevados de Maca (latitude 45 degrees 19 minutes), of Cuptano (latitude 44 degrees 58 minutes), of Yanteles (latitude 43 degrees 52 minutes), of Corcovado, Chayapirca (latitude 42 degrees 52 minutes) and of Llebean (latitude 41 degrees 49 minutes). The peak of Cuptana rises like the peak of Teneriffe, from the bosom of the sea; but being scarcely visible at thirty-six or forty leagues distance, it cannot be more than 1500 toises high. Corcovado, situated on the coast of the continent, opposite the southern point of the island of Chiloe, appears to be more than 1950 toises high; it is perhaps the loftiest summit of the whole globe, south of the parallel of 42 degrees south latitude. On the north of San Carlos de Chiloe, in the whole length of Chile to the desert of Atacama, the low western regions not having been overwhelmed by floods, the Andes there appear farther from the coast. The Abbe Molina affirms that the Cordilleras of Chile form three parallel chains, of which the intermediary is the most elevated; but to prove that this division is far from general, it suffices to recollect the barometric survey made by MM. Bauza and Espinosa, in 1794, between Mendoza and Santiago de Chile. The road leading from one of those towns to the other, rises gradually from 700 to 1987 toises; and after passing the Col des Andes (La Cumbre, between the houses of refuge called Las Calaveras and Las Cuevas), it descends continually as far as the temperate valley of Santiago de Chile, of which the bottom is only 409 toises above the level of the sea. The same survey has made known the minimum of height at Chile of the lower limit of snow, in 33 degrees south latitude. The

limit does not lower in summer to 2000 toises.* (* On the southern declivity of the
Himalayas snow begins (3 degrees nearer the equator) at 1970 toises.) I think we may
conclude according to the analogy of the Snowy Mountains of Mexico and southern Europe,
and considering the difference of the summer temperature of the two hemispheres, that the
real Nevadas at Chile, in the parallel of Valdivia (latitude 40 degrees), cannot be below 1300
toises; in Valparaiso (latitude 33 degrees) not lower than 2000 toises, and in that of Copiapo
(latitude 27 degrees) not below 2200 toises of height. These are the limit-numbers, the
minimum of elevation, which the ridge of the Andes of Chile must attain in different degrees
of latitude, to enable their summits to rise above the line of perpetual snow. The numerical
results which I have just marked and which are founded on the laws of distribution of heat,
have still the same importance which they possessed at the time of my travels in America;
for there does not exist in the immense extent of the Andes, from 8 degrees south latitude to
the Straits of Magellan, one Nevada of which the height above the sea-level has been
determined, either by a simple geometric measure, or by the combined means of barometric
and geodesic measurements.

Between 33 and 18 degrees south latitude, between the parallels of Valparaiso and
Arica, the Andes present towards the east three remarkable spurs, the Sierra de Cordova, the
Sierra de Salta, and the Nevados de Cochabamba. Travellers partly cross and partly go along
the side of the Sierra de Cordova (between 33 and 31 degrees of latitude) in their way from
Buenos Ayres to Mendoza; it may be said to be the most southern promontory which
advances, in the Pampas, towards the meridian of 65 degrees; it gives birth to the great river
known by the name of Desaguadero de Mendoza and extends from San Juan de la Frontera
and San Juan de la Punta to the town of Cordova. The second spur, called the Sierra de Salta
and the Jujui, of which the greatest breadth is 25 degrees of latitude, widens from the valley
of Catamarca and San Miguel del Tucuman, in the direction of the Rio Vermejo (longitude
64 degrees). Finally, the third and most majestic spur, the Sierra Nevada de Cochabamba and
Santa Cruz (from 22 to 17 1/2 degrees of latitude), is linked with the knot of the mountains
of Porco. It forms the points of partition (divortia aquarum, between the basin of the
Amazon and that of the Rio de la Plata. The Cachimayo and the Pilcomayo, which rise
between Potosi, Talavera de la Puna, and La Plata or Chuquisaca, run in the direction of
south-east, while the Parapiti and the Guapey (Guapaiz, or Rio de Mizque) pour their waters
into the Mamori, to north-east. The ridge of partition being near Chayanta, south of Mizque,
Tomina and Pomabamba, nearly on the southern declivity of the Sierra de Cochabamba in
latitude 19 and 20 degrees, the Rio Guapey flows round the whole group, before it reaches
the plains of the Amazon, as in Europe the Poprad, a tributary of the Vistula, makes a circuit
in its course from the southern part of the Carpathians to the plains of Poland. I have
already observed above, that where the mountains cease (west* of the meridian of 66 1/2
degrees (* I agree with Captain Basil Hall, in fixing the port of Valparaiso in 71 degrees 31
minutes west of Greenwich, and I place Cordova 8 degrees 40 minutes, and Santa Cruz de la
Sierra 7 degrees 4 minutes east of Valparaiso. The longitudes mentioned in the text refer
always to the meridian of the Observatory of Paris.)) the partition ridge of Cochabamba goes
up towards the north-east, to 16 degrees of latitude, forming, by the intersection of two
slightly inclined planes, only one ridge amidst the savannahs, and separating the waters of the
Guapore, a tributary of the Madeira, from those of the Aguapehy and Jauru, tributaries of
the Rio Paraguay. This vast country between Santa Cruz de la Sierra, Villabella, and
Matogrosso, is one of the least known parts of South America. The two spurs of Cordova
and Salta present only a mountainous territory of small elevation, and linked to the foot of
the Andes of Chile. Cochabamba, on the contrary, attains the limit of perpetual snow (2300
toises) and forms in some sort a lateral branch of the Cordilleras, diverging even from their
tops between La Paz and Oruro. The mountains composing this branch (the Cordillera de
Chiriguanaes, de los Sauces and Yuracarees) extend regularly from west to east; their eastern
declivity* is very rapid, and their loftiest summits are not in the centre, but in the northern
part of the group. (* For much information concerning the Sierra de Cochabamba I am
indebted to the manuscripts of my countryman, the celebrated botanist Taddeus Haenke,
which a monk of the congregation of the Escurial, Father Cisneros, kindly communicated to
me at Lima. Mr. Haenke, after having followed the expedition of Alexander Malaspina,

settled at Cochabamba in 1798. A part of the immense herbal of this botanist is now at Prague.)

The principal Cordillera of Chile and Upper Peru is, for the first time, ramified very distinctly into two branches, in the group of Porco and Potosi, between latitude 19 and 20 degrees. These two branches comprehend the table-land extending from Carangas to Lamba (latitude 19 3/4 to 15 degrees) and in which is situated the small mountain lake of Paria, the Desaguadero, and the great Laguna of Titicaca or Chucuito, of which the western part bears the name of Vinamarca. To afford an idea of the colossal dimensions of the Andes, I may here observe that the surface of the lake of Titicaca alone (448 square sea leagues) is twenty times greater than that of the Lake of Geneva, and twice the average extent of a department of France. On the banks of this lake, near Tiahuanacu, and in the high plains of Callao, ruins are found which bear evidence of a state of civilization anterior to that which the Peruvians assign to the reign of the Inca Manco Capac. The eastern Cordillera, that of La Paz, Palca, Ancuma, and Pelechuco, join, north-west of Apolobamba, the western Cordillera, which is the most extensive of the whole chain of the Andes, between the parallels 14 and 15 degrees. The imperial city of Cuzco is situated near the eastern extremity of this knot, which comprehends, in an area of 3000 square leagues, the mountains of Vilcanota, Carabaya, Abancai, Huando, Parinacochas, and Andahuaylas. Though here, as in general, in every considerable widening of the Cordillera, the grouped summits do not follow the principal axis in uniform and parallel directions, a phenomenon observable in the general disposition of the chain of the Andes, from latitude 18 degrees, is well worthy the attention of geologists. The whole mass of the Cordilleras of Chile and Upper Peru, from the Straits of Magellan to the parallel of the port of Arica (18 degrees 28 minutes 35 seconds), runs from south to north, in the direction of a meridian at most 5 degrees north-east; but from the parallel of Arica, the coast and the two Cordilleras east and west of the Alpine lake of Titicaca, abruptly change their direction and incline to north-west. The Cordilleras of Ancuma and Moquehua, and the longitudinal valley, or rather the basin of Titicaca, which they inclose, take a direction north 42 degrees west. Further on, the two branches again unite in the group of the mountains of Cuzco, and thence their direction is north 80 degrees west. This group of which the table-land inclines to the north-east, forms a curve, nearly from east to west, so that the part of the Andes north of Castrovireyna is thrown back more than 242,000 toises westward. This singular geological phenomenon resembles the variation of dip of the veins, and especially of the two parts of the chain of the Pyrenees, parallel to each other, and linked by an almost rectangular elbow, 16,000 toises long, near the source of the Garonne;* (* Between the mountain of Tentenade and the Port d'Espot.); but in the Andes, the axes of the chain, south and north of the curve, do not preserve parallelism. On the north of Castrovireyna and Andahuaylas (latitude 14 degrees), the direction is north 22 degrees west, while south of 15 degrees, it is north 42 degrees west. The inflexions of the coast follow these changes. The shore separated from the Cordillera by a plain 15 leagues in breadth, stretches from Camapo to Arica, between 27 1/2 and 18 1/2 degrees latitude north 5 degrees east; from Arica to Pisco, between 18 1/2 and 14 degrees latitude at first north 42 degrees west, afterwards north 65 degrees west; and from Pisco to Truxillo, between 14 and 8 degrees of latitude north 27 degrees west. The parallelism between the coast and the Cordillera of the Andes is a phenomenon the more worthy of attention, as it occurs in several parts of the globe where the mountains do not in the same manner form the shore.

After the great knot of mountains of Cuzco and Parinacochas, in 14 degrees south latitude, the Andes present a second bifurcation, on the east and west of the Rio Jauja, which throws itself into the Mantaro, a tributary stream of the Apurimac. The eastern chain stretches on the east of Huanta, the convent of Ocopa and Tarma; the western chain, on the west of Castrovireyna, Huancavelica, Huarocheri, and Yauli. The basin, or rather the lofty table-land which is inclosed by these chains, is nearly half the length of the basin of Chucuito or Titicaca. Two mountains covered with eternal snow, seen from the town of Lima, and which the inhabitants name Toldo de la Nieve, belong to the western chain, that of Huarocheri.

North-west of the valleys of Salcabamba, in the parallel of the ports of Huaura and Guarmey, between 11 and 10 degrees latitude, the two chains unite in the knot of the

147

Huanuco and the Pasco, celebrated for the mines of Yauricocha or Santa Rosa. There rise two peaks of colossal height, the Nevados of Sasaguanca and of La Viuda. The table-land of this knot of mountains appears in the Pambas de Bombon to be more than 1800 toises above the level of the ocean. From this point, on the north of the parallel of Huanuco (latitude 11 degrees), the Andes are divided into three chains: the first, and most eastern, rises between Pozuzu and Muna, between the Rio Huallaga, and the Rio Pachitea, a tributary of the Ucayali; the second, or central, is between the Huallaga, and the Upper Maranon; the third, or western, between the Upper Maranon and the coast of Truxillo and Payta. The eastern chain is a small lateral branch which lowers into a range of hills: its direction is first north-north-east, bordering the Pampas del Sacramento, afterwards it turns west-north-west, where it is broken by the Rio Huallaga, in the Pongo, above the confluence of Chipurana, and then it loses itself in latitude 6 1/4 degrees, on the north-west of Lamas. A transversal ridge seems to connect it with the central chain, south of Paramo de Piscoguanuna (or Piscuaguna), west of Chachapoyas. The intermediary or central chain stretches from the knot of Pasco and Huanuco, towards north-north-west, between Xican and Chicoplaya, Huacurachuco and the sources of the Rio Monzan, between Pataz and Pajatan, Caxamarquilla and Moyobamba. It widens greatly in the parallel of Chachapoyas, and forms a mountainous territory, traversed by deep and extremely hot valleys. On the north of the Paramo de Piscoguanuna (latitude 6 degrees) the central chain throws two branches in the direction of La Vellaca and San Borja. We shall soon see that this latter branch forms, below the Rio Neva a tributary stream of the Amazon, the rocks that border the famous Pongo de Manseriche. In this zone, where North Peru approximates to the confines of New Grenada in latitude 10 and 5 degrees, no summit of the eastern and central chains rises as high as the region of perpetual snow; the only snowy summits are in the western chain. The central chain, that of the Paramos de Callacalla, and Piscoguanuna, scarcely attains 1800 toises, and lowers gently to 800 toises; so that the mountainous and temperate tract of country which extends on the north of Chachapoyas towards Pomacocha, La Vellaca and the source of the Rio Nieva is rich in fine cinchona trees. After having passed the Rio Huallaga and the Pachitea, which with the Beni forms the Ucayali, we find, in advancing towards the east, only ranges of hills. The western chain of the Andes, which is the most elevated and nearest to the coast, runs almost parallel with the shore north 22 degrees west, between Caxatambo and Huary, Conchucos and Guamachuco, by Caxamarca, the Paramo de Yanaguanga, and Montan, towards the Rio de Guancabamba. It comprises (between 9 and 7 1/2 degrees) the three Nevados de Pelagatos, Moyopata and Huaylillas. This last snowy summit, situated near Guamachuco (in 7 degrees 55 minutes latitude), is the more remarkable, since from thence on the north, as far as Chimborazo, on a length of 140 leagues, there is not one mountain that enters the region of perpetual snow. This depression, or absence of snow, extends in the same interval, over all the lateral chains; while, on the south of the Nevado de Huaylillas, it always happens that when one chain is very low, the summits of the other exceed the height of 2460 toises. It was on the south of Micuipampa (latitude 7 degrees 1 minute) that I found the magnetic equator.

The Amazon, or as it is customary to say in those regions, the Upper Maranon, flows through the western part of the longitudinal valley lying between the Cordilleras of Chachapayas and Caxamarca. Comprehending in one point of view, this valley, and that of the Rio Jauja, bounded by the Cordilleras of Tarma and Huarocheri, we are inclined to consider them as one immense basin 180 leagues long, and crossed in the first third of its length, by a dyke, or ridge 18,000 toises broad. In fact, the two alpine lakes of Lauricocha and Chinchaycocha, where the river Amazon and the Rio de Jauja take their rise, are situated south and north of this rocky dyke, which is a prolongation of the knot of Huanuco and Pasco. The Amazon, on issuing from the longitudinal valley which bounds the chains of Caxamarca and Chachacocha, breaks the latter chain; and the point where the great river penetrates the mountains, is very remarkable. Entering the Amazon by the Rio Chamaya or Guancabamba, I found opposite the confluence, the picturesque mountain of Patachuana; but the rocks on both banks of the Amazon begin only between Tambillo and Tomependa (latitude 5 degrees 31 minutes, longitude 80 degrees 56 minutes). From thence to the Pongo de Rentema, a long succession of rocks follow, of which the last is the Pongo de

Tayouchouc, between the strait of Manseriche and the village of San Borja. The course of the Amazon, which is first directed north, then east, changes near Puyaya, three leagues north-east of Tomependa. Throughout the whole distance between Tambillo and San Borja, the waters force a way, more or less narrow, across the sandstones of the Cordillera of Chachapoyas. The mountains are lofty near the Embarcadero, at the confluence of the Imasa, where large trees of cinchona, which might be easily transplanted to Cayenne, or the Canaries, approach the Amazon. The rocks in the famous strait of Manseriche are scarcely 40 toises high; and further eastward the last hills rise near Xeberos, towards the mouth of the Rio Huallaga.

I have not yet noticed the extraordinary widening of the Andes near the Apolobamba. The sources of the Rio Beni being found in the spur which stretches northward beyond the confluence of that river with the Apurimac, I shall give to the whole group the name of "the spur of Beni." The following is the most certain information I have obtained respecting those countries, from persons who had long inhabited Apolobamba, the Real das Minas of Pasco, and the convent of Ocopa. Along the whole eastern chain of Titicaca, from La Paz to the knot of Huanuco (latitude 17 1/2 to 10 1/2 degrees) a very wide mountainous land is situated eastward, at the back of the declivity of the Andes. It is not a widening of the eastern chain itself, but rather of the small heights that surround the foot of the Andes like a penumbra, filling the whole space between the Beni and the Pachitca. A chain of hills bounds the eastern bank of the Beni to latitude 8 degrees; for the rivers Coanache and Magua, tributaries of the Ucayali (flowing in latitude 6 and 7 degrees) come from a mountainous tract between the Ucayali and the Javari. The existence of this tract in so eastern a longitude (probably longitude 74 degrees), is the more remarkable, as we find at four degrees of latitude further north, neither a rock nor a hill on the east of Xeberos, or the mouth of the Huallaga (longitude 77 degrees 56 minutes).

We have just seen that the spur of Beni, a sort of lateral branch, loses itself about latitude 8 degrees; the chain between the Ucayali and the Huallaga terminates at the parallel of 7 degrees, in joining, on the west of Lamas, the chain of Chachapayas, stretching between the Huallaga and the Amazon. Finally, the latter chain, to which I have given the designation of central, after forming the rapids and cataracts of the Amazon, between Tomependa and San Borja, turns to north-north-west, and joins the western chain, that of Caxamarca, or the Nevados of Pelagatos and Huaylillas, and forms the great knot of the mountains of Loxa. The mean height of this knot is only from 1000 to 1200 toises: its mild climate renders it peculiarly favourable to the growth of the cinchona trees, the finest kinds of which are found in the celebrated forest of Caxanuma and Uritusinga, between the Rio Zamora and the Cachiyacu, and between Tavacona and Guancabamba. Before the cinchona of Popayan and Santa Fe de Bogota (north latitude 2 1/2 to 5 degrees), of Huacarachuco, Huamalies and Huanuco (south latitude 9 to 11 degrees) became known, the group of the mountains of Loxa had for ages been regarded as the sole region whence the febrifuge bark of cinchona could be obtained. This group occupies the vast territory between Guancabamba, Avayaca, Ona and the ruined towns of Zamora and Loyola, between latitude 5 1/2 and 3 1/4 degrees. Some of the summits (the Paramos of Alpachaca, Saraguru, Savanilla, Gueringa, Chulucanas, Guamani, and Yamoca, which I measured) rise from 1580 to 1720 toises, but are not even sporadically covered with snow, which in this latitude falls only above 1860 to 1900 toises of absolute height. Eastward, in the direction of the Rio Santiago and the Rio de Chamaya, two tributary streams of the Amazon, the mountains lower rapidly: between San Felipe, Matara, and Jaen de Bracamoros, they are not more than 500 or 300 toises.

As we advance from the mica-slate mountain of Loxa towards the north, between the Paramos of Alpachaca and Sara (in latitude 3 degrees 15 minutes) the knot of mountains ramifies into two branches which comprehend the longitudinal valley of Cuenca. This separation continues for a length of only 12 leagues; for in latitude 2 degrees 27 minutes the two Cordilleras again re-unite in the knot of Assuy, a trachytic group, of which the table-land near Cadlud (2428 toises high) nearly enters the region of perpetual snow.

The group of the mountains of Assuy, which affords a very frequented pass of the Andes between Cuenca and Quito (latitude 2 1/2 to 0 degrees 40 minutes south) is succeeded by another division of the Cordilleras, celebrated by the labours of Bouguer and

La Condamine, who placed their signals sometimes on one, sometimes on the other of the two chains. The eastern chain is that of Chimborazo (3350 toises) and Carguairazo; the western is the chain of the volcano Sangay, the Collanes, and of Llanganate. The latter is broken by the Rio Pastaza. The bottom of the longitudinal basin that bounds those two chains, from Alausi to Llactacunga, is somewhat higher than the bottom of the basin of Cuenca. North of Llactacanga, 0 degrees 40 minutes latitude, between the tops of Yliniza (2717 toises) and Cotopaxi (2950 toises), of which the former belongs to the chain of Chimborazo, and the latter to that of Sangay, is situated the knot of Chisinche; a kind of narrow dyke that closes the basin, and divides the waters between the Atlantic and the Pacific. The Alto de Chisinche is only 80 toises above the surrounding table-lands. The waters of its northern declivity form the Rio de San Pedro, which, joining the Rio Pita, throws itself into the Gualabamba, or Rio de las Esmeraldas. The waters of the southern declivity, called Cerro de Tiopullo, run into the Rio San Felipe and the Pastaza, a tributary stream of the Amazon.

The bipartition of the Cordilleras re-commences and continues from 0 degrees 40 minutes latitude south to 0 degrees 20 minutes latitude north; that is, as far as the volcano of Imbabura near the villa of Ibarra. The eastern Cordillera presents the snowy summits of Antisana (2992 toises), of Guamani, Cayambe (3070 toises) and of Imbabura; the western Cordillera, those of Corazon, Atacazo, Pichinca (2491 toises) and Catocache (2570 toises). Between these two chains, which may be regarded as the classic soil of the astronomy of the 18th century, is a valley, part of which is again divided longitudinally by the hills of Ichimbio and Poignasi. The table-lands of Puembo and Chillo are situated eastward of those hills; and those of Quito, Inaquito and Turubamba lie westward. The equator crosses the summit of the Nevado de Cayambe and the valley of Quito, in the village of San Antonio de Lulumbamba. When we consider the small mass of the knot of Assuy, and above all, of that of Chisinche, we are inclined to regard the three basins of Cuenca, Hambato and Quito as one valley (from the Paramo de Sarar to the Villa de Ibarra) 73 sea leagues long, from 4 to 5 leagues broad, having a general direction north 8 degrees east, and divided by two transverse dykes one between Alausi and Cuenca (2 degrees 27 minutes south latitude), and the other between Machache and Tambilbo (0 degrees 40 minutes). Nowhere in the Cordillera of the Andes are there more colossal mountains heaped together than on the east and west of this vast basin of the province of Quito, one degree and a half south, and a quarter of a degree north of the equator. This basin which, next to the basin of Titicaca, is the centre of the most ancient native civilization, touches, southward, the knot of the mountains of Loxa, and northward the tableland of the province of Los Pastos.

In this province, a little beyond the villa of Ibarra, between the snowy summits of Cotocache and Imbabura, the two Cordilleras of Quito unite, and form one mass, extending to Meneses and Voisaco, from 0 degrees 21 minutes north latitude to 1 degree 13 minutes. I call this mass, on which are situated the volcanoes of Cumbal and Chiles, the knot of the mountains of Los Pastos, from the name of the province that forms the centre. The volcano of Pasto, the last eruption of which took place in the year 1727, is on the south of Yenoi, near the northern limit of this group, of which the inhabited table-lands are more than 1600 toises above sea-level. It is the Thibet of the equinoctial regions of the New World.

On the north of the town of Pasto (latitude 1 degree 13 minutes north; longitude 79 degrees 41 minutes) the Andes again divide into two branches and surround the table-land of Mamendoy and Almaguer. The eastern Cordillera contains the Sienega of Sebondoy (an alpine lake which gives birth to the Putumayo), the sources of the Jupura or Caqueta, and the Paramos of Aponte and Iscanse. The western Cordillera, that of Mamacondy, called in the country Cordillera de la Costa, on account of its proximity to the shore of the Pacific, is broken by the great Rio de Patias, which receives the Guativa, the Guachicon and the Quilquase. The table-land or intermediary basin has great inequalities; it is partly filled by the Paramos of Pitatumba and Paraguay, and the separation of the two chains appeared to me indistinct as far as the parallel of Almaguer (latitude 1 degree 54 minutes; longitude 79 degrees 15 minutes). The general direction of the Andes, from the extremity of the basin of the province of Quito to the vicinity of Popayan, changes from north 8 degrees east to north 36 degrees east; and follows the direction of the coast of Esmeralda and Barbacoas.

On the parallel of Almaguer, or rather a little north-east of that town, the geological structure of the ground displays very remarkable changes. The Cordillera, to which we have given the name of eastern, that of the lake of Sebondoy, widens considerably between Pansitara and Ceja. The knot of the Paramo de las Papas and of Socoboni gives birth to the great rivers of Cauca and Magdalena, and is divided into two chains, latitude 2 degrees 5 minutes east and west of La Plata, Vieja and Timana. These two chains continue nearly parallel as far as 5 degrees of latitude, and they bound the longitudinal valley through which winds the Rio Magdalena. We shall give the name of the eastern Cordillera of New Grenada to that chain which stretches towards Santa Fe de Bogota, and the Sierra Nevada de Merida, east of Magdalena; the chain which lies between the Magdalena and the Cauca, in the direction of Mariquita, we will call the central Cordillera of New Grenada; and the chain which continues the Cordillera de la Costa from the basin of Almaguer, and separates the bed of the Rio Cauca from the platiniferous territory of Choco, we will designate the western Cordillera of New Grenada. For additional clearness, we may also name the chain, that of Suma Paz, after the colossal group of mountains on the south of Santa Fe de Bogota, which empties the waters of its eastern declivity into the Rio Meta. The second chain may bear the name of the chain of Guanacas or Quindiu, after the two celebrated passages of the Andes, on the road from Santa Fe de Bogota to Popayan. The third chain may be called the chain of Choco, or of the shore. Some leagues south of Popayan (latitude 2 degrees 21 minutes north), west of Paramo de Palitara and the volcano of Purace, a ridge of mica-slate runs from the knot of the mountains of Sacoboni to north-west, and divides the waters between the Pacific and the Caribbean Sea; they flow from the northern declivity into the Rio Cauca, and from the southern declivity, into the Rio de Patias.

The tripartition of the Andes (north latitude 1 3/4 to 2 1/4 degrees) resembles that which takes place at the source of the Amazon in the knot of the mountains of Huanuco and Pasco (latitude 11 degrees south); but the most western of the three chains that bound the basins of the Amazon and the Huallaga, is the loftiest; while that of Choco, or the shore, is the least elevated of the three chains of New Grenada. Ignorance of this tripartition of the Andes in that part of South America near the Rio Atrato and the isthmus of Panama, has led to many erroneous opinions respecting the possibility of a canal that should connect the two seas.

The eastern chain of the Andes of New Grenada* preserves its parallelism during some time with the two other chains, those of Quindiu and Choco; but beyond Tunja (latitude 5 1/2 degrees) it inclines more towards the north-east, passing somewhat abruptly from the direction north 25 degrees east to that of north 45 degrees east. (* I employ a systematic denomination, for the name of the Andes is unknown in the countries situated north of the equator.) It is like a vein that changes its direction; and it rejoins the coast after being greatly enlarged by the grouping of the snowy mountains of Merida. The tripartition of the Cordilleras, and above all, the spreading of their branches, have a vast influence on the prosperity of the nations of New Grenada. The diversity of the superposed table-lands and climates varies the agricultural productions as well as the character of the inhabitants. It gives activity to the exchange of productions, and renews over a vast surface, north of the equator, the picture of the sultry valleys and cool and temperate plains of Peru. It is also worthy of remark that, by the separation of one of the branches of the Cordilleras of Cundinamarca and by the deviation of the chain of Bogota towards the north-east, the colossal group of the mountains of Merida is enclosed in the territory of the ancient Capitania-general of Venezuela, and that the continuity of the same mountainous land from Pamplona to Barquisimeto and Nirgua may be said to have facilitated the political union of the Columbian territory. As long as the central chain (that of Quindiu) presents its snowy summits, no peak of the eastern chain (that of La Suma Paz) rises, in the same parallels, to the limit of perpetual snow. Between latitude 2 and 5 1/2 degrees neither the Paramos situated on the east of Gigante and Neiva, nor the tops of La Suma Paz, Chingasa, Guachaneque, and Zoraca, exceed the height of 1900 to 2000 toises; while on the north of the parallel of Paramo d'Erve (latitude 5 degrees 5 minutes), the last of the Nevados of the central Cordillera, we discover in the eastern chain the snowy summits of Chita (latitude 5 degrees 50 minutes), and of Mucuchies (latitude 8 degrees 12 minutes). Hence it results that from latitude 5 degrees the only mountains covered with snow during the whole year are the Cordilleras of the east; and although the Sierra Nevada of Santa Marta is not, properly speaking, a continuation of the Nevados of Chita and Mucuchies (west of Patute and east of Merida), it is at least very near their meridian.

Having now arrived at the northern extremity of the Cordilleras, comprehended between Cape Horn and the isthmus of Panama, we shall proceed to notice the loftiest summits of the three chains which separate in the knot of the mountains of Socoboni, and the ridge of Roble (latitude 1 degree 50 minutes to 2 degrees 20 minutes). I begin with the most eastern chain, that of Timana and Suma Paz, which divides the tributary streams of the Magdalena and the Meta: it runs by the Paramos de Chingasu, Guachaneque, Zoraca, Toquillo (near Labranza Grande), Chita, Almorsadero, Laura, Cacota, Zumbador and Porqueras, in the direction of the Sierra Nevada de Merida. These Paramos indicate ten partial risings of the back of the Cordilleras. The declivity of the eastern chain is extremely rapid on the eastern side, where it bounds the basin of the Meta and the Orinoco; it is widened on the west by the spurs on which are situated the towns of Santa Fe de Bogota, Tunja, Sogamoso and Leiva. They are like tablelands fixed to the western declivity, and are from 1300 to 1400 toises high; that of Bogota (the bottom of an ancient lake) contains fossil bones of the mastodon, in the plain called (from them) the Campo de Gigantes, near Suacha.

The intermediary, or central chain, runs east of Popayan, by the high plains of Mabasa, the Paramos of Guanacas, Huila, Savelillo, Iraca, Baraguan, Tolima, Ruiz and Herveo, towards the province of Antioquia. In 5 degrees 15 minutes of latitude this chain, the only one that shows traces of recent volcanic fire, in the summits of Sotara and Purace, widens considerably towards the west, and joins the western chain, which we have called the chain of Choco, because the platiniferous land of that province lies on the slope opposite the Pacific ocean. By the union of the two chains, the basin of the province of Popayan is close on the north of Cartago Viejo; and the river of Cauca, issuing from the plain of Buga, is forced, from the Salto de San Antonio, to La Boca del Espiritu Santo, to open its way across the mountains, along a course of from 40 to 50 leagues. The difference of the level is very remarkable in the bottom of the two parallel basins of Cauca and Magdalena. The former, between Cali and Cantago, is from 500 to 404 toises; the latter, from Neiva to Ambalema, is from 265 to 150 toises high. According to different geological hypotheses, it may be said either that the secondary formations have not accumulated to the same thickness between the eastern and central, as between the central and western chains; or, that the deposits have been made on the base of primitive rocks, unequally upheaved on the east and west of the Andes of Quindiu. The average difference of the thickness of these formations is 300 toises. The rocky ridge of the Angostura of Carare branches from the south-east, from the spur of Muzo, through which winds the Rio Negro. By this spur, and by those that come from the west, the eastern and central chains approach between Nares, Honda, and Mendales. In fact, the bed of the Rio Magdalena is narrowed in 5 and 5 degrees 18 minutes, on the east by the mountains of Sergento, and on the west by the spurs that are linked with the granitic mountains of Maraquito and Santa Ana. This narrowing of the bed of the river is in the same parallel with that of the Cauca, near the Salto de San Antonio; but, in the knot of the mountains of Antioquia the central and western chains join each other, while between Honda and Mendales, the tops of the central and eastern chains are so far removed that it is only the spurs of each system that draw near and are confounded together. It is also worthy of remark that the central Cordillera of New Grenada displays the loftiest summit of the Andes in the northern hemisphere. The peak of Tolima (latitude 4 degrees 46 minutes) which is almost unknown even by name in Europe, and which I measured in 1801, is at least 2865 toises high. It consequently surpasses Imbabura and Cotocache in the province of Quito, the Chiles of the table-lands of Los Pastos, the two volcanoes of Popayan and even the Nevados of Mexico and Mount Saint Elias of Russian America. The peak of Tolima, which in form resembles Cotapaxi, is perhaps inferior in height only to the ridge of the Sierra Nevada de Santa Marta, which may be considered as an insulated system of mountains.

The eastern chain, also called the chain of Choco and the east coast (of the Pacific), separates the provinces of Popayan and Antioquia from those of Barbacoas, Raposo and

Choco. It is in general but little elevated, compared to the height of the central and eastern chains; it however presents great obstacles to the communications between the valley of Cauca and the shore. On its western slope lies the famous auriferous and platiniferous land,* which has during ages yielded more than 13,000 marks of gold annually. (* Choco, Barbacoas and Brazil are the only countries in which the existence of grains of platinum and palladium has hitherto been fully ascertained. The small town of Barbacoas is situated on the left bank of the Rio Telembi (a tributary of Patias or the Rio del Castigo) a little above the confluence of Telembi and the Guagi or Guaxi, nearly in latitude 1 degree 48 minutes. The ancient Provincia, or rather the Partido del Raposo, comprehends the insalubrious land extending from the Rio Dagua, or San Buenaventura, to the Rio Iscuande, the southern limit of Choco.) This alluvial zone is from ten to twelve leagues broad; its maximum of productiveness lies between the parallels of 2 and 6 degrees latitude; it sensibly impoverishes towards the north and south, and almost entirely disappears between 1 1/4 degree north latitude and the equator. The auriferous soil fills the basin of Cauca, as well as the ravines and plains west of the Cordillera of Choco; it rises sometimes nearly 600 toises above the level of the sea, and descends at least 40 toises.* (* M. Caldas assigns to the upper limit of the zone of gold-washings, only the height of 350 toises. Semanario tome 1 page 18; but I found the Seraderos[?] of Quilichao, on the north of Popayan, to be 565 toises high.) Platinum (and this fact is worthy of attention) has hitherto been found only on the west of the Cordillera of Choco, and not on the east, notwithstanding the analogy of the fragments of rocks of greenstone, phonolite, trachyte, and ferruginous quartz, of which the soil of the two slopes is composed. From the ridge of Los Robles, which separates the table-land of Almaguer from the basin of Cauca, the western chain forms, first, in the Cerros de Carpinteria, east of the Rio San Juan de Micay, the continuation of the Cordillera of Sindagua, broken by the Rio Patias; then, lowering northward, between Cali and Las Juntas de Dagua, and at the elevation of 800 to 900 toises, it sends out considerable spurs (latitude 4 1/4 to 5 degrees) towards the source of the Calima, the Tamana and the Andagueda. The two former of these auriferous rivers are tributary streams of the Rio San Juan del Choco; the second empties its waters into the Atrato. This widening of the western chain forms the mountainous part of Choco: here, between the Tado and Zitara, called also Francisco de Quibdo, lies the isthmus of Raspadura, across which a monk traced a navigable line of communication between the two oceans. The culminant point of this system of mountains appears to be the Peak of Torra, situated south-east of Novita.

The northern extremity of this enlargement of the Cordillera of Choco, which I have just described, corresponds with the junction formed on the east, between the same Cordillera and the central chain, that of Quindiu. The mountains of Antioquia, on which we have the excellent observations of Mr. Restrepo, may be called a knot of mountains, and on the northern limit of the plains of Buga, or the basin of Cauca, they join the central and western chains. The ridge of the eastern Cordillera is at the distance of thirty-five leagues from this knot, so that the contraction of the bed of the Rio Magdalena, between Honda and Ambalema, is caused only by the approximation of the spurs of Mariquita and Guaduas. There is not, therefore, properly speaking, a group of mountains between latitude 5 and 5 1/4 degrees, uniting the three chains at once. In the group of the province of Antioquia, which forms the junction of the central and western Cordilleras, we may distinguish two great masses; one between the Magdalena and the Cauca, and the other between the Cauca and the Atrato. The first of these masses, which is linked most immediately to the snowy summits of Herveo, gives birth on the east to the Rio de la Miel and the Nare; and on the north to Porce and Nechi; its average height is only from 1200 to 1350 toises. The culminant point appears to be near Santa Rosa, south-west of the celebrated Valley of Bears (Valle de Osos). The towns of Rio Negro and Marinilla are built on table-lands 1060 toises high. The western mass of the knot of the mountains of Antioquia, between the Cauca and the Atrato, gives rise, on its western descent, to the Rio San Juan, Bevara, and Murri. It attains its greatest height in the Alto del Viento, north of Urrao, known to the first conquistadores by the name of the Cordilleras of Abide or Dabeida. This height (latitude 7 degrees 15 minutes) does not, however, exceed 1500 toises. Following the western slope of this system of mountains of Antioquia, we find that the point of partition of the waters that flow towards the Pacific and the Caribbean Sea (latitude 5 1/2 and 6 degrees) nearly corresponds with the parallel of the isthmus of Raspadura, between the Rio San Juan and the Atrato. It is remarkable that in this group, more than 30 leagues broad, without sharp summits, between latitude 5 1/4 and 7 degrees, the highest masses rise towards the west; while, further south, before the union of the two chains of Quindiu and Choco, we saw them on the east of Cauca.

The ramifications of the knot of Antioquia, on the north of the parallel 7 degrees, are very imperfectly known; it is observed only that their lowering is in general more rapid and complete towards the north-west, in the direction of the ancient province of Biruquete and Darien, than towards the north and north-east, on the side of Zaragoza and Simiti. From the northern bank of the Rio Nare, near its confluence with the Samana, a spur stretches out, known by the name of La Simitarra, and the Mountains of San Lucar. We may call it the first branch of the group of Antioquia. I saw it, in going up the Rio Magdalena, on the west, from the Regidor and the mouth of the Rio Simiti, as far as San Bartolome (on the south of the mouth of the Rio Sogamozo); while, eastward, in latitude 7 3/4 and 8 1/4 degrees, the spur of the mountains of Ocana appear in the distance; they are inhabited by some tribes of Molitone Indians. The second branch of the group of Antioquia (west of Samitarra) commences at the mountains of Santa Rosa, stretches out between Zaragoza and Caceres, and terminates abruptly at the confluence of the Rio Nechi (latitude 8 degrees 33 minutes): at least if the hills, often conical, between the mouth of the Rio Sinu and the small town of Tolu, or even the calcareous heights of Turbaco and Popa, near Carthagena, may not be regarded as the most northern prolongation of this second branch. A third advances towards the gulf of Uraba or Darien, between the Rio San Jorge and the Atrato. It is linked southward with the Alto del Viento, or Sierra de Abide, and is rapidly lost, advancing as far as the parallel of 8 degrees. Finally, the fourth branch of the Andes of Antioquia, situated westward of Zitara and the Rio Atrato, undergoes, long before it enters the isthmus of Panama, such a depression, that between the Gulf of Cupica and the embarcadero of the Rio Naipipi, we find only a plain across which M. Gogueneche has projected a canal for the junction of the two seas. It would be interesting to know the configuration of the strata between Cape Garachine, or the Gulf of St. Miguel, and Cape Tiburon, especially towards the source of the Rio Tuyra and Chucunaque or Chucunque, so as to determine with precision where the mountains of the isthmus of Panama begin to rise; mountains whose elevation does not appear to be more than 100 toises. The interior of Darfur is not more unknown to geographers than the humid, insalubrious forest-land which extends on the north-west of Betoi and the confluence of the Bevara with the Atrato, towards the isthmus of Panama. All that we positively know of it hitherto is that between Cupica and the left bank of the Atrato there is either a land-strait, or a total absence of the Cordillera. The mountains of the isthmus of Panama, by their direction and their geographical position, may be considered as a continuation of the mountains of Antioquia and Choco; but on the west of Bas-Atrato, there is scarcely a ridge in the plain. We do not find in this country a group of interposed mountains like that which links (between Barquisimeto, Nirgua and Valencia) the eastern chain of New Grenada (that of Suma Paz and the Sierra Nevada de Merida) to the Cordillera of the shore of Venezuela.

The Cordillera of the Andes, considered in its whole extent, from the rocky wall of the island of Diego Ramirez to the isthmus of Panama, is sometimes ramified into chains more or less parallel, and sometimes articulated by immense knots of mountains. We distinguish nine of those knots, and consequently an equal number of branching-points and ramifications. The latter are generally bifurcations. The Andes are twice only divided into three chains; in the knot of Huanuco, near the source of the Amazon, and the Huallaga (latitude 10 to 11 degrees) and in the knot of the Paramo de las Papas (latitude 2 degrees), near the source of the Magdalena and the Cauca. Basins, almost shut in at their extremities, parallel with the axis of the Cordillera and bounded by two knots and two lateral chains, are characteristic features of the structure of the Andes. Among these knots of mountains some, for instance those of Cuzco, Loxa and Los Pastos, comprise 3300, 1500 and 1130 square leagues, while others no less important in the eye of the geologist are confined to ridges or transversal dykes. To the latter belong the Altos de Chisinche (latitude 0 degrees 40 minutes south) and the Los Robles (latitude 2 degrees 20 minutes north), on the south of Quito and Popayan. The knot of Cuzco, so celebrated in the annals of Peruvian civilization, presents an average height of from 1200 to 1400 toises, and a surface nearly three times greater than the whole of Switzerland. The ridge of Chisinche, which separates the basins of Tacunga and Quito, is 1580 toises high, but scarcely a mile broad. The knots or groups which unite several partial chains have not the highest summits, either in the Andes or, for the most part, in the great mountain ranges of the old continent; it is not even certain that there is always in those knots a widening of the chain. The greatness of the mass, and the height so long attributed to points whence several considerable branches issue, was founded either on theoretic ideas or on false measures. The Cordilleras were compared to rivers that swell as they receive a number of tributary streams.

Among the basins which the Andes present, and which form probably as many lakes or small inland seas, those of Titicaca, Rio Jauja and the Upper Maranon, comprise respectively 3500, 1300, and 2400 square leagues of surface.* (* I here subjoin some measures interesting to geologists. Area of the Andes, from Tierra del Fuego to the Paramo de las Rosas (latitude 9 1/4 degrees north), where the mountainous land of Tocuyo and Barquesimeto begins, part of the Cordillera of the shore of Venezuela, 58,900 square leagues, (20 to a degree) the four spurs of Cordova, Salta, Cochabamba and Beni alone, occupy 23,300 square leagues of this surface, and the three basins contained between latitude 6 and 20 degrees south measure 7200 square leagues. Deducting 33,200 square leagues for the whole of the enclosed basins and spurs, we find, in latitude 65 degrees, the area of the Cordilleras elevated in the form of walls, to be 25,700 square leagues, whence results (comprehending the knots, and allowing for the inflexion of the chains) an average breadth of the Andes of 18 to 20 leagues. The valleys of Huallaga and the Rio Magdalena are not comprehended in these 58,900 square leagues, on account of the diverging direction of the chain, east of Cipoplaya and Santa Fe de Bogota.) The first is so encompassed that no drop of water can escape except by evaporation; it is like the enclosed valley of Mexico,* (* We consider it in its primitive state, without respect to the gap or cleft of the mountains, known by the name of Desaghue de Huehuetoca.) and of those numerous circular basins which have been discerned in the moon, and which are surrounded by lofty mountains. An immense alpine lake characterizes the basin of Tiahuanaco or Titicaca; this phenomenon is the more worthy of attention, as in South America there are scarcely any of those reservoirs of fresh water which are found at the foot of the European Alps, on the northern and southern slopes, and which are permanent during the season of drought. The other basins of the Andes, for instance, those of Jauja, the Upper Maranon and Cauca, pour their waters into natural canals, which may be considered as so many crevices situated either at one of the extremities of the basin, or on its banks, nearly in the middle of the lateral chain. I dwell on this articulated form of the Andes, on those knots or transverse ridges, because, in the continuation of the Andes called the Cordilleras of the shore of Venezuela, we shall find the same transverse dykes, and the same phenomena.

The ramification of the Andes and of all the great masses of mountains into several chains merits particular consideration in reference to the height more or less considerable of the bottom of the enclosed basins, or longitudinal valleys. Geologists have hitherto directed more attention to the successive narrowing of these basins, their depth compared with the walls of rock that surround them, and the correspondence between the re-entering and the salient angles, than to the level of the bottom of the valleys. No precise measure has yet fixed the absolute height of the three basins of Titicaca, Jauja and the Upper Maranon;* (* I am inclined to believe that the southern part of the basin of the Upper Maranon, between Huary and Huacarachuco, exceeds 350 toises.) but I was fortunate enough to be able to determine the six other basins, or longitudinal valleys, which succeed each other, as if by steps, towards the north. The bottom of the valley of Cuenca, between the knots of Loxa and Assuay, is 1350 toises; the valley of Allansi and of Hambato, between the knot of the Assuay and the ridge of Chisinche, 1320 toises; the valley of Quito in the eastern part, 1340 toises, and in the western part, 1490 toises; the basin of Almaguer, 1160 toises; the basin of the Rio Cauca, between the lofty plains of Cali, Buga, and Cartago, 500 toises; the valley of Magdalena, first between Neiva and Honda, 200 toises; and further on, between Honda and

Mompox, 100 toises of average height above the level of the sea.* (* In the region of the Andes comprehended between 4 degrees of south latitude and 2 degrees of north, the longitudinal valleys or basins inclosed by parallel chains are regularly between 1200 and 1500 toises high; while the transversal valleys are remarkable for their depression, or rather the rapid lowering of their bottom. The valley of Patias, for instance, running from north-east to south-west is only 350 toises of absolute height, even above the junction of the Rio Guachion with the Quilquasi, according to the barometric measures of M. Caldas; and yet it is surrounded by the highest summits, the Paramos de Puntaurcu and Mamacondy. Going from the plains of Lombardy, and penetrating into the Alps of the Tyrol, by a line perpendicular to the axis of the chain, we advance more than 20 marine leagues towards the north, yet we find the bottom of the valley of the Adige and of Eysack near Botzen, to be only 182 toises of absolute height, an elevation which exceeds but 117 toises that of Milan. From Botzen however, to the ridge of Brenner (culminant point 746 toises) is only 11 leagues. The Valais is a longitudinal valley; and in a barometric measurement which I made very recently from Paris to Naples and Berlin, I was surprised to find that from Sion to Brigg, the bottom of the valley rises only to from 225 to 350 toises of absolute height; nearly the level of the plains of Switzerland, which, between the Alps and the Jura, are only from 274 to 300 toises.) In this region, which has been carefully measured, the different basins lower very sensibly from the equator northward. The elevation of the bottom of enclosed basins merits great attention in connection with the causes of the formation of the valleys. I do not deny that the depressions in the plains may be sometimes the effect of ancient pelagic currents, or slow erosions. I am inclined to believe that the transversal valleys, resembling crevices, have been widened by running waters; but these hypotheses of successive erosions cannot well be applied to the completely enclosed basins of Titicaca and Mexico. These basins, as well as those of Jauja, Cuenca and Almaguer, which lose their waters only by a lateral and narrow issue, owe their origin to a cause more instantaneous, more closely linked with the upheaving of the whole chain. It may be said that the phenomenon of the narrow declivities of the Sarenthal and of the valley of Eysack in the Tyrol, is repeated at every step, and on a grander scale, in the Cordilleras of equinoctial America. We seem to recognize in the Cordilleras those longitudinal sinkings, those rocky vaults, which, to use the expression of a great geologist,* "are broken when extended over a great space, and leave deep and almost perpendicular rents." (* Von Buch, Tableau du Tyrol meridional page 8 1823.)

If, to complete the sketch of the structure of the Andes from Tierra del Fuego to the northern Polar Sea, we pass the boundaries of South America, we find that the western Cordillera of New Grenada, after a great depression between the mouth of the Atrato and the gulf of Cupica, again rises in the isthmus of Panama to 80 or 100 toises high, augmenting towards the west, in the Cordilleras of Veragua and Salamanca,* and extending by Guatimala as far as the confines of Mexico. (* If it be true, as some navigators affirm, that the mountains at the north-western extremity of the republic of Columbia, known by the names of Silla de Veragua, and Castillo del Choco, be visible at 36 leagues distance, the elevation of their summits must be nearly 1400 toises, little lower than the Silla of Caracas.) Within this space it extends along the coast of the Pacific where, from the gulf of Nicoya to Soconusco (latitude 9 1/2 to 16 degrees) is found a long series of volcanoes,* most frequently insulated, and sometimes linked to spurs or lateral branches. (* See the list of twenty-one volcanoes of Guatimala, partly extinct and partly still burning, given by Arago and myself, in the Annuaire du Bureau des Longitudes pour 1824 page 175. No mountain of Guatimala having been hitherto measured, it is the more important to fix approximately the height of the Volcan de Agua, or the Volcano of Pacaya, and the Volcan de Fuego, called also Volcano of Guatimala. Mr. Juarros expressly says that this volcano which, by torrents of water and stones, destroyed, on the 11th September, 1541, the Ciudad Vieja, or Almolonga (the ancient capital of the country, which must not be confounded with the ancient Guatimala), is covered with snow, during several months of the year. This phenomenon would seem to indicate a height of more than 1750 toises.) Passing the isthmus of Tehuantepecor Huasacualco, on the Mexican territory, the Cordillera of central America extends on toward the intendancia of Oaxaca, at an equal distance from the two oceans; then from 18 1/2 to 21 degrees latitude, from Misteca to the mines of Zimapan, it approximates to the eastern coast. Nearly in the

156

parallel of the city of Mexico, between Toluca, Xalapa and Cordoba, it attains its maximum height; several colossal summits rising to 2400 and 2770 toises. Farther north the chain called Sierra Madre runs north 40 degrees west towards San Miguel el Grande and Guanaxuato. Near the latter town (latitude 21 degrees 0 minutes 15 seconds) where the richest silver mines of the known world are situated, it widens in an extraordinary degree and separates into three branches. The most eastern branch advances towards Charcas and the Real de Catorce, and lowers progressively (turning to north-east) in the ancient kingdom of Leon, in the province of Cohahuila and Texas. That branch is prolonged from the Rio Colorado de Texas, crossing the Arkansas near the confluence of the Mississippi and the Missouri (latitude 38 degrees 51 minutes). In those countries it bears the name of the Mountains of Ozark,* and attains 300 toises of height. (* Ozark is at once the ancient name of Arkansas and of the tribe of Quawpaw Indians who inhabit the banks of that great river. The culminant point of the Mountains of Ozark is in latitude 37 1/2 degrees, between the sources of the White and Osage rivers.) It has been supposed that on the east of the Mississippi (latitude 44 to 46 degrees) the Wisconsin Hills, which stretch out to north-north-east in the direction of Lake Superior, may be a continuation of the mountains of Ozark. Their metallic wealth seems to denote that they are a prolongation of the eastern Cordillera of Mexico. The western branch or Cordillera occupies a part of the province of Guadalajara and stretches by Culiacan, Aripe and the auriferous lands of the Pimeria Alta and La Sonora, as far as the banks of the Rio Gila (latitude 33 to 34 degrees), one of the most ancient dwellings of the Aztek nations. We shall soon see that this western chain appears to be linked by the spurs that advance to the west, with the maritime Alps of California. Finally, the central Cordillera of Anahuac, which is the most elevated, runs first from south-east to north-west, by Zacatecas towards Durango, and afterwards from south to north, by Chihuahua, towards New Mexico. It takes successively the names of Sierra de Acha, Sierra de Los Mimbres, Sierra Verde, and Sierra de las Grullas, and about the 29 and 39 degrees of latitude, it is connected by spurs with two lateral chains, those of the Texas and La Sonora, which renders the separation of the chains more imperfect than the trifurcations of the Andes in South America.

That part of the Cordilleras of Mexico which is richest in silver beds and veins, is comprehended between the parallels of Oaxaca and Cosiquiriachi (latitude 16 1/2 to 29 degrees); the alluvial soil that contains disseminated gold extends some degrees still further northwards. It is a very striking phenomenon that the gold-washing of Cinaloa and Sonora, like that of Barbacoas and Choco on the south and north of the isthmus of Panama, is uniformly situated on the west of the central chain, on the descent opposite the Pacific. The traces of a still-burning volcanic fire which was no longer seen, on a length of 200 leagues, from Pasto and Popayan to the gulf of Nicoya (latitude 1 1/4 to 9 1/2 degrees), become very frequent on the western coast of Guatimala (latitude 9 1/2 to 16 degrees); these traces of fire again cease in the gneiss-granite mountains of Oaxaca, and re-appear, perhaps for the last time, towards the north, in the central Cordillera of Anahuac, between latitude 18 1/4 and 19 1/2 degrees, where the volcanoes of Taxtla, Orizaba, Popocatepetl, Toluca, Jorullo and Colima appear to be situated in a crevice* extending from east-south-east to west-north-west, from one ocean to the other. (* On this zone of volcanoes is the parallel of the greatest heights of New Spain. If the survey of Captain Basil Hall afford results alike certain in latitude and in longitude, the volcano of Colima is north of the parallel of Puerto de Navidad in latitude 19 degrees 36 minutes; and, like the volcano of Tuxtla, if not beyond the zone, at least beyond the average parallel of the volcanic fire of Mexico, which parallel seems to be between 18 degrees 59 minutes and 19 degrees 12 minutes.) This line of summits, several of which enter the limit of perpetual snow, and which are the loftiest of the Cordilleras from the peak of Tolima (latitude 40 degrees 46 minutes north), is almost perpendicular to the great axis of the chain of Guatimala and Anahuac, advancing to the 27th parallel, uniformly north 42 degrees east. A characteristic feature of every knot, or widening of the Cordilleras, is that the grouping of the summits is independent of the general direction of the axis. The backs of the mountains in New Spain form very elevated plains, along which carriages can roll for an extent of 400 leagues, from the capital to Santa-Fe and Taos, near the sources of Rio del Norte. This immense table-land, in 19 and 24 1/2 degrees, is constantly at the height of from 950 to 1200 toises, that is, at the elevation of the passes of the Great Saint Bernard and the Splugen. We find on the back of the Cordilleras of Anahuac, which lower progressively from the city of Mexico towards Taos, a succession of basins: they are separated by hills little striking to the eye of the traveller because they rise only from 250 to 400 toises above the surrounding plains. The basins are sometimes closed, like the valley of Tenochtitlan, where lie the great Alpine lakes, and sometimes they exhibit traces of ancient ejections, destitute of water.

Between latitude 33 and 38 degrees, the Rio del Norte forms, in its upper course, a great longitudinal valley; and the central chain seems here to be divided into several parallel ranges. This distribution continues northward, in the Rocky Mountains,* where, between the parallels of 37 and 41 degrees, several summits covered with eternal snow (Spanish Peak, James Peak and Big Horn) are from 1600 to 1870 toises of absolute height. (* The Rocky Mountains have been at different periods designated by the names of Chypewyan, Missouri, Columbian, Caous, Stony, Shining and Sandy Mountains.) Towards latitude 40 degrees south of the sources of the Paduca, a tributary of the Rio de la Plata, a branch known by the name of the Black Hills, detaches itself towards the north-east from the central chain. The Rocky Mountains at first seem to lower considerably in 46 and 48 degrees; and then rise to 48 and 49 degrees, where their tops are from 1200 to 1300 toises, and their ridge near 950 toises. Between the sources of the Missouri and the River Lewis, one of the tributaries of the Oregon or Columbia, the Cordilleras form in widening, an elbow resembling the knot of Cuzco. There, also, on the eastern declivity of the Rocky Mountains, is the partition of water between the Caribbean Sea and the Polar Sea. This point corresponds with those in the Andes of South America, at the spur of Cochabamba, on the east, latitude 19 degrees 20 minutes south; and in the Alto de los Robles (latitude 2 degrees 20 minutes north), on the west. The ridge that separates the Rocky Mountains extends from west to east, towards Lake Superior, between the basins of the Missouri and those of Lake Winnipeg and the Slave Lake. The central Cordillera of Mexico and the Rocky Mountains follow the direction north 10 degrees west, from latitude 25 to 38 degrees; the chain from that point to the Polar Sea prolongs in the direction north 24 degrees west, and ends in the parallel 69 degrees, at the mouth of the Mackenzie River.*

(* The eastern boundary of the Rocky Mountains lies:—
In 38 degrees latitude : 107 degrees 20 minutes longitude.
In 40 degrees latitude : 108 degrees 30 minutes longitude.
In 63 degrees latitude : 124 degrees 40 minutes longitude.
In 68 degrees latitude : 130 degrees 30 minutes longitude.)

In thus developing the structure of the Cordilleras of the Andes from 56 degrees south to beyond the Arctic circle, we see that its northern extremity (longitude 130 degrees 30 minutes) is nearly 61 degrees of longitude west of its southern extremity (longitude 60 degrees 40 minutes); this is the effect of the long-continued direction from south-east to north-west north of the isthmus of Panama. By the extraordinary breadth of the New Continent, in the 30 and 60 degrees north latitude, the Cordillera of the Andes, continually approaching nearer to the western coast in the southern hemisphere, is removed 400 leagues on the north from the source of the Rio de la Paz. The Andes of Chile may be considered as maritime Alps,* (* Geognostically speaking, a littoral chain is not a range of mountains forming of itself the coast; this name is extended to a chain separated from the coast by a narrow plain.) while, in their most northern continuation, the Rocky Mountains are a chain in the interior of a continent. There is, no doubt, between latitude 23 and 60 degrees from Cape Saint Lucas in California, to Alaska on the western coast of the Sea of Kamschatka, a real littoral Cordillera; but it forms a system of mountains almost entirely distinct from the Andes of Mexico and Canada. This system, which we shall call the Cordillera of California, or of New Albion, is linked between latitude 33 and 34 degrees with the Pimeria alta, and the western branch of the Cordilleras of Anahuac; and between latitude 45 and 53 degrees, with the Rocky Mountains, by transversal ridges and spurs that widen towards the east. Travellers who may at some future time pass over the unknown land between Cape Mendocino and the source of the Rio Colorado, may perhaps inform us whether the connexion of the maritime Alps of California or New Albion, with the western branch of the Cordilleras of

Mexico, resembles that which, notwithstanding the depression, or rather total interruption observed on the west of the Rio Atrato, is admitted by geographers to exist between the mountains of the isthmus of Panama and the western branch of the Andes of New Grenada. The maritime Alps, in the peninsula of Old California, rise progressively towards the north in the Sierra of Santa Lucia (latitude 34 1/2 degrees), in the Sierra of San Marcos (latitude 37 to 38 degrees) and in the Snowy Mountains near Cape Mendocino (latitude 39 degrees 41 minutes); the last seem to attain at least the height of 1500 toises. From Cape Mendocino the chain follows the coast of the Pacific, but at the distance of from twenty to twenty-five leagues. Between the lofty summits of Mount Hood and Mount Saint Helen, in latitude 45 3/4 degrees, the chain is broken by the River Columbia. In New Hanover, New Cornwall and New Norfolk these rents of a rocky coast are repeated, these geologic phenomena of the fjords that characterize western Patagonia and Norway. At the point where the Cordillera turns towards the west (latitude 58 3/4 degrees, longitude 139 degrees 40 minutes) there are two volcanic peaks, one of which (Mount Saint Elias) perhaps equals Cotopaxi in height; the other (Fair-Weather Mountain) equals the height of Mount Rosa. The elevation of the former exceeds all the summits of the Cordilleras of Mexico and the Rocky Mountains, north of the parallel 19 1/4 degrees; it is even the culminant point in the northern hemisphere, of the whole known world north of 50 degrees of latitude. North-west of the peaks of Saint Elias and Fair-Weather the chain of California widens considerably in the interior of Russian America. Volcanoes multiply in number as we advance westward, in the peninsula of Alaska and the Fox Islands, where the volcano Ajagedan rises to the height of 1175 toises above the level of the sea. Thus the chain of the maritime Alps of California appears to be undermined by subterraneous fires at its two extremities; on the north in 60 degrees of latitude, and on the south, in 28 degrees, in the volcanoes of the Virgins.* (* Volcanes de las Virgenes. The highest summit of Old California, the Cerro de la Giganta (700 toises), appears to be also an extinguished volcano.) If it were certain that the mountains of California belong to the western branch of the Andes of Anahuac, it might be said that the volcanic fire, still burning, abandons the central Cordillera when it recedes from the coast, that is, from the volcano of Colima; and that the fire is borne on the north-west by the peninsula of Old California, Mount Saint Elias, and the peninsula of Alaska, towards the Aleutian Islands and Kamschatka.

I shall terminate this sketch of the structure of the Andes by recapitulating the principal features that characterize the Cordilleras, north-west of Darien.

Latitude 8 to 11 degrees. Mountains of the isthmus of Panama, Veragua and Costa Rica, slightly linked to the western chain of New Grenada, which is that of Choco.

Latitude 11 to 16 degrees. Mountains of Nicaragua and Guatimala; line of volcanoes north 50 degrees west, for the most part still burning, from the gulf of Nicoya to the volcano of Soconusco.

Latitude 16 to 18 degrees. Mountains of gneiss-granite in the province of Oaxaca.

Latitude 18 1/2 to 19 1/2 degrees. Trachytic knot of Anahuac, parallel with the Nevados and the burning volcanoes of Mexico.

Latitude 19 1/2 to 20 degrees. Knot of the metaliferous mountains of Guanaxuato and Zacatecas.

Latitude 21 3/4 to 22 degrees. Division of the Andes of Anahuac into three chains:

Eastern chain (that of Potosi and Texas), continued by the Ozark and Wisconsin mountains, as far as Lake Superior.

Central chain (of Durango, New Mexico and the Rocky Mountains), sending on the north of the source of the river Platte (latitude 42 degrees) a branch (the Black hills) to north-east, widening greatly between the parallels 46 and 50 degrees, and lowering progressively as it approaches the mouth of Mackenzie River (latitude 68 degrees).

Western chain (of Cinaloa and Sonora). Linked by spurs to the maritime Alps, or mountains of California.

We have yet no means of judging with precision the elevation of the Andes south of the knot of the mountains of Loxa (south latitude 3 degrees 5), but we know that on the north of that knot the Cordilleras rise five times higher than the majestic elevation of 2600 toises:

In the group of Quito, 0 to 2 degrees south latitude (Chimborazo, Antisano, Cayambe, Cotopaxi, Collanes, Yliniza, Sangay, Tungurahua.)

In the group of Cundinamarca, latitude 4 3/4 degrees north (peak of Tolima, north of the Andes of Quindiu).

In the group of Anahuac, from latitude 18 degrees 59 minutes to 19 degrees 12 minutes (Popocatepetl or the Great Volcano of Mexico, and Peak of Orizaba). If we consider the maritime Alps or mountains of California and New Norfolk, either as a continuation of the western chain of Mexico, that of Sonora, or as being linked by spurs to the central chain, that of the Rocky Mountains, we may add to the three preceding groups:

The group of Russian America, from latitude 60 to 70 degrees (Mount Saint Elias). Over an extent of 63 degrees of latitude, I know only twelve summits of the Andes which reach the height of 2600 toises, and consequently exceed by 140 toises, the height of Mont Blanc. Only three of these twelve summits are situated north of the isthmus of Panama.

2. INSULATED GROUP OF THE SNOWY MOUNTAINS OF SANTA MARTA.

In the enumeration of the different systems of mountains, I place this group before the littoral chain of Venezuela, though the latter, being a northern prolongation of the Cordillera of Cundinamarca, is immediately linked with the chain of the Andes. The Sierra Nevado of Santa Marta is encompassed within two divergent branches of the Andes, that of Bogota, and that of the isthmus of Panama. It rises abruptly like a fortified castle, amidst the plains extending from the gulf of Darien, by the mouth of the Magdalena, to the lake of Maracaybo. The old geographers erroneously considered this insulated group of mountains covered with eternal snow, as the extremity of the high Cordilleras of Chita and Pamplona. The loftiest ridge of the Sierra Nevada de Santa Marta is only three or four leagues in length from east to west; it is bounded (at nine leagues distance from the coast) by the meridians of the capes of San Diego and San Augustin. The culminant points, called El Picacho and Horqueta, are near the western border of the group; they are entirely separated from the peak of San Lorenzo, also covered with eternal snow, but only four leagues distant from the port of Santa Marta, towards the south-east. I saw this latter peak from the heights that surrounded the village of Turbaco, south of Carthagena. No precise measurement has hitherto given us the height of the Sierra Nevada, which Dampier affirms to be one of the highest mountains of the northern hemisphere. Calculations founded on the maximum of distance at which the group is discerned at sea, give a height of more than 3004 toises. That the group of the mountains of Santa Marta is insulated is proved by the hot climate of the lands (tierras calientes) that surround it. Low ridges and a succession of hills indicate, perhaps, an ancient connection between the Sierra Nevada de Santa Marta on one side, by the Alto de las Minas, with the phonolitic and granitic rocks of the Penon and Banca, and on the other, by the Sierra de Perija, with the mountains of Chiliguana and Ocana, which are the spurs of the eastern chain of the Andes of New Grenada. In this latter chain, the febrifuge species of cinchona (corollis hirsutis, staminibus inclusis) are found in the Sierra Nevada de Merida; but the real cinchona, the most northern of South America, is found in the temperate region of the Sierra Nevada de Santa Marta.

3. LITTORAL CHAIN OF VENEZUELA.

This is the system of mountains the configuration and direction of which have excited so powerful an influence on the cultivation and commerce of the ancient Capitania General of Venezuela. It bears different names, as the mountains of Coro, of Caracas, of the Bergantin, of Barcelona, of Cumana, and of Paria; but all these names belong to the same chain, of which the northern part runs along the coast of the Caribbean Sea. This system of mountains, which is 160 leagues long,* is a prolongation of the eastern Cordillera of the Andes of Cundinamarca. (* It is more than double the length of the Pyrenees, from Cape Creux to the point of Figuera.) There is an immediate connection of the littoral chain with the Andes, like that of the Pyrenees with the mountains of Asturia and Galicia; it is not the effect of transversal ridges, like the connection of the Pyrenees with the Swiss Alps, by the Black Mountain and the Cevennes. The points of junction are between Truxillo and the lake of Valencia.

The eastern chain of New Grenada stretches north-east by the Sierra Nevada de Merida, as well as by the four Paramos of Timotes, Niquitao, Bocono and Las Rosas, of which the absolute height cannot be less than from 1400 to 1600 toises. After the Paramo of Las Rosas, which is more elevated than the two preceding, there is a great depression, and we no longer see a distinct chain or ridge, but merely hills, and high table-lands surrounding the towns of Tocuyo and Barquisimeto. We know not the height even of Cerro del Altar, between Tocuyo and Caranacatu; but we know by recent measures that the most inhabited spots are from 300 to 350 toises above sea-level. The limits of the mountainous land between Tocuyo and the valleys of Aragua are, the plains of San Carlos on the south, and the Rio Tocuyo on the north; the Rio Siquisique flows into that river. From the Cerro del Altar on the north-east towards Guigue and Valencia, succeed, as culminant points, the mountains of Santa Maria (between Buria and Nirgua); then the Picacho de Nirgua, supposed to be 600 toises high; and finally Las Palomeras and El Torito (between Valencia and Nirgua). The line of water-partition runs from west to east, from Quibor to the lofty savannahs of London, near Santa Rosa. The waters flow on the north, towards the Golfo triste of the Caribbean Sea; and on the south, towards the basins of the Apure and the Orinoco. The whole of this mountainous country, by which the littoral chain of Caracas is linked to the Cordilleras of Cundinamarca, was celebrated in Europe in the middle of the nineteenth century; for that part of the territory formed of gneiss-granite, and lying between the Rio Tocuyo and the Rio Yaracui, contains the auriferous veins of Buria, and the copper-mine of Aroa which is worked at the present day. If, across the knot of the mountains of Barquisimeto, we trace the meridians of Aroa, Nirgua and San Carlos, we find that on the north-west that knot is linked with the Sierra de Coro, and on the north-east with the mountains of Capadare, Porto Cabello and the Villa de Cura. It may be said to form the eastern wall of that vast circular depression of which the lake of Maracaybo is the centre and which is bounded on the south and west by the mountains of Merida, Ocana, Perija and Santa Marta.

The littoral chain of Venezuela presents towards the centre and the east the same phenomena of structure as those observed in the Andes of Peru and New Grenada; namely, the division into several parallel ranges and the frequency of longitudinal basins or valleys. But the irruptions of the Caribbean Sea having apparently overwhelmed, at a very remote period, a part of the mountains of the shore, the ranges or partial chains are interrupted and some basins have become oceanic gulfs. To comprehend the Cordillera of Venezuela in mass we must carefully study the direction and windings of the coast from Punta Tucacas (west of Porto Cabello) as far as Punta de la Galera of the island of Trinidad. That island, those of Los Testigos, Marguerita and Tortuga constitute, with the mica-slates of the peninsula of Araya, one and the same system of mountains. The granitic rocks which appear between Buria, Duaca and Aroa cross the valley of the Rio Yaracui and draw near the shore, whence they extend, like a continuous wall, from Porto Cabello to Cape Codera. This prolongation forms the northern chain of the Cordillera of Venezuela and is traversed in going from south to north, either from Valencia and the valleys of Aragua, to Burburata and Turiamo, or from Caracas to La Guayra. Hot springs* issue from those mountains (* The other hot springs of the Cordillera of the shore are those of San Juan, Provisor, Brigantin, the gulf of Cariaco, Cumucatar and Irapa. MM. Rivero and Boussingault, who visited the thermal waters of Mariara in February, 1823, during their journey from Caracas to Santa Fe de Bogota, found their maximum to be 64 degrees centigrade. I found it at the same season only 59.2 degrees. Has the great earthquake of the 26th March, 1812, had an influence on the temperature of these springs? The able chemists above mentioned were, like myself, struck with the extreme purity of the hot waters that issue from the primitive rocks of the basin of Aragua. Those of Onoto, which flow at the height of 360 toises above the level of the sea, have no smell of sulphuretted hydrogen; they are without taste, and cannot be precipitated, either by nitrate of silver or any other re-agent. When evaporated they have an inappreciable residue which consists of a little silica and a trace of alkali; their temperature is only 44.5 degrees, and the bubbles of air which are disengaged at intervals are at Onoto, as well as in the thermal waters of Mariara, pure nitrogen. The waters of Mariara (244 toises) have a faint smell of sulphuretted hydrogen; they leave, by evaporation, a slight residuum, that yields carbonic acid, sulphuric acid, soda, magnesia and lime. The quantities are so small that the water is altogether without taste. In the course of my journey I found only the springs of Cumangillas hotter than the thermal waters of Las Trincheras: they are situated on the south of Porto Cabello. The waters of Comangillas are at the height of 1040 toises and are alike remarkable for their purity and their temperature of 96.3 degrees centigrade.), those of Las Trincheras (90.4 degrees) on its southern slope and those of Onoto and Mariara on its southern slope. The former issue from a granite with large grains, very regularly stratified; the latter from a rock of gneiss. What especially characterizes the northern chain is a summit which is not only the loftiest of the system of the mountains of Venezuela, but of all South America, on the east of the Andes. The eastern summit of the Silla of Caracas, according to my barometric measurement made in 1800, is 1350 toises high,* (* The Silla of Caracas is only 80 toises lower than the Canigou in the Pyrenees.) and notwithstanding the commotion which took place on the Silla during the great earthquake of Caracas, that mountain did not sink 50 or 60 toises, as some North American journals asserted. Four or five leagues south of the northern chain (that of Mariara, La Silla and Cape Codera) the mountains of Guiripa, Ocumare and Panaquire form the southern chain of the coast, which stretches in a parallel direction from Guigue to the mouth of the Rio Tuy, by the Guesta of Yusma and the Guacimo. The latitudes of the Villa de Cura and San Juan, so erroneously marked on our maps, enabled me to ascertain the mean breadth of the whole Cordillera of Venezuela. Ten or twelve leagues may be reckoned as the distance from the descent of the northern chain which bounds the Caribbean Sea, to the descent of the southern chain bounding the immense basin of the Llanos. This latter chain, which also bears the name of the Inland Mountains, is much lower than the northern chain; and I can hardly believe that the Sierra de Guayraima attains the height of 1200 toises.

The two partial chains, that of the interior, and that which runs along the coast, are linked by a ridge or knot of mountains known by the names of Altos de las Cocuyzas (845 toises) and the Higuerote (835 toises between Los Teques and La Victoria) in longitude 69 degrees 30 minutes and 69 degrees 50 minutes. On the west of this ridge lies the enclosed basin* of the lake of Valencia or the Valles de Aragua (* This basin contains a small system of inland rivers which do not communicate with the ocean. The southern chain of the litteral Cordillera of Venezuela is so depressed on the south-west that the Rio Pao is separated from the tributary streams of the lake of Tacarigua or Valencia. Towards the east the Rio Tuy, which takes its rise on the western declivity of the knot of mountains of Las Cocuyzas, appears at first to empty itself into the valleys of Aragua; but hills of calcareous tufa, forming a ridge between Consejo and Victoria, force it to take its course south-east.); and on the east the basin of Caracas and of the Rio Tuy. The bottom of the first-mentioned basins is between 220 and 250 toises high; the bottom of the latter is 460 toises above the level of the Caribbean Sea. It follows from these measures that the most western of the two longitudinal valleys enclosed by the littoral Cordillera is the deepest; while in the plains near the Apure and the Orinoco the declivity is from west to east; but we must not forget that the peculiar disposition of the bottom of the two basins, which are bounded by two parallel chains, is a local phenomenon altogether separate from the causes on which the general structure of the country depends. The eastern basin of the Cordillera of Venezuela is not shut up like the basin of Valencia. It is in the knot of the mountains of Las Cocuyzas, and of Higuerote, that the Serrania de los Teques and Oripoto, stretching eastward, form two valleys, those of the Rio Guayre and Rio Tuy; the former contains the town of Caracas and both unite below the Caurimare. The Rio Tuy runs through the rest of the basin, from west to east, as far as its mouth which is situated on the north of the mountains of Panaquire.

Cape Codera seems to terminate the northern range of the littoral mountains of Venezuela but this termination is only apparent. The coast forms a vast nook, thirty-five sea leagues in length, at the bottom of which is the mouth of the Rio Unare and the road of Nueva Barcelona. Stretching first from west to east, in the parallel of 10 degrees 37 minutes, this coast recedes at the parallel 10 degrees 6 minutes, and resumes its original direction (10 degrees 37 minutes to 10 degrees 44 minutes) from the western extremity of the peninsula of Araya to the eastern extremities of Montana de Paria and the island of Trinidad. From this dissection of the coast it follows that the range of mountains bordering the shore of the provinces of Caracas and Barcelona, between the meridian 66 degrees 32 minutes and 68 degrees 29 minutes (which I saw on the south of the bay of Higuerote and on the north of the Llanos of Pao and Cachipo), must be considered as the continuation of the southern chain of Venezuela and as being linked on the west with the Sierras de Panaquire and Ocumare. It may therefore be said that between Cape Codera and Cariaco the inland chain itself forms the coast. This range of very low mountains, often interrupted from the mouth of the Rio Tuy to that of the Rio Neveri, rises abruptly on the east of Nueva Barcelona, first in the rocky island of Chimanas, and then in the Cerro del Bergantin, elevated probably more than 800 toises, but of which the astronomical position and the precise height are yet alike unknown. On the meridian of Cumana the northern chain (that of Cape Codera and the Silla of Caracas) again appears. The micaceous slate of the peninsula of Araya and Maniquarez joins by the ridge or knot of mountains of Meapire the southern chain, that of Panaquire the Bergantin, Turimiquiri, Caripe and Guacharo. This ridge, not more than 200 toises of absolute height, has, in the ancient revolutions of our planet, prevented the irruption of the ocean, and the union of the gulfs of Paria and Cariaco. On the west of Cape Codera the northern chain, composed of primitive granitic rocks, presents the loftiest summits of the whole Cordillera of Venezuela; but the culminant points east of that cape are composed in the southern chain of secondary calcareous rocks. We have seen above that the peak of Turimiquiri, at the back of the Cocollar, is 1050 toises, while the bottoms of the high valleys of the convent of Caripe and of Guardia de San Augustin are 412 and 533 toises of absolute height. On the east of the ridge of Meapire the southern chain sinks abruptly towards the Rio Arco and the Guarapiche; but, on quitting the main land, we again see it rising on the southern coast of the island of Trinidad which is but a detached portion of the continent, and of which the northern side unquestionably presents the vestiges of the northern chain of Venezuela, that is, of the Montana de Paria (the Paradise of Christopher Columbus), the peninsula of Araya and the Silla of Caracas. The observations of latitude I made at the Villa de Cura (10 degrees 2 minutes 47 seconds), the farm of Cocollar (10 degrees 9 minutes 37 seconds) and the convent of Caripe (10 degrees 10 minutes 14 seconds), compared with the more anciently known position of the south coast of Trinidad (latitude 10 degrees 6 minutes), prove that the southern chain, south of the basins of Valencia and of Tuy* (* The bottom of the first of these four basins bounded by parallel chains is from 230 to 460 toises above, and that of the two latter from 30 to 40 toises below the present sea-level. Hot springs gush from the bottom of the gulf of the basin of Cariaco, as from the bottom of the basin of Valencia on the continent.) and of the gulfs of Cariaco and Paria, is still more uniform in the direction from west to east than the northern chain from Porto Cabello to Punta Galera. It is highly important to know the southern limit of the littoral Cordillera of Venezuela because it determines the parallel at which the Llanos or the savannahs of Caracas, Barcelona and Cumana begin. On some well-known maps we find erroneously marked between the meridians of Caracas and Cumana two Cordilleras stretching from north to south, as far as latitude 8 3/4 degrees, under the names of Cerros de Alta Gracia and del Bergantin, thus describing as mountainous a territory of 25 leagues broad, where we should seek in vain a hillock of a few feet in height.

Turning to the island of Marguerita, composed, like the peninsula of Araya, of micaceous slate, and anciently linked with that peninsula by the Morro de Chacopata and the islands of Coche and Cubagua, we seem to recognize in the two mountainous groups of Macanao and La Vega de San Juan traces of a third coast-chain of the Cordillera of Venezuela. Do these two groups of Marguerita, of which the most westerly is above 600 toises high, belong to a submarine chain stretching by the isle of Tortuga, towards the Sierra de Santa Lucia de Coro, on the parallel of 11 degrees? Must we admit that in latitude 11 1/4 and 12 1/2 degrees a fourth chain, the most northerly of all, formerly stretched out in the direction of the island of Hermanos, by Blanquilla, Los Roques, Orchila, Aves, Buen Ayre, Curacao and Oruba, towards Cape Chichivacoa? These important problems can only be solved when the chain of islands parallel with the coast has been properly examined. It must not be forgotten that a great irruption of the ocean appears to have taken place between Trinidad and Grenada,* and that no where else in the long series of the Lesser Antilles are two neighbouring islands so far removed from each other. (* It is affirmed that the island of

Trinidad is traversed in the northern part by a chain of primitive slate, and that Grenada furnishes basalt. It would be important to examine of what rock the island of Tobago is composed; it appeared to me of dazzling whiteness; and on what point, in going from Trinidad northward, the trachytic and trappean system of the Lesser Antilles begins.) We observe the effect of the rotatory current in the direction of the coast of Trinidad, as in the coasts of the provinces of Cumana and Caracas, between Cape Paria and Punta Araya and between Cape Codera and Porto Cabello. If a part of the continent has been overwhelmed by the ocean on the north of the peninsula of Araya it is probable that the enormous shoal which surrounds Cubagua, Coche the island of Marguerita, Los Frailes, La Sola and the Testigos marks the extent and outline of the submerged land. This shoal or placer, which is of the extent of 200 square leagues, is well known only to the tribe of the Guayqueries; it is frequented by these Indians on account of its abundant fishery in calm weather. The Gran Placer is believed to be separated only by some canals or deep furrows of the bank of Grenada from the sand-bank that extends like a narrow dyke from Tobago to Grenada, and which is known by the lowering of the temperature of the water and from the sand-banks of Los Roques and Aves. The Guayquerie Indians and, generally speaking, all the inhabitants of the coast of Cumana and Barcelona, are imbued with an idea that the water of the shoals of Marguerita and the Testigos diminishes from year to year; they believe that, in the lapse of ages, the Morro do Chacopata on the peninsula of Araya will be joined by a neck of land to the islands of Lobos and Coche. The partial retreat of the waters on the coast of Cumana is undeniable and the bottom of the sea has been upheaved at various times by earthquakes; but these local phenomena, which it is so difficult to account for by the action of volcanic force, the changes in the direction of currents, and the consequent swelling of the waters, are very different from the effects manifested at once over the space of several hundred square leagues.

4. GROUP OF THE MOUNTAINS OF PARIME.

It is essential to mineralogical geography to designate by one name all the mountains that form one system. To attain this end, a denomination belonging to a partial group only may be extended over the whole chain; or a name may be employed which, by reason of its novelty, is not likely to give rise to homogenic mistakes. Mountaineers designate every group by a special denomination; and a chain is generally considered as forming a whole only when it is seen from afar bounding the horizon of the plains. We find the name of snowy mountains (Himalaya, Imaus) repeated in every zone, white (Alpes, Alb), black and blue. The greater part of the Sierra Parime is, as it were, edged round by the Orinoco. I have, however, avoided a denomination having reference to this circumstance, because the group of mountains to which I am about to direct attention extends far beyond the banks of the Orinoco. It stretches south-east, towards the banks of the Rio Negro and the Rio Branco, to the parallel of 1 1/2 degrees north latitude. The geographical name of Parime has the advantage of reviving recollections of the fable of El Dorado, and the lofty mountains which, in the sixteenth century, were supposed to surround the lake Rupunuwini, or the Laguna de Parime. The missionaries of the Orinoco still give the name of Parime to the whole of the vast mountainous country comprehended between the sources of the Erevato, the Orinoco, the Caroni, the Rio Parime* (a tributary of the Rio Branco) and the Rupunuri or Rupunuwini, a tributary of the Rio Essequibo. (* The Rio Parime, after receiving the waters of the Uraricuera, joins the Tacutu, and forms, near the fort of San Joacquim, the Rio Branco, one of the tributary streams of the Rio Negro.) This country is one of the least known parts of South America and is covered with thick forests and savannahs; it is inhabited by independent Indians and is intersected by rivers of dangerous navigation, owing to the frequency of shoals and cataracts.

The system of the mountains of Parime separates the plains of the Lower Orinoco from those of the Rio Negro and the Amazon; it occupies a territory of trapezoidal form, comprehended between the parallels of 3 and 8 degrees, and the meridians of 61 and 70 1/2 degrees. I here indicate only the elements of the loftiest group, for we shall soon see that towards south-east the mountainous country, in lowering, draws near the equator, as well as to French and Portuguese Guiana. The Sierra Parime extends most in the direction north 85 degrees west and the partial chains into which it separates on the westward generally follow the same direction. It is less a Cordillera or a continuous chain in the sense given to those denominations when applied to the Andes and Caucasus than an irregular grouping of mountains separated the one from the other by plains and savannahs. I visited the northern, western and southern parts of the Sierra Parime, which is remarkable by its position and its extent of more than 25,000 square leagues. From the confluence of the Apure, as far as the delta of the Orinoco, it is uniformly three or four leagues removed from the right bank of the great river; only some rocks of gneiss-granite, amphibolic slate and greenstone advance as far as the bed of the Orinoco and create the rapids of Torno and of La Boca del Infierno.* (* To this series of advanced rocks also belong those which pierce the soil between the Rio Aquire and the Rio Barima; the granitic and amphibolic rocks of the Vieja Guayana and of the town of Angostura; the Cerro de Mono on the south-east of Muitaco or Real Corono; the Cerro of Taramuto near the Alta Gracia, etc.) I shall name successively, from north-north-east to south-south-west, the different chains seen by M. Bonpland and myself as we approached the equator and the river Amazon. First. The most northern chain of the whole system of the mountains of Parime appeared to us to be that which stretches (latitude 7 degrees 50 minutes) from the Rio Arui, in the meridian of the rapids of Camiseta, at the back of the town of Angostura, towards the great cataracts of the Rio Carony and the sources of the Imataca. In the missions of the Catalonian Capuchins this chain, which is not 300 toises high, separates the tributary streams of the Orinoco and those of the Rio Cuyuni, between the town of Upata, Cupapui and Santa Marta. Westward of the meridian of the rapids of Camiseta (longitude 67 degrees 10 minutes) the high mountains in the basin of the Rio Caura only commence at 7 degrees 20 minutes of latitude, on the south of the mission of San Luis Guaraguaraico, where they occasion the rapids of Mura. This chain stretches westward by the sources of the Rio Cuchivero, the Cerros del Mato, the Cerbatana and Maniapure, as far as Tepupano, a group of strangely-formed granitic rocks surrounding the Encaramada. The culminant points of this chain (latitude 7 degrees 10 minutes to 7 degrees 28 minutes) are, according to the information I gathered from the Indians, situated near the sources of Cano de la Tortuga. In the chain of the Encaramada there are some traces of gold. This chain is also celebrated in the mythology of the Tamanacs; for the painted rocks it contains are associated with ancient local traditions. The Orinoco changes its direction at the confluence of the Apure, breaking a part of the chain of the Encaramada. The latter mountains and scattered rocks in the plain of the Capuchino and on the north of Cabruta may be considered either as the vestiges of a destroyed spur or (on the hypothesis of the igneous origin of granite) as partial eruptions and upheavings. I shall not here discuss the question whether the most northerly chain, that of Angostura and of the great fall of Carony, be a continuation of the chain of Encaramada. Third. In navigating the Orinoco from north to south we observe, alternately, on the east, small plains and chains of mountains of which we cannot distinguish the profiles, that is, the sections perpendicular to their longitudinal axes. From the mission of the Encaramada to the mouth of the Rio Qama I counted seven recurrences of this alternation of savannahs and high mountains. First, on the south of the isle Cucuruparu rises the chain of Chaviripe (latitude 7 degrees 10 minutes); it stretches, inclining towards the south (latitude 6 degrees 20 minutes to 6 degrees 40 minutes), by the Cerros del Corozal, the Amoco, and the Murcielago, as far as the Erevato, a tributary of the Caura. It there forms the rapids of Paru and is linked with the summits of Matacuna. Fourth. The chain of Chaviripe is succeeded by that of the Baraguan (latitude 6 degrees 50 minutes to 7 degrees 5 minutes), celebrated for the strait of the Orinoco, to which it gives its name. The Saraguaca, or mountain of Uruana, composed of detached blocks of granite, may be regarded as a northern spur of the chain of the Baraguan, stretching south-west towards Siamacu and the mountains (latitude 5 degrees 50 minutes) that separate the sources of the Erevato and the Caura from those of the Ventuari. Fifth. The chain of Carichana and of Paruaci (latitude 6 degrees 25 minutes), wild in aspect, but surrounded by charming meadows. Piles of granite crowned with trees and insulated rocks of prismatic form (the Mogote of Cocuyza and the Marimaruta or Castillito of the Jesuits) belong to this chain. Sixth. On the western bank of the Orinoco, which is low and flat, the Peak of Uniana rises abruptly more than 3000 feet high. The spurs (latitude 5 degrees 35 minutes to 5 degrees 40 minutes) which this peak sends eastward are crossed by the Orinoco in the first Great

Cataract (that of Mapura or the Atures); further on they unite together and, rising in a chain, stretch towards the sources of the Cataniapo, the rapids of Ventuari, situated on the north of the confluence of the Asisi (latitude 5 degrees 10 minutes) and the Cerro Cunevo. Seventh. Five leagues south of the Atures is the chain of Quittuna, or of Maypures (latitude 15 degrees 13 minutes), which forms the bar of the Second Great Cataract. None of those lofty summits are situated on the west of the Orinoco; on the east of that river rises the Cunavami, the truncated peak of Calitamini and the Jujamari, to which Father Gili attributes an extraordinary height. Eighth. The last chain of the south-west part of the Sierra Parime is separated by woody plains from the chain of Maypures; it is the chain of the Cerros de Sipapo (latitude 4 degrees 50 minutes); an enormous wall behind which the powerful chief of the Guaypunabi Indians intrenched himself during the expedition of Solano. The chain of Sipapo may be considered as the beginning of the range of lofty mountains which bound, at the distance of some leagues, the right bank of the Orinoco, where that river runs from south-east to north-west, between the mouth of the Ventuari, the Jao and the Padamo (latitude 3 degrees 15 minutes). In ascending the Orinoco, above the cataract of Maypures, we find, long before we reach the point where it turns, near San Fernando del Atabapo, the mountains disappearing from the bed of the river, and from the mouth of the Zama there are only insulated rocks in the plains. The chain of Sipapo forms the south-west limit of the system of mountains of Parime, between 70 1/2 and 68 degrees of longitude. Modern geologists have observed that the culminant points of a group are less frequently found at its centre than towards one of its extremities, preceding, and announcing in some sort, a great depression* of the chain. (* As seen in Mont Blanc and Chimborazo.) This phenomenon is again observed in the group of the Parime, the loftiest summits of which, the Duida and the Maraguaca, are in the most southerly range of mountains, where the plains of the Cassiquiare and the Rio Negro begin.

These plains or savannahs which are covered with forests only in the vicinity of the rivers do not, however, exhibit the same uniform continuity as the Llanos of the Lower Orinoco, of the Meta and of Buenos Ayres. They are interrupted by groups of hills (Cerros de Daribapa) and by insulated rocks of grotesque form which pierce the soil and from a distance fix the attention of the traveller. These granitic and often stratified masses resemble the ruins of pillars or edifices. The same force which upheaved the whole group of the Sierra Parime has acted here and there in the plains as far as beyond the equator. The existence of these steeps and sporadic hills renders it difficult to determine the precise limits of a system in which the mountains are not longitudinally ranged as in a vein. As we advance towards the frontier of the Portuguese province of the Rio Negro the high rocks become more rare and we no longer find the shelves or dykes of gneiss-granite which cause rapids and cataracts in the rivers.

Such is the surface of the soil between 68 1/2 and 70 1/2 degrees of longitude, between the meridian of the bifurcation of the Orinoco and that of San Fernando de Atabapo; further on, westward of the Upper Rio Negro, towards the source of that river, and its tributary streams the Xie and the Uaupes (latitude 1 to 2 1/4 degrees, longitude 72 to 74 degrees) lies a small mountainous tableland, in which Indian traditions place a Laguna de oro, that is, a lake surrounded with beds of auriferous earth.* (* According to the journals of Acunha and Fritz the Manao Indians (Manoas) obtained from the banks of the Yquiari (Iguiare or Iguare) gold of which they made thin plates. The manuscript notes of Don Apollinario also mention the gold of the Rio Uaupes. La Condamine, Voyage a l'Amazone. We must not confound the Laguna de Oro, which is said to be found in going up the Uaupes (north latitude 0 degrees 40 minutes) with another gold lake (south latitude 1 degree 10 minutes) which La Condamine calls Marahi or Morachi (water), and which is merely a tract often inundated between the sources of the Jurubech (Urubaxi) and the Rio Marahi, a tributary stream of the Caqueta.) At Maroa, the most westerly mission of the Rio Negro, the Indians assured me that that river as well as the Inirida (a tributary of the Guavare) rises at the distance of five days' march, in a country bristled with hills and rocks. The natives of San Marcellino speak of a Sierra Tunuhy, nearly thirty leagues west of their village, between the Xie and the Icanna. La Condamine learned also from the Indians of the Amazon that the Quiquiari comes from a country of mountains and mines. Now, the Iquiari is placed by the

French astronomer between the equator and the mouth of the Xie (Ijie), which identifies it with the Iguiare that falls into the Icanna. We cannot advance in the geologic knowledge of America without having continually recourse to the researches of comparative geography. The small system of mountains, which we may provisionally call that of the sources of the Rio Negro and the Uaupes, and the culminant points of which are not probably more than 100 or 120 toises high, appears to extend southward to the basin of Rio Yupura, where rocky ridges form the cataracts of the Rio de los Enganos and the Salto Grande de Yupura (south latitude 0 degrees 40 minutes to north latitude 0 degrees 28 minutes), and the basin of the Upper Guaviare towards the west. We find in the course of this river, from 60 to 70 leagues west of San Fernando del Atabapo, two walls of rocks bounding the strait (nearly 3 degrees 10 minutes north latitude and 73 3/4 degrees longitude) where father Maiella terminated his excursion. That missionary told me that, in going up the Guaviare, he perceived near the strait (angostura) a chain of mountains bounding the horizon on the south. It is not known whether those mountains traverse the Guaviare more to the west, and join the spurs which advance from the eastern Cordillera of New Grenada, between the Rio Umadea and the Rio Ariari, in the direction of the savannahs of San Juan de los Llanos. I doubt the existence of this junction. If it really existed, the plains of the Lower Orinoco would communicate with those of the Amazon only by a very narrow land-strait, on the east of the mountainous country which surrounds the source of the Rio Negro: but it is more probable that this mountainous country (a small system of mountains, geognostically dependent on the Sierra Parime) forms as it were an island in the Llanos of Guaviare and Yupura. Father Pugnet, Principal of the Franciscan convent at Popayan, assured me, that when he went from the missions settled on the Rio Caguan to Aramo, a village situated on the Rio Guayavero, he found only treeless savannahs, extending as far as the eye could reach. The chain of mountains placed by several modern geographers, between the Meta and the Vichada, and which appears to link the Andes of New Grenada with the Sierra Parime, is altogether imaginary.

We have now examined the prolongation of the Sierra Parime on the west, towards the source of the Rio Negro: it remains for us to follow the same group in its eastern direction. The mountains of the Upper Orinoco, eastward of the Raudal of the Guaharibos (north latitude 1 degree 15 minutes longitude 67 degrees 38 minutes), join the chain of Pacaraina, which divides the waters of the Carony and the Rio Branco, and of which the micaceous schist, resplendent with silvery lustre, figures so conspicuously in Raleigh's El Dorado. The part of that chain containing the sources of the Orinoco has not yet been explored; but its prolongation more to the east, between the meridian of the military post of Guirior and the Rupunuri, a tributary of the Essequibo, is known to me through the travels of the Spaniards Antonio Santos and Nicolas Rodriguez, and also by the geodesic labours of two Portuguese, Pontes and Almeida. Two portages but little frequented* are situated between the Rio Branco and the Rio Essequibo, south of the chain of Pacaraina; they shorten the land-road leading from the Villa del Rio Negro to Dutch Guiana. (* The portages of Sarauru and the lake Amucu.) On the contrary, the portage between the basin of the Rio Branco and that of the Carony crosses the summit of the chain of Pacaraina. On the northern slope of this chain rises the Anocapra, a tributary of the Paraguamusi or Paravamusi; and on the southern slope, the Araicuque, which, with the Uraricapara, forms the famous Valley of Inundations, above the destroyed mission of Santa Rosa (latitude 3 degrees 46 minutes, longitude 65 degrees 10 minutes). The principal Cordillera, which appears of little breadth, stretches on a length of 80 leagues, from the portage of Anocapra (longitude 65 degrees 35 minutes), to the left bank of the Rupunuri (longitude 61 degrees 50 minutes), following the parallels of 4 degrees 4 minutes and 4 degrees 12 minutes. We there distinguish from west to east the mountains of Pacaraina, Tipique, Tauyana, among which rises the Rio Parime (a tributary of the Uraricuera), Tubachi, Christaux (latitude 3 degrees 56 minutes, longitude 62 degrees 52 minutes) and Canopiri. The Spanish traveller, Rodriguez, marks the eastern part of the chain by the name of Quimiropaca; but preferring to adopt general names, I continue to give the name of Pacaraina to the whole of this Cordillera which links the mountains of the Orinoco to the interior of Dutch and French Guiana, and which Raleigh and Keymis made known in Europe at the end of the 16th century. This chain is broken by the Rupunuri and the Essequibo, so that one of their tributary streams, the Tavaricuru, takes its rise on the southern declivity, and the other, the Sibarona, on the northern. On approaching the Essequibo, the mountains are more developed towards the south-east, and extend beyond 2 1/2 degrees north latitude. From this eastern branch of the chain of Pacaraina the Rio Rupunuri rises near the Cerro Uassari. On the right bank of the Rio Branco, in a still more southern latitude (between 1 and 2 degrees north) is a mountainous territory in which the Caritamini, the Padaviri, the Cababuri (Cavaburis) and the Pacimoni take their source, from east to west. This western branch of the mountains of Pacaraina separates the basin of Rio Branco from that of the Upper Orinoco, the sources of which are probably not found east of the meridian of 66 15 minutes: it is linked with the mountains of Unturan and Yumariquin, situated south-east of the mission of Esmeralda. Thence it results that, while on the west of the Cassiquiare, between that river, the Atabapo, and the Rio Negro, we find only vast plains, in which rise some little hills and insulated rocks; real spurs stretch eastward of the Cassiquiare, from north-west to south-east, and form a continued mountainous territory as far as 2 degrees north latitude. The basin only, or rather the transversal valley of the Rio Branco, forms a kind of gulf, a succession of plains and savannahs (campos) several of which penetrate from south to north, into the mountainous land between the eastern and western branches of the chain of Pacaraina, to the distance of eight leagues north of the parallel of San Joaquin.

We have just examined the southern part of the vast system of the mountains of Parime, between 2 and 4 degrees of latitude, and between the meridians of the sources of the Orinoco and the Essequibo. The development of this system of mountains northward between the chain of Pacaraina and Rio Cuyuni, and between the meridians 66 and 61 3/4 degrees, is still less known. The only road frequented by white men is that of the river Paragua, which receives the Paraguamusi, near the Guirior. We find indeed, in the journal of Nicolas Rodriguez, that he was constantly obliged to have his canoe carried by men (arrastrando) past the cataracts which intercept the navigation; but we must not forget a circumstance of which my own experience furnished me with frequent proofs—that the cataracts in this part of South America are often caused only by ridges of rocks which do not form mountains. Rodriguez names but two between Barceloneta and the mission of San Jose; while the missionaries place more to the east, in 6 degrees latitude, between the Rio Caroni and the Cuyuni, the Serranias of Usupama and Rinocote. The latter crosses the Mazaruni, and forms thirty-nine cataracts in the Essequibo, from the military post of Arinda (latitude 5 degrees 30 minutes) to the mouth of Rupunuri.

With respect to the continuation of the system of the mountains of Parime, south-east of the meridian of the Essequibo, the materials are entirely wanting for tracing it with precision. The whole interior of Dutch, French and Portuguese Guiana is a terra incognita; and the astronomical geography of those countries has scarcely made any progress during the space of thirty years. If the American limits recently fixed between France and Portugal should one day cease to be mere diplomatic illusions and acquire reality in being traced on the territory by means of astronomical observations (as was projected in 1817), this undertaking would lead geographical engineers to that unknown region which, at 3 1/2 degrees west of Cayenne, divides the waters between the coast of Guiana and the Amazon. Till that period, which the political state of Brazil seems to retard, the geognostic table of the group of Parime can only be completed by scattered notions collected in the Portuguese and Dutch colonies. In going from the Uassari mountains (latitude 2 degrees 25 minutes, longitude 61 degrees 50 minutes) which form a part of the eastern branch of the Cordillera of Pacaraina, we find towards the east a chain of mountains, called by the missionaries Acaray and Tumucuraque. Those two names are found on our maps between 1/2 and 3 degrees north latitude. Raleigh first made known, in 1596, the system of the mountains of Parime, between the sources of the Rio Carony and the Essequibo, by the name of Wacarima (Pacarima), and the Jesuits Acunha and Artedia furnished, in 1639, the first precise notions of that part of this system which extends from the meridian of Essequibo to that of Oyapoc. There they place the mountains of Yguaracuru and Paraguaxo, the former of which gives birth to a gold river (Rio de oro), a tributary of the Curupatuba;* (* When we know that in Tamanac gold is called caricuri; in Carib, caricura: in Peruvian, cori (curi), we easily recognize in the names of the mountains and rivers (Yguara-curu, Cura-patuba) which we have just marked, the indication of auriferous soil. Such is the analogy of the imported roots in the American tongues, which otherwise differ altogether from each other, that 300 leagues west of the mountain Ygaracuru, on the banks of the Caqueta, Pedro de Ursua heard of the province of Caricuri, rich in gold washings. The Curupatuba falls into the Amazon near the Villa of Monte Alegre, north-east of the mouth of the Rio Topayos.); and according to the assertion of the natives, subterraneous noises are sometimes heard from the latter. The ridge of this chain of mountains, which runs in a direction south 85 degrees east from the peak of Duida near the Esmeralda (latitude 3 degrees 19 minutes), to the rapids of the Rio Manaye near Cape Nord (latitude 1 degree 50 minutes), divides, in the parallel of 2 degrees, the northern sources of the Essequibo, the Maroni and the Oyapoc, from the southern sources of the Rio Trombetas, Curupatuba and Paru. The most southern spurs of this chain approach nearer to the Amazon, at the distance of fifteen leagues. These are the first heights which we perceived after having left Xeberos and the mouth of the Huallaga. They are constantly seen in navigating from the mouth of the Rio Topayo towards that of Paru, from the town of Santarem to Almeirim. The peak Tripoupou is nearly in the meridian of the former of those towns and is celebrated among the Indians of Upper Maroni. It is said that farther eastward, at Melgaco, the Serras do Velho and do Paru are still distinguished in the horizon. The real boundaries of this series of sources of the Rio Trombetas are better known southward than northward, where a mountainous country appears to advance in Dutch and French Guiana, as far as within twenty to twenty-five leagues of the coast. The numerous cataracts of the rivers of Surinam, Maroni and Oyapoc, prove the extent and the prolongation of rocky ridges; but in those regions nothing indicates the existence of continued plains or table-lands some hundred toises high, fitted for the cultivation of the plants of the temperate zone.

The system of the mountains of Parime surpasses in extent nineteen times that of the whole of Switzerland. Even considering the mountainous group of the sources of the Rio Negro and the Xie as independent or insulated amidst the plains, we still find the Sierra Parime (between Maypures and the sources of the Oyapoc) to be 340 leagues in length; its greatest breadth (the rocks of Imataca, near the delta of the Orinoco, at the sources of the Rio Paru) is 140 leagues. In the group of the Parime, as well as in the group of the mountains of central Asia, between the Himalaya and the Altai, the partial chains are often interrupted and have no uniform parallelism. Towards the south-west, however (between the strait of Baraguan, the mouth of the Rio Zama and the Esmeralda), the line of the mountains is generally in the direction of north 70 degrees west. Such is also the position of a distant coast, that of Portuguese, French, Dutch and English Guiana, from Cape North to the mouth of the Orinoco; such is the mean direction of the course of the Rio Negro and Yupura. It is desirable to fix our attention on the angles formed by the partial chains, in different regions of America, with the meridians, because on less extended surfaces, for instance in Germany, we find also this singular co-existence of groups of neighbouring mountains following laws of direction altogether different, though every separate group exhibits the greatest uniformity in the line of chains.

The soil on which the mountains of Parime rise, is slightly convex. By barometric measures I found that, between 3 and 4 degrees north latitude, the plains are elevated from 160 to 180 toises above sea-level. This height will appear considerable if we reflect that at the foot of the Andes of Peru, at Tomependa, 900 leagues from the coast of the Atlantic Ocean, the Llanos or plains of the Amazon rise only to the height of 194 toises. The distinctive characteristics of the group of the mountains of Parime are the rocks of granite and gneiss-granite, the total absence of calcareous secondary formations, and the shelves of bare rock (the tsy of the Chinese deserts), which occupy immense spaces in the savannahs.

5. GROUP OF THE BRAZIL MOUNTAINS.

This group has hitherto been marked on the maps in a very erroneous way. The temperate table-lands and real chains of 300 to 500 toises high have been confounded with countries of exceedingly hot temperature, and of which the undulating surface presents only ranges of hills variously grouped. But the observations of scientific travellers have recently thrown great light on the orography of Portuguese America. The mountainous region of

Brazil, of which the mean height rises at least to 400 toises, is comprehended within very narrow limits, nearly between 18 and 28 degrees south latitude; it does not appear to extend, between the provinces of Goyaz and Matogrosso, beyond longitude 53 degrees west of the meridian of Paris.

When we regard in one view the eastern configuration of North and South America, we perceive that the coast of Brazil and Guiana, from Cape Saint Roque to the mouth of the Orinoco (stretching from south-east to north-west), corresponds with that of Labrador, as the coast from Cape Saint Roque to the Rio de la Plata corresponds with that of the United States (stretching from south-west to north-east). The chain of the Alleghenies is opposite to the latter coast, as the principal Cordilleras of Brazil are nearly parallel to the shore of the provinces of Porto Seguro, Rio Janeiro and Rio Grande. The Alleghenies, generally composed of grauwacke and transition rocks, are somewhat loftier than the almost primitive mountains (of granite, gneiss and mica-slate) of the Brazilian group; they are also of a far more simple structure, their chains lying nearer to each other and preserving, as in the Jura, a more uniform parallelism.

If, instead of comparing those parts of the new continent situated north and south of the equator, we confine ourselves to South America, we find on the western and northern coasts in their whole length, a continued chain near the shore (the Andes and the Cordillera of Venezuela), while the eastern coast presents masses of more or less lofty mountains only between the 12 and 30 degrees south latitude. In this space, 360 leagues in length, the system of the Brazil mountains corresponds geologically in form and position with the Andes of Chile and Peru. Its most considerable portion lies between the parallels 15 and 22 degrees, opposite the Andes of Potosi and La Paz, but its mean height is five toises less, and cannot even be compared with that of the mountains of Parime, Jura and Auvergne. The principal direction of the Brazilian chains, where they attain the height of from four to five hundred toises, is from south to north, and from south-south-west to north-north-east; but, between 13 and 19 degrees the chains are considerably enlarged, and at the same time lowered towards the west. Ridges and ranges of hills seem to advance beyond the land-straits which separate the sources of the Rio Araguay, Parana, Topayos, Paraguay, Guapore and Aguapehy, in 63 degrees longitude. As the western widening of the Brazilian group, or rather the undulations of the soil in the Campos Parecis, correspond with the spurs of Santa Cruz de la Sierra, and Beni, which the Andes send out eastward, it was formerly concluded that the system of the mountains of Brazil was linked with that of the Andes of Upper Peru. I myself laboured under this error in my first geologic studies.

A coast chain (Serra do Mar) runs nearly parallel with the coast, north-east of Rio Janeiro, lowering considerably towards Rio Doce, and losing itself almost entirely near Bahia (latitude 12 degrees 58 minutes). According to M. Eschwege* some small ridges reach Cape Saint Roque (latitude 5 degrees 12 minutes). (* Geognostiches Gemulde von Brasilien, 1822. The limestone of Bahia abounds in fossil wood.) South-east of Rio Janeiro the Serra do Mar follows the coast behind the island of Saint Catherine as far as Torres (latitude 29 degrees 20 minutes); it there turns westward and forms an elbow stretching by the Campos of Vacaria towards the banks of the Jacuy.

Another chain is situated westward of the shore-chain of Brazil. This is the most lofty and considerable of all and is called the chain of Villarica. Mr. Eschwege distinguishes it by the name of Serra do Espinhaco and considers it as the principal part of the whole structure of the mountains of Brazil. This Cordillera loses itself northward,* between Minas Novas and the southern extremity of the Capitania of Bahia, in 16 degrees latitude. (* The rocky ridges that form the cataract of Paulo Affonso, in the Rio San Francisco, are supposed to belong to the northern prolongation of the Serra do Espinhaco, as a series of heights in the province of Seara (fetid calcareous rocks containing a quantity of petrified fish) belong to the Serra dos Vertentes.) It is there more than 60 leagues removed from the coast of Porto Seguro; but southward, between the parallels of Rio Janeiro and Saint Paul (latitude 22 to 23 degrees), in the knot of the mountains of Serra da Mantiquiera, it draws so near to the Cordillera of the shore (Serra do Mar), that they are almost confounded together. In the same manner the Serra do Espinhaco follows constantly the direction of a meridian, towards the north; while towards the south it runs south-east, and terminates about 25 degrees

170

latitude. The chain reaches its highest elevation between 18 and 21 degrees; and there the spurs and table-lands at its back are of sufficient extent to furnish lands for cultivation where, at successive heights, there are temperate climates comparable to the delicious climates of Xalapa, Guaduas, Caracas and Caripe. This advantage, which depends at once on the widening of the mass of the chain and of its spurs, is nowhere found in the same degree east of the Andes, not even in chains of more considerable absolute height, as those of Venezuela and the Orinoco. The culminant points of the Serra do Espinhaco, in the Capitania of Minas Geraes, are the Itambe (932 toises), the Serra da Piedade, near Sabara (910 toises), the Itacolumi, properly Itacunumi (900 toises), the Pico of Itabira (816 toises), the Serras of Caraca, Ibitipoca and Papagayo. Saint Hilaire felt piercing cold in the month of November (therefore in summer) in the whole Cordillera of Lapa, from the Villa do Principe to the Morro de Gaspar Suares.

We have just noticed two chains of mountains nearly parallel but of which the most extensive (the littoral chain) is the least lofty. The capital of Brazil is situated at the point where the two chains draw nearest together and are linked together on the east of the Serra de Mantiqueira, if not by a transversal ridge, at least by a mountainous territory. Old systematic ideas respecting the rising of mountains in proportion as we advance into a country, would have warranted the belief that there existed, in the Capitania of Mato Grosso, a central Cordillera much loftier than that of Villarica or do Espinhaco; but we now know (and this is confirmed by climateric circumstances) that there exists no continued chain, properly speaking, westward of Rio San Francisco, on the frontiers of Minas Geraes and Goyaz. We find only a group of mountains, of which the culminant points are the Serras da Canastra (south-west of Paracatu) and da Marcella (latitude 18 1/2 and 19.10 degrees), and, further north, the Pyrenees stretching from east to west (latitude 16 degrees 10 minutes) between Villaboa and Mejaponte). M. Eschwege has named the group of mountains of Goyaz the Serra dos Vertentes, because it divides the waters between the southern tributary streams of the Rio Grande or Parana, and the northern tributary streams of Rio Tucantines. It runs southward beyond the Rio Grande (Parana), and approaches the chain of Espinpapo in 23 degrees latitude, by the Serra do Franca. It attains only the height of 300 or 400 toises, with the exception of some summits north-west of Paracatu, and is consequently much lower than the chain of Villarica.

Further on, west of the meridian of Villaboa, there are only ridges and a series of low hills which, on a length of 12 degrees, form the division of water (latitude 13 to 17 degrees) between the Araguay and the Paranaiba (a tributary of the Parana), between the Rio Topayos and the Paraguay, between the Guapore and the Aguapehy. The Serra of San Marta (longitude 15 1/2 degrees) is somewhat lofty, but maps have vastly exaggerated the height of the Serras or Campos Parecis north of the towns of Cuyaba and Villabella (latitude 13 to 14 degrees, longitude 58 to 62 degrees). These Campos, which take their name from that of a tribe of wild Indians, are vast, barren table-lands, entirely destitute of vegetation; and in them the sources of the tributary streams of three great rivers, the Topayos, the Madeira and the Paraguay, take their rise.

According to the measures and geologic observations of M. Eschwege, the high summits of the Serra do Mar (the coast-chain) scarcely attain 660 toises; those of the Serra do Espinhaco (chain of Villarica), 950 toises; those of Serra de los Vertentes (group of Canastra and the Brazilian Pyrenees), 450 toises. Further west the surface of the soil seems to present but slight undulations; but no measure of height has been made beyond the meridian of Villaboa. Considering the system of the mountains of Brazil in their real limits, we find, except some conglomerates, the same absence of secondary formations as in the system of the mountains of the Orinoco (group of Parime). These secondary formations, which rise to considerable heights in the Cordillera of Venezuela and Cumana, belong only to the low regions of Brazil.

B. PLAINS (LLANOS) OR BASINS.

In that part of South America situated on the east of the Andes we have successively examined three systems of mountains, those of the shore of Venezuela, of the Parime and Brazil: we have seen that this mountainous region, which equals the Cordillera of the Andes, not in mass, but in area and horizontal section of surface, is three times less elevated, much less rich in precious metals adhering to the rock, destitute of recent traces of volcanic fire and, with the exception of the coast of Venezuela, little exposed to the violence of earthquakes. The average height of the three systems diminishes from north to south, from 750 to 400 toises; those of the culminant points (maxima of the height of each group) from 1350 to 1000 or 900 toises. Hence it results that the loftiest chain, with the exception of the small insulated system of the Sierra Nevada of Santa Marta, is the Cordillera of the shore of Venezuela, which is itself but a continuation of the Andes. Directing our attention northward, we find in Central America (latitude 12 to 30 degrees) and North America (latitude 30 to 70 degrees), on the east of the Andes of Guatimala, Mexico and Upper Louisiana, the same regular lowering which struck us towards the south. In this vast extent of land, from the Cordillera of Venezuela to the polar circle, eastern America presents two distinct systems, the group of the mountains of the West Indies (which in its eastern part is volcanic) and the chain of the Alleghenies. The former of these systems, partly covered by the ocean, may be compared, with respect to its relative position and form, to the Sierra Parime; the latter, to the Brazil chains, running also from south-west to north-east. The culminant points of those two systems rise to 1138 and 1040 toises. Such are the elements of this curve, of which the convex summit is in the littoral chain of Venezuela:

AMERICA, EAST OF THE ANDES.
COLUMN 1 : SYSTEMS OF MOUNTAINS.
COLUMN 2 : MAXIMA OF HEIGHTS IN TOISES.
Brazil Group : Itacolumi 900 (south latitude 20 1/2 degrees).
Parime Group : Duida 1300 (north latitude 3 1/4 degrees).
Littoral Chain of Venezuela : Silla of Caracas 1350 (north latitude 10 1/2 degrees).
Group of the West Indies : Blue Mountains 1138 (north latitude 18 1/5 degrees).
Chain of the Alleghenies : Mount Washington 1040 (north latitude 44 1/4 degrees).
I have preferred indicating in this table the culminant points of each system to the mean height of the line of elevation; the culminant points are the results of direct measures, while the mean height is an abstract idea somewhat vague, particularly when there is only one group of mountains, as in Brazil, Parime and the West Indies, and not a continued chain. Although it cannot be doubted that, among the five systems of mountains on the east of the Andes, of which one only belongs to the southern hemisphere, the littoral chain of Venezuela is the most elevated (having a culminant point of 1350 toises, and a mean height from the line of elevation of 750), we yet recognise with surprise that the mountains of eastern America (whether continental or insular) differ very inconsiderably in their height above the level of the sea. The five groups are all nearly of an average height of from 500 to 700 toises; and the culminant points (maxima of the lines of elevation) from 1000 to 1300 toises. That uniformity of structure, in an extent twice as great as Europe, appears to me a very remarkable phenomenon. No summit east of the Andes of Peru, Mexico and Upper Louisiana rises beyond the limit of perpetual snow.* (* Not even the White Mountains of the state of New Hampshire, to which Mount Washington belongs. Long before the accurate measurement of Captain Partridge I had proved (in 1804), by the laws of the decrement of heat, that no summit of the White Mountains could attain the height assigned to them by Mr. Cutler, of 1600 toises.) It may be added that, with the exception of the Alleghenies, no snow falls sporadically in any of the eastern systems which we have just examined. From these considerations it results, and above all, from the comparison of the New Continent with those parts of the old world which we know best, with Europe and Asia, that America, thrown into the aquatic hemisphere* of our planet, is still more remarkable for the continuity and extent of the depressions of its surface, than for the height and continuity of its longitudinal ridge. Beyond and within the isthmus of Panama, but eastward of the Cordillera of the Andes, the mountains scarcely attain, over an extent of 600,000 square leagues, the height of the Scandinavian Alps, the Carpathians, the Monts-Dores (in Auvergne) and the Jura. (* The southern hemisphere, owing to the unequal distribution of seas and continents, has long been marked as eminently aquatic; but the same inequality is found when we consider the globe as divided not according to the equator but by meridians. The great masses of land are stinted between the meridian of 10 degrees west, and 150 degrees east of Paris, while the hemisphere eminently aquatic begins westward of the

meridian of the coast of Greenland, and ends on the east of the meridian of the eastern coast of New Holland and the Kurile Isles. This unequal distribution of land and water has the greatest influence on the distribution of heat over the surface of the globe, on the inflexions of the isothermal lines, and the climateric phenomena in general. For the inhabitants of the central parts of Europe the aquatic hemisphere may be called western, and the land hemisphere eastern; because in going to the west we reach the former sooner than the latter. It is the division according to the meridians, which is intended in the text. Till the end of the 15th century the western hemisphere was as much unknown to the nations of the eastern hemisphere, as one half of the lunar globe is to us at present, and will probably always remain.) One system only, that of the Andes, comprises in America, over a long and narrow zone of 3000 leagues, all the summits exceeding 1400 toises high. In Europe, on the contrary, even considering the Alps and the Pyrenees as one sole line of elevation, we still find summits far from this line or principal ridge, in the Sierra Nevada of Grenada, Sicily, Greece, the Apennines, perhaps also in Portugal, from 1500 to 1800 toises high.* (* Culminant points; Malhacen of Grenada, 1826 toises; Etna, according to Captain William Henry Smith, 1700 toises; Monte Corno of the Apennines, 1489 toises. If Mount Tomoros in Greece and the Serra Gaviarra of Portugal enter, as is alleged, into the limit of perpetual snow, those summits, according to their position in latitude, should attain from 1400 to 1600 toises. Yet on the loftiest mountains of Greece, Tomoros, Olympus in Thessaly, Polyanos in Dolope and Mount Parnassus, M. Pouqueville saw, in the month of August, snow lying only in patches, and in cavities sheltered from the rays of the sun.) The contrast between America and Europe, with respect to distribution of the culminant points, which attain from 1300 to 1500 toises, is the more striking, as the low eastern mountains of South America, of which the maximum of elevation is only from 1300 to 1400 toises, are situated beside a Cordillera of which the mean height exceeds 1800 toises, while the secondary system of the mountains of Europe rises to maxima of elevation of 1500 to 1800 toises, near a principal chain of at least 1200 toises of average height.

MAXIMA OF THE LINE OF ELEVATION IN THE SAME PARALLELS.

Andes of Chile, Upper Peru. Knots of the mountains of Porco and Cuzco, 2500 toises. : Group of the Brazil Mountains; a little lower than the Cevennes 900 to 1000 toises.

Andes of Popayan and Cundinamarca. Chain of Guacas, Quindiu, and Antioquia. More than 2800 toises. : Group of Parime Mountains; little lower than the Carpathians; 1300 toises.

Insulated group of the Snowy Mountains of Santa Marta. It is believed to be 3000 toises high. : Littoral Chain of Venezuela; 80 toises lower than the Scandinavian Alps; 1350 toises.

Volcanic Andes of Guatimala, and primitive Andes of Oaxaca, from 1700 to 1800 toises. : Group of the West Indies, 170 toises higher than the mountains of Auvergne, 1140 toises.

Andes of New Mexico and Upper Louisiana (Rocky Mountains) and further west. The Maritime Alps of New Albion, 1600 to 1900 toises. : Chain of the Alleghenies; 160 toises higher than the chains of Jura and the Gates of Malabar; 1040 toises.

This table contains the whole system of mountains of the New Continent; namely: the Andes, the maritime Alps of California or New Albion and the five groups of the east.

I may subjoin to the facts I have just stated an observation equally striking; in Europe the maxima of secondary systems, which exceed 1500 toises, are found solely on the south of the Alps and Pyrenees, that is, on the south of the principal continental ridge. They are situated on the side where that ridge approaches nearest the shore, and where the Mediterranean has not overwhelmed the land. On the north of the Alps and Pyrenees, on the contrary, the most elevated secondary systems, the Carpathian and the Scandinavian mountains* do not attain the height of 1300 toises. (* The Lomnitzer Spitz of the Carpathians is, according to M. Wahlenberg, 1245 toises; Sneehattan, in the chain of Dovrefjeld in Norway (the highest summit of the old continent, north of the parallel of 55 degrees), is 1270.) The depression of the line of elevation of the second order is consequently found in Europe as well as in America, where the principal ridge is farthest removed from the shore. If we did not fear to subject great phenomena to too small a scale, we might compare the difference of the height of the Alps and the mountains of eastern America, with the difference of height observable between the Alps or the Pyrenees, and the Monts Dores, the Jura, the Vosges or the Black Forest.

We have just seen that the causes which upheaved the oxidated crust of the globe in ridges, or in groups of mountains, have not acted very powerfully in the vast extent of country stretching from the eastern part of the Andes towards the Old World; that depression and that continuity of plains are geologic facts, the more remarkable, as they extend nowhere else in other latitudes. The five mountain systems of eastern America, of which we have stated the limits, divide that part of the continent into an equal number of basins of which only that of the Caribbean Sea remains submerged. From north to south, from the polar circle to the Straits of Magellan, we see in succession:

1. THE BASIN OF THE MISSISSIPPI AND OF CANADA.

An able geologist, Mr. Edwin James, has recently shown that this basin is comprehended between the Andes of New Mexico, or Upper Louisiana, and the chains of the Alleghenies which stretch northward in crossing the rapids of Quebec. It being quite as open northward as southward, it may be designated by the collective name of the basin of the Mississippi, the Missouri, the river St. Lawrence, the great lakes of Canada, the Mackenzie river, the Saskatchewan and the coast of Hudson's Bay. The tributary streams of the lakes and those of the Mississippi are not separated by a chain of mountains running from east to west, as traced on several maps; the line of partition of the waters is marked by a slight ridge, a rising of two counter-slopes in the plain. There is no chain between the sources of the Missouri and the Assineboine, which is a branch of the Red River and of Hudson's Bay. The surface of these plains, almost all savannah, between the polar sea and the gulf of Mexico, is more than 270,000 square sea leagues, nearly equal to the area of the whole of Europe. On the north of the parallel of 42 degrees the general slope of the land runs eastward; on the south of that parallel it inclines southward. To form a precise idea how little abrupt are these slopes we must recollect that the level of Lake Superior is 100 toises; that of Lake Erie, 88 toises, and that of Lake Ontario, 36 toises above the level of the sea. The plains around Cincinnati (latitude 39 degrees 6 minutes) are scarcely, according to Mr. Drake, 80 toises of absolute height. Towards the west, between the Ozark mountains and the foot of the Andes of Upper Louisiana (Rocky Mountains, latitude 35 to 38 degrees), the basin of the Mississippi is considerably elevated in the vast desert described by Mr. Nuttal. It presents a series of small table-lands, gradually rising one above another, and of which the most westerly (that nearest the Rocky Mountains, between the Arkansas and the Padouca), is more than 450 toises high. Major Long measured a base to determine the position and height of James Peak. In the great basin of the Mississippi the line that separates the forests and the savannahs runs, not, as may be supposed, in the manner of a parallel, but like the Atlantic coast, and the Allegheny mountains themselves, from north-east to south-west, from Pittsburg towards Saint Louis, and the Red River of Nachitoches, so that the northern part only of the state of Illinois is covered with gramina. This line of demarcation is not only interesting for the geography of plants, but exerts, as we have said above, great influence in retarding culture and population north-west of the Lower Mississippi. In the United States the prairie countries are more slowly colonized; and even the tribes of independent Indians are forced by the rigour of the climate to pass the winter on the banks of rivers, where poplars and willows are found. The basins of the Mississippi, of the lakes of Canada and the St. Lawrence, are the largest in America; and though the total population does not rise at present beyond three millions, it may be considered as that in which, between latitude 29 and 45 degrees (longitude 74 to 94 degrees), civilization has made the greatest progress. It may even be said that in the other basins (of the Orinoco, the Amazon and Buenos Ayres) agricultural life scarcely exists; it begins, on a small number of points only, to supersede pastoral life, and that of fishing and hunting nations. The plains between the Alleghenies and the Andes of Upper Louisiana are of such vast extent that, like the Pampas of Choco and Buenos Ayres, bamboos (Ludolfia miega) and palm-trees grow at one extremity, while the other, during a great part of the year, is covered with ice and snow.

174

2. THE BASIN OF THE GULF OF MEXICO, AND OF THE CARIBBEAN SEA.

This is a continuation of the basin of the Mississippi, Louisiana and Hudson's Bay. It may be said that all the low lands on the coast of Venezuela situated north of the littoral chain and of the Sierra Nevada de Merida belong to the submerged part of this basin. If I treat here separately of the basin of the Caribbean Sea, it is to avoid confounding what, in the present state of the globe, is partly above and partly below the ocean. The recent coincidence of the periods of earthquakes observed at Caracas and on the banks of the Mississippi, the Arkansas and the Ohio, justifies the geologic theories which regard as one basin the plains bounded on the south, by the littoral Cordillera of Venezuela; on the east, by the Alleghenies and the series of the volcanoes of the West Indies; and on the west, by the Rocky Mountains (Mexican Andes) and by the series of the volcanoes of Guatimala. The basin of the West Indies forms, as we have already observed, a Mediterranean with several issues, the influence of which on the political destinies of the New Continent depends at once on its central position and the great fertility of its islands. The outlets of the basin, of which the four largest* are 75 miles broad, are all on the eastern side, open towards Europe, and agitated by the current of the tropics. (* Between Tobago and Grenada; Saint Martin and the Virgin Isles; Porto Rico and Saint Domingo; and between the Little Bank of Bahama and Cape Canaveral of Florida.) In the same manner as we recognize, in our Mediterranean, the vestiges of three ancient basins by the proximity of Rhodes, Scarpanto, Candia, and Cerigo, as well as by that of Cape Sorello of Sicily, the island of Pantelaria and Cape Bon, in Africa; so the basin of the West India Islands, which exceeds the Mediterranean in extent, seems to present the remains of ancient dykes which join* Cape Catoche of Yucatan to Cape San Atonio of the island of Cuba (* I do not pretend that this hypothesis of the rupture and the ancient continuity of lands can be extended to the eastern foot of the basin of the West Indies, that is, to the series of the volcanic islands in a line from Trinidad to Porto Rico.); and that island to Cape Tiburon of St. Domingo; Jamaica, the Bank of La Vibora and the rock of Serranilla to Cape Gracias a Dios on the Mosquito Shore. From this situation of the most prominent islands and capes of the continent, there results a division into three partial basins. The most northerly has long been distinguished by a particular denomination, that of the Gulf of Mexico; the intermediary or central basin may be called the Sea of Honduras, on account of the gulf of that name which makes a part of it; and the southern basin, comprehended between the Caribbean Islands and the coast of Venezuela, the isthmus of Panama, and the country of the Mosquito Indians, would form the Caribbean Sea. The modern volcanic rocks distributed on the two opposite banks of the basin of the West Indies on the east and west, but not on the north and south, is also a phenomenon worthy of attention. In the Caribbean Islands, a group of volcanoes, partly extinct and partly burning, stretches from 12 to 18 degrees; and in the Cordilleras of Guatimala and Mexico from latitude 9 to 19 1/2 degrees. I noticed on the north west extremity of the basin of the West Indies that the secondary formations dip towards south-east; along the coast of Venezuela rocks of gneiss and primitive mica-slate dip to north-west. The basalts, amygdaloids, and trachytes, which are often surmounted by tertiary limestones, appear only towards the eastern and western banks.

3. THE BASIN OF THE LOWER ORINOCO, OR THE PLAINS OF VENEZUELA.

This basin, like the plains of Lombardy, is open to the east. Its limits are the littoral chain of Venezuela on the north, the eastern Cordillera of New Grenada on the west, and the Sierra Parime on the south; but as the latter group extends on the west only to the meridian of the cataracts of Maypures (longitude 70 degrees 37 minutes), there remains an opening or land-strait, running from north to south, by which the Llanos of Venezuela communicate with the basin of the Amazon and the Rio Negro. We must distinguish between the basin of the Lower Orinoco, properly so called (north of that river and the Rio Apure), and the plains of Meta and Guaviare. The latter occupy the space between the mountains of Parime and New Grenada. The two parts of this basin have an opposite direction; but being alike covered with gramina, they are usually comprehended in the country under the same denomination. Those Llanos extend, in the form of an arch, from

175

the mouth of the Orinoco, by San Fernando de Apure, to the confluence of the Rio Caguan with the Jupura, consequently along a length of more than 360 leagues.

(3a.) PART OF THE BASIN OF VENEZUELA RUNNING FROM EAST TO WEST.

The general slope is eastward, and the mean height from 40 to 50 toises. The western bank of that great sea of verdure (mar de yerbas) is formed by a group of mountains, several of which equal or exceed in height the Peak of Teneriffe and Mont Blanc. Of this number are the Paramos del Almorzadero, Cacota, Laura, Porquera, Mucuchies, Timotes, and Las Rosas. The height of the northern and southern banks is generally less than 500 or 600 toises. It is somewhat extraordinary that the maximum of the depression of the basin is not in its centre, but on its southern limit, at the Sierra Parime. It is only between the meridians of Cape Codera and Cumana, where a great part of the littoral Cordillera of Venezuela has been destroyed, that the waters of the Llanos (the Rio Unare and the Rio Neveri) reach the northern coast. The partition ridge of this basin is formed by small table-lands, known by the names of Mesas de Amana, Guanipa and Jonoro. In the eastern part, between the meridians 63 and 66 degrees, the plains or savannahs run southward beyond the bed of the Orinoco and the Imataca, and form (as they approach the Cujuni and the Essequibo) a kind of gulf along the Sierra Pacaraina.

(3b.) PART OF THE BASIN OF VENEZUELA RUNNING FROM SOUTH TO NORTH.

The great breadth of this zone of savannahs (from 100 to 120 leagues) renders the denomination of land-strait somewhat improper, at least if it be not geognostically applied to every communication of basins bounded by high Cordilleras. Perhaps this denomination more properly belongs to that part in which is situated the group of almost unknown mountains that surround the sources of the Rio Negro. In the basin comprehended between the eastern declivity of the Andes of New Grenada and the western part of the Sierra Parime, the savannahs, as we have observed above, stretch far beyond the equator; but their extent does not determine the southern limits of the basin here under consideration. These limits are marked by a ridge which divides the waters between the Orinoco and the Rio Negro, a tributary stream of the Amazon. The rising of a counter-slope almost imperceptible to the eye, forms a ridge that seems to join the eastern Cordillera of the Andes to the group of the Parime. This ridge runs from Ceja (latitude 1 degree 45 minutes), or the eastern slope of the Andes of Timana, between the sources of the Guayavero and the Rio Caguan, towards the isthmus that separates the Tuamini from Pimichin. In the Llanos, consequently, it follows the parallels of 20 degrees 30 minutes and 2 degrees 45 minutes. It is remarkable that we find the divortia aquarum further westward on the back of the Andes, in the knot of mountains containing the sources of the Magdalena, at a height of 900 toises above the level of the Llanos, between the Caribbean Sea and the Pacific ocean, and almost in the same latitude (1 degree 45 minutes to 2 degrees 20 minutes). From the isthmus of Javita towards the east, the line of the partition of waters is formed by the mountains of the Parime group; it first rises a little on the north-east towards the sources of the Orinoco (latitude 3 degrees 45 minutes ?) and the chain of Pacaraina (latitude 4 degrees 4 minutes to 4 degrees 12 minutes); then, during a course of 80 leagues, between the portage of the Anocapra and the banks of the Rupunuri, it runs very regularly from west to east; and finally, beyond the meridian 61 degrees 50 minutes, it again deviates towards lower latitudes, passing between the northern sources of the Rio Suriname, the Maroni, the Oyapoc and the southern sources of Rio Trombetas, Curupatuba, and Paru (latitude 2 degrees to 1 degree 50 minutes). These facts suffice to prove that this first line of partition of the waters of South America (that of the northern hemisphere) traverses the whole continent between the parallels of 2 and 4 degrees. The Cassiquiare alone has cut its way across the ridge just described. The hydraulic system of the Orinoco displays the singular phenomenon of a bifurcation where the limit of two basins (those of the Orinoco and the Rio Negro) crosses the bed of the principal recipient. In that part of the basin of the Orinoco which runs in the direction of from south to north, as well as in that running from west to east, the maxima of depression are found at the foot of the Sierra Parime, we may even say, on its outline.

4. THE BASIN OF THE RIO NEGRO AND THE AMAZON.

176

This is the central and largest basin of South America. It is exposed to frequent equatorial rains, and the hot and humid climate develops a force of vegetation to which nothing in the two continents can be compared. The central basin, bounded on the north by the Parime group, and on the south by the mountains of Brazil, is entirely covered by thick forests, while the two basins at the extremities of the continent (the Llanos of Venezuela and the Lower Orinoco, and the Pampas of Buenos Ayres or the Rio de la Plata) are savannahs or prairies, plains without trees and covered with gramina. This symmetric distribution of savannahs bounded by impenetrable forests, must be connected with physical revolutions which have operated simultaneously over great surfaces.

(4a.) PART OF THE BASIN OF THE AMAZON, RUNNING FROM EAST TO WEST, BETWEEN 2 DEGREES NORTH AND 12 DEGREES SOUTH; 880 LEAGUES IN LENGTH.

The western shore of this basin is formed by the chain of the Andes, from the knot of the mountains of Huanuco to the sources of the Magdalena. It is enlarged by the spurs of the Rio Beni,* (* The real name of this great river, respecting the course of which geographers have been so long divided, is Uchaparu, probably water (para) of Ucha; Peni also signifies river or water; for the language of the Maypures has very many analogies with that of the Moxos; and veni (oueni) signifies water in Maypure, as una in Moxo. Perhaps the river retained the name of Maypure, after the Indians who spoke that language had emigrated northward in the direction of the banks of the Orinoco.) rich in gem-salt, and composed of several ranges of hills (latitude 8 degrees 11 minutes south) which advance into the plains on the eastern bank of the Paro. These hills are transformed on our maps into Upper Cordilleras and Andes of Cuchao. Towards the north the basin of the Amazon, of which the area (244,000 square leagues) is only one-sixth less than the area of all Europe, rises in a gentle slope towards the Sierra Parime. At 68 degrees of west longitude the elevated part of this Sierra terminates at 3 1/2 degrees north latitude. The group of little mountains surrounding the source of the Rio Negro, the Inirida and the Xie (latitude 2 degrees) the scattered rocks between the Atabapo and the Cassiquiare, appear like groups of islands and rocks in the middle of the plain. Some of those rocks are covered with signs or symbolical sculpture. Nations, very different from those who now inhabit the banks of the Cassiquiare, penetrated into the savannahs; and the zone of painted rocks, extending more than 150 leagues in breadth, bears traces of ancient civilization. On the east of the sporadic groups of rocks (between the meridian of the bifurcation of the Orinoco and that of the confluence of the Essequibo with the Rupunuri) the lofty mountains of the Parime commence only in 3 degrees north latitude; where the plains of the Amazon terminate.

The limits of the plains of the Amazon are still less known towards the south than towards the north. The mountains that exceed 400 toises of absolute height do not appear to extend in Brazil northward of the parallels 14 or 15 degrees of south latitude, and west of the meridian of 52 degrees; but it is not known how far the mountainous country extends, if we may call by that name a territory bristled with hills of one hundred or two hundred toises high. Between the Rio dos Vertentes and the Rio de Tres Barras (tributary streams of the Araguay and the Topayos) several ridges of the Monts Parecis run northward. On the right bank of the Topayos a series of little hills advance as far as the parallel of 5 degrees south latitude, to the fall (cachoeira) of Maracana; while further west, in the Rio Madeira, the course of which is nearly parallel with that of the Topayos, the rapids and cataracts indicate no rocky ridges beyond the parallel of 8 degrees. The principal depression of the basin of which we have just examined the outline, is not near one of its banks, as in the basin of the Lower Orinoco, but at the centre, where the great recipient of the Amazon forms a longitudinal furrow inclining from west to east, under an angle of at least 25 degrees. The barometric measurements which I made at Javita on the banks of the Tuamini, at Vasivia on the banks of the Cassiquiare and at the cataract of Rentema, in the Upper Maranon, seem to prove that the rising of the Llanos of the Amazon northward (at the foot of the Sierra Parime) is 150 toises, and westward (at the foot of the Cordillera of the Andes of Loxa), 190 toises above the sea-level.

(4b.) PART OF THE BASIN OF THE AMAZON STRETCHING FROM SOUTH TO NORTH.

This is the zone or land-strait by which, between 12 and 20 degrees of south latitude, the plains of the Amazon communicate with the Pampas of Buenos Ayres. The western bank of this zone is formed by the Andes, between the knot of Porco and Potosi, and that of Huanuco and Pasco. Part of the spurs of the Rio Beni, which is but a widening of the Cordilleras of Apolobamba and Cuzco and the whole promontory of Cochabamba, advance eastward into the plains of the Amazon. The prolongation of this promontory has given rise to the idea that the Andes are linked with a series of hills which the Serras dos Parecis, the Serra Melgueira, and the supposed Cordillera of San Fernando, throw out towards the west. This almost unknown part of the frontiers of Brazil and Upper Peru merits the attention of travellers. It is understood that the ancient mission of San Jose de Chiquitos (nearly latitude 17 degrees, longitude 67 degrees 10 minutes, supposing Santa Cruz de la Sierra, in latitude 17 degrees 25 minutes, longitude 66 degrees 47 minutes) is situated in the plains, and that the mountains of the spur of Cochabamba terminate between the Guapaix (Rio de Mizque) and the Parapiti, which lower down takes the names of Rio San Miguel and Rio Sara. The savannahs of the province of Chiquitos communicate on the north with those of Moxos, and on the south with those of Chaco; but a ridge or line of partition of the waters is formed by the intersection of two gently sloping plains. This ridge takes its origin on the north of La Plata (Chuquisaca) between the sources of the Guapaix and the Cachimayo, and it ascends from the parallel of 20 degrees to that of 15 1/2 degrees south latitude, consequently on the north-east, towards the isthmus of Villabella. From this point, one of the most important of the whole hydrography of America, we may follow the line of the partition of the water to the Cordillera of the shore (Serra do Mar). It is seen winding (latitude 17 to 20 degrees) between the northern sources of the Araguay, the Maranhao or Tocantines, the Rio San Francisco and the southern sources of the Parana. This second line of partition which enters the group of the Brazil mountains on the frontier of Capitania of Goyaz separates the flowings of the basin of the Amazon from those of the Rio de la Plata, and corresponds, south of the equator, with the line we have indicated in the northern hemisphere (latitude 2 to 4 degrees), on the limits of the basins of the Amazon and the Lower Orinoco.

If the plains of the Amazon (taking that denomination in the geognostic sense we have given it) are in general distinguished from the Llanos of Venezuela and the Pampas of Buenos Ayres, by the extent and thickness of their forests, we are the more struck by the continuity of the savannahs in that part running from south to north. It would seem as though this sea of verdure stretched forth an arm from the basin of Buenos Ayres, by the Llanos of Tucuman, Manso, Chuco, the Chiquitos, and the Moxos, to the Pampas del Sacramento and the savannahs of Napo, Guaviare, Meta and Apure. This arm crosses, between 7 and 3 degrees south latitude, the basin of the forests of the Amazon; and the absence of trees on so great an extent of territory, together with the preponderance which the small monocotyledonous plants have acquired, is a phenomenon of the geography of plants which belongs perhaps to the action of ancient pelagic currents or other partial revolutions of our planet.

5. PLAINS OF THE RIO DE LA PLATA, AND OF PATAGONIA, FROM THE SOUTH-WESTERN SLOPE OF THE GROUP OF THE BRAZIL MOUNTAINS TO THE STRAIT OF MAGELLAN; FROM 20 TO 53 DEGREES OF LATITUDE.

These plains correspond with those of the Mississippi and of Canada in the northern hemisphere. If one of their extremities approaches less nearly to the polar regions, the other enters much further into the region of palm-trees. That part of this vast basin extending from the eastern coast towards the Rio Paraguay does not present a surface so perfectly smooth as the part situated on the west and the south-east of the Rio de la Plata, and which has been known for ages by the name of Pampas, derived from the Peruvian or Quichua language.* (* Hatan Pampa signifies in that language, a great plain. We find the word Pampa also in Riobamba and Guallabamba; the Spaniards, in order to soften the geographical names, changing the p into b.) Geognostically speaking these two regions of east and west form only one basin, bounded on the east by the Sierra de Villarica or do Espinhaco, which loses itself in the Capitania of San Paul, near the parallel of 24 degrees; issuing on the north-east by little hills, from the Serra da Canastra and the Campos Parecis towards the province

178

of Paraguay; on the west by the Andes of Upper Peru and Chile; and on the north-west by the ridge of the partition of the waters which runs from the spur of Santa Cruz de la Sierra, across the plains of the Chiquitos, towards the Serras of Albuquerque (latitude 19 degrees 2 minutes) and San Fernando. That part only of this basin lying on the west of the Rio Paraguay, and which is entirely covered with gramina, is 70,000 square leagues. This surface of the Pampas or Llanos of Manse, Tucuman, Buenos Ayres and eastern Patagonia is consequently four times greater than the surface of the whole of France. The Andes of Chile narrow the Pampas by the two spurs of Salta and Cordova; the latter promontory forms so projecting a point that there remains (latitude 31 to 32 degrees) a plain only 45 leagues broad between the eastern extremity of the Sierra de Cordova and the right bank of the river Paraguay, stretching in the direction of a meridian, from the town of Nueva Coimbra to Rosario, below Santa Fe. Far beyond the southern frontiers of the old viceroyalty of Buenos Ayres, between the Rio Colorado and the Rio Negro (latitude 38 to 39 degrees) groups of mountains seem to rise in the form of islands in the middle of a muriatiferous plain. A tribe of Indians of the south (Tehuellet) have there long borne the characteristic name of men of the mountains (Callilehet) or Serranos. From the parallel of the mouth of the Rio Negro to that of Cabo Blanco (latitude 41 to 47 degrees) scattered mountains on the eastern Patagonian coast denote more considerable inequalities inland. All that part, however, of the Straits of Magellan, from the Virgins' Cape to the North Cape, on the breadth of more than 30 leagues, is surrounded by savannahs or Pampas; and the Andes of western Patagonia only begin to rise near the latter cape, exercising a marked influence on the direction of that part of the strait nearest the Pacific, proceeding from south-east to north-west.

If we have given the plains or great basins of South America the names of the rivers that flow in their longitudinal furrows, we have not meant by so-doing to compare them to mere valleys. In the plains of the Lower Orinoco and the Amazon all the lines of the declivity doubtless reach a principal recipient, and the tributaries of tributary streams, that is the basins of different orders, penetrate far into the group of the mountains. The upper parts or high valleys of the tributary streams must be considered in a geological table as belonging to the mountainous region of the country, and beyond the plains of the Lower Orinoco and the Amazon. The views of the geologist are not identical with those of the hydrographer. In the basin of the Rio de la Plata and Patagonia the waters that follow the lines of the greatest declivities have many issues. The same basin contains several valleys of rivers; and when we examine nearly the polyedric surface of the Pampas and the portion of their waters which, like the waters of the steppes of Asia, do not go to the sea, we conceive that these plains are divided by small ridges or lines of elevation, and have alternate slopes, inclined, with reference to the horizon, in opposite directions. In order to point out more clearly the difference between geological and hydrographic views, and to prove that in the former, abstracting the course of the waters which meet in one recipient, we obtain a far more general point of view, I shall here again recur to the hydrographic basin of the Orinoco. That immense river rises on the southern slope of the Sierra Parime. It is bounded by plains on the left bank, from the Cassiquiare to the mouth of the Atabapo, and flows in a basin which, geologically speaking, according to one great division of the surface of South America into three basins, we have called the basin of the Rio Negro and the Amazon. The low regions, which are bounded by the southern and northern declivities of the Parime and Brazil mountains, and which the geologist ought to mark by one name, contain, according to the no less precise language of hydrography, two basins of rivers, those of the Upper Orinoco and the Amazon, separated by a ridge that runs from Javita towards Esmeralda. From these considerations it results that a geological basin (sit venia verbo) may have several recipients and several emissaries, divided by small ridges almost imperceptible; it may at the same time contain waters that flow to the sea by different furrows independent of each other, and the systems of inland rivers flowing into lakes more or less charged with saline matter. A basin of a river, or hydrographic basin, has but one recipient, one emissary; if, by a bifurcation, it gives a part of its waters to another hydrographic basin, it is because the bed of the river, or the principal recipient, approaches so near the banks of the basin or the ridge of partition that the ridge partly crosses it.

The distribution of the inequalities of the surface of the globe does not present any strongly marked limits between the mountainous country and the low regions, or geologic basins. Even where real chains of mountains rise like rocky dykes issuing from a crevice, spurs more or less considerable, seem to indicate a lateral upheaving. While I admit the difficulty of properly defining the groups of mountains and the basins or continuous plains, I have attempted to calculate their surfaces according to the statements contained in the preceding sheets.

TABLE OF AREAS FOR SOUTH AMERICA.
COLUMN 1 : GEOGRAPHICAL LOCATION.
COLUMN 2 : AREA IN SQUARE MARINE LEAGUES.
1. MOUNTAINOUS PART:

Andes : 58,900.
Littoral Chain of Venezuela : 1,900.
Sierra Nevada de Merida : 200.
Group of the Parime : 25,800.
System of the Brazil mountains : 27,600.
TOTAL : 114,400.
2. PLAINS:

Llanos of the Lower Orinoco, the Meta, : 29,000.
and the Guaviare
Plains of the Amazon : 260,400.
Pampas of Rio de la Plata and Patagonia : 135,200.
Plains between the eastern chain of the
Andes of Cundinamarca and the chain of Choco : 12,300.
Plains of the shore on the west of the Andes : 20,000.
TOTAL : 456,900.

The whole surface of South America contains 571,300 square leagues (20 to a degree), and the proportion of the mountainous country to the region of the plains is as 1 to 3.9. The latter region, on the east of the Andes, comprises more than 424,600 square leagues, half of which consists of savannahs; that is to say, it is covered with gramina.

SECTION 2.
GENERAL PARTITION OF GROUND. DIRECTION AND INCLINATION OF THE STRATA. RELATIVE HEIGHT OF THE FORMATIONS ABOVE THE LEVEL OF THE OCEAN.

In the preceding section we have examined the inequalities of the surface of the soil, that is to say, the general structure of the mountains and the form of the basins rising between those variously grouped mountains. These mountains are sometimes longitudinal, running in narrow bands or chains, similar to the veins that preserve their directions at great distances, as the Andes, the littoral chain of Venezuela, the Serra do Mar of Brazil, and the Alleghenies of the United States. Sometimes they are in masses with irregular forms, in which upheavings seem to have taken place as on a labyrinth of crevices or a heap of veins, as for example in the Sierra Parime and the Serra dos Vertentes. These modes of formation are linked with a geognostic hypothesis, which has at least the recommendation of being founded on facts observed in remote times, and which strongly characterize the chains and groups of mountains. Considerations on the aspect of a country are independent of those which indicate the nature of the soil, the heterogeneity of matter, the superposition of rocks and the direction and inclination of strata.

In taking a general view of the geological constitution of a chain of mountains, we may distinguish five elements of direction too often confounded in works of geognosy and physical geography. These elements are:—

      1. The longitudinal axis of the whole chain.
  2. The line that divides the waters (divortia aquarum).
  3. The line of ridges or elevation passing along the maxima of height.
  4. The line that separates two contiguous formations into horizontal
      sections.
  5. The line that follows the fissures of stratification.

This distinction is the more necessary, there existing probably no chain on the globe that furnishes a perfect parallelism of all these directing lines. In the Pyrenees, for instance, 1, 2, 3, do not coincide, but 4 and 5 (that is, the different formations which come to light successively, and the direction of the strata) are obviously parallel to 1, or to the direction of the whole chain. We find so often in the most distant parts of the globe, a perfect parallelism between 1 and 5, that it may be supposed that the causes which determine the direction of the axis (the angle under which that axis cuts the meridian) are generally linked with causes that determine the direction and inclination of the strata. This direction of the strata is independent of the line of the formations, or their visible limits at the surface of the soil; the lines 4 and 5 sometimes cross each other, even when one of them coincides with 1, or with the direction of the longitudinal axis of the whole chain. The RELIEF of a country cannot be precisely explained on a map, nor can the most erroneous opinions on the locality and superposition of the strata be avoided, if we do not apprehend with clearness the relation of the directing lines just mentioned.

In that part of South America to which this memoir principally relates, and which is bounded by the Amazon on the south, and on the west by the meridian of the Snowy Mountains (Sierra Nevada) of Merida, the different bands or zones of formations (4) are sensibly parallel with the longitudinal axis (1) of the chains of mountains, basins or interposed plains. It may be said in general that the granitic zone (including under that denomination the rocks of granite, gneiss and mica-slate) follows the direction of the Cordillera of the shore of Venezuela, and belongs exclusively to that Cordillera and the group of the Parime mountains; since it nowhere pierces the secondary and tertiary strata in the Llanos or basin of the Lower Orinoco. Thence it results that the same formations do not constitute the region of plains and that of mountains.

If we may be allowed to judge of the structure of the whole Sierra Parime, from the part which I examined in 6 degrees of longitude, and 4 degrees of latitude, we may believe it to be entirely composed of gneiss-granite; I saw some beds of greenstone and amphibolic slate, but neither mica-slate, clay-slate, nor banks of green limestone, although many phenomena render the presence of mica-slate probable on the east of the Maypures and in the chain of Pacaraina. The geological formation of the Parime group is consequently still more simple than that of the Brazilian group, in which granites, gneiss and mica-slate are covered with thonschiefer, chloritic quartz (Itacolumite), grauwacke and transition-limestone; but those two groups exhibit in common the absence of a real system of secondary rocks; we find in both only some fragments of sandstone or silicious conglomerate. In the littoral Cordillera of Venezuela the granitic formations predominate; but they are wanting towards the east, and especially in the southern chain, where we observe (in the missions of Caripe and around the gulf of Cariaco) a great accumulation of secondary and tertiary calcareous rocks. From the point where the littoral Cordillera is linked with the Andes of New Grenada (longitude 71 1/2 degrees) we observe first the granitic mountains of Aroa and San Felipe, between the rivers Yaracui and Tocuyo; these granitic formations extend on the east of the two coasts of the basin of the Valleys of Aragua, in the northern chain, as far as Cape Codera; and in the southern as far as the mountains (altas savanas) of Ocumare. After the remarkable interruption of the littoral Cordillera in the province of Barcelona, granitic rocks begin to appear in the island of Marguerita and in the isthmus of Araya, and continue, perhaps, towards the Boca del Drago; but on the east of the meridian of Cape Codera the northern chain only is granitic (of micaceous slate); the southern chain is entirely composed of secondary limestone and sandstone.

If, in the granitic series, where a very complex formation, we would distinguish mineralogically between the rocks of granite, gneiss, and mica-slate, it must be borne in mind that coarse-grained granite, not passing to gneiss, is very rare in this country. It belongs peculiarly to the mountains that bound the basin of the lake of Valencia towards the north; for in the islands of that lake, in the mountains near the Villa de Cura, and in the whole northern chain, between the meridian of Vittoria and Cape Codera, gneiss predominates, sometimes alternating with granite, or passing to mica-slate. Mica-slate is the most frequent rock in the peninsula of Araya and the group of Macanao, which forms the western part of the island of Marguerita. On the west of Maniquarez the mica-slate of the peninsula of Araya loses by degrees its semi-metallic lustre; it is charged with carbon, and becomes a clay-slate (thonschiefer) even an ampelite (alaunschiefer). Beds of granular limestone are most common in the primitive northern chain; and it is somewhat remarkable that they are found in gneiss, and not in mica-slate.

We find at the back of this granitic, or rather mica-slate-gneiss soil of the southern chain, on the south of the Villa de Cura, a transition stratum, composed of greenstone, amphibolic serpentine, micaceous limestone, and green and carburetted slate. The most southern limit of this district is marked by volcanic rocks. Between Parapara, Ortiz and the Cerro de Flores (latitude 9 degrees 28 minutes to 9 degrees 34 minutes; longitude 70 degrees 2 minutes to 70 degrees 15 minutes) phonolites and amygdaloids are found on the very border of the basin of the Llanos, that vast inland sea which once filled the whole space between the Cordilleras of Venezuela and Parime. According to the observations of Major Long and Dr. James, trap-formations (bulleuses dolerites and amygdaloids with pyroxene) also border the plains or basin of the Mississippi, towards the west, at the declivity of the Rocky Mountains. The ancient pyrogenic rocks which I found near Parapara where they rise in mounds with rounded summits, are the more remarkable as no others have hitherto been discovered in the whole eastern part of South America. The close connection observed in the strata of Parapara, between greenstone, amphibolic serpentine, and amygdaloids containing crystals of pyroxene; the form of the Morros of San Juan, which rise like cylinders above the table-land; the granular texture of their limestone, surrounded by trap rocks, are objects worthy the attention of the geologist who has studied in the southern Tyrol the effects produced by the contact of poroxenic porphyries.* (* Leopold von Buch. Tableau geologique du Tyrol page 17. M. Boussingault states that these singular Morros de San Juan, which furnish a limestone with crystalline grains, and thermal springs, are hollow, and contain immense grottos filled with stalactites, which appear to have been anciently inhabited by the natives.)

The calcareous soil of the littoral Cordillera prevails most on the east of Cape Unare, in the southern chain; it extends to the gulf of Paria, opposite the island of Trinidad, where we find gypsum of Guire, containing sulphur. I have been informed that in the northern chain also, in the Montana de Paria, and near Carupana, secondary calcareous formations are found, and that they only begin to show themselves on the east of the ridge of rock called the Cerro de Meapire, which joins the calcareous group of Guacharo to the mica-slate group of the peninsula of Araya; but I have not had an opportunity of ascertaining the accuracy of this information. The calcareous stratum of the southern chain is composed of two formations which appear to be very distinct the one from the other: namely limestone of Cumanacoa and that of Caripe. When I was on the spot the former appeared to me to have some analogy with zechstein, or Alpine limestone; the latter with Jura limestone; I even thought that the granular gypsum of Guire might be that which belongs in Europe to zechstein, or is placed between zechstein and variegated sandstone. Strata of quartzose sandstone, alternating with slaty clay, cover the limestone of Cumanacoa, Cerro del Imposible, Turimiquiri, Guarda de San Agustin, and the Jura limestone in the province of Barcelona (Aguas Calientes). According to their position these sandstones may be considered as belonging to the formation of green sandstone, or sandstone with lignites below chalk. But if, as I thought I observed at Cocollar, sandstone forms strata in the Alpine limestone before it is superposed, it appears doubtful whether the sandstone of the Imposible, and of Aguas Calientes, constitute one series. Muriatiferous clay (with petroleum and lamellar gypsum) covers the western part of the peninsula of Araya, opposite to the town of Cumana, and in the centre of the island of Marguerita. This clay appears to lie immediately over the mica-slate, and under the calcareous breccia of the tertiary strata. I cannot decide whether Araya, which is rich in disseminated muriate of soda, belongs to the sandstone formation of the Imposible, which from its position may be compared to variegated sandstone (red marl).

There is no doubt that fragments of tertiary strata surround the castle and town of Cumana (Castillo de San Antonio) and they also appear at the south-western extremity of the peninsula of Araya (Cerro de la Vela et del Barigon); at the ridge of the Cerro de Meapire, near Cariaco; at Cabo Blanco, on the west of La Guayra, and on the shore of Porto Cabello;

they are consequently found at the foot of the two slopes of the northern chain of the Cordillera of Venezuela. This tertiary stratum is composed of alternate beds of calcareous conglomerate, compact limestone, marl, and clay, containing selenite and lamellar gypsum. The whole system (of very recent beds) appears to me to constitute but one formation, which is found at the Cerro de la Popa, near Carthagena, and in the islands of Guadaloupe and Martinico.

Such is the geological distribution of strata in the mountainous part of Venezuela, in the group of the Parime and in the littoral Cordillera. We have now to characterize the formations of the Llanos (or of the basin of the Lower Orinoco and the Apure); but it is not easy to determine the order of their superposition, because in this region ravines or beds of torrents and deep wells dug by the hands of man are entirely wanting. The formations of the Llanos are, first, a sandstone or conglomerate, with rounded fragments of quartz, Lydian stone, and kieselschiefer, united by a ferruginous clayey cement, extremely tenacious, olive-brown, sometimes of a vivid red; second, a compact limestone (between Tisnao and Calabozo) which, by its smooth fracture and lithographic aspect, approaches the Jura limestone: third, alternate strata of marl and lamellar gypsum (Mesa de San Diego, Ortiz, Cachipo). These three formations appeared to me to succeed each other in the order I have just described, the sandstone inclining in a concave position, northward, on the transition-slates of Malpasso, and southward, on the gneiss-granite of Parime. As the gypsum often immediately covers the sandstone of Calabozo, which appeared to me, on the spot, to be identical with our red sandstone, I am uncertain of the age of its formation. The secondary rocks of the Llanos of Cumana, Barcelona and Caracas occupy a space of more than 5000 square leagues. Their continuity is the more remarkable, as they appear to have no existence, at least on the east of the meridian of Porto Cabello (70 degrees 37 minutes) in the whole basin of the Amazon not covered by granitic sands. The causes which have favoured the accumulation of calcareous matter in the eastern region of the coast chain, in the Llanos of Venezuela (from 10 1/2 to 8 degrees north), cannot have operated nearer the equator, in the group of the mountains of the Parime and in the plains of the Rio Negro and the Amazon (latitude 1 degree north to 1 degree south). The latter plains, however, furnish some ledges of fragmentary rocks on the south-west of San Fernando de Atabapo, as well as on the south-east, in the lower part of the Rio Negro and the Rio Branco. I saw in the plains of Jaen de Bracamoros a sandstone which alternates with ledges of sand and conglomerate nodules of porphyry and Lydian stone. MM. Spix and Martius affirm that the banks of the Rio Negro on the south of the equator are composed of variegated sandstone; those of the Rio Branco, Jupura and Apoporis of quadersandstein; and those of the Amazon, on several points, of ferruginous sandstone.* (* Braunes eisenschussiges Sandstein-Conglomerat (Iron-sand of the English geologists, between the Jura limestone and green sandstone.) MM. Spix and Martius found on rocks of quadersandstein, between the Apoporis and the Japura, the same sculptures which we have pointed out from the Essequibo to the plains of Cassiquiare, and which seem to prove the migrations of a people more advanced in civilization than the Indians who now inhabit those countries.) It remains to examine if (as I am inclined to suppose) the limestone and gypsum formations of the eastern part of the littoral Cordillera of Venezuela differ entirely from those of the Llanos, and to what series belongs that rocky wall* named the Galera, which bounds the steppes of Calabozo towards the north? (* Is this wall a succession of rocks of dolomite or a dyke of quadersandstein, like the Devil's Wall (Teufelsmauer), at the foot of the Hartz? Calcareous shelves (coral banks), either ledges of sandstone (effects of the revulsion of the waves) or volcanic eruptions, are commonly found on the borders of great plains, that is, on the shores of ancient inland seas. The Llanos of Venezuela furnish examples of such eruptions near Para(?) like Harudje (Mons Ater, Plin.) on the northern boundary of the African desert (the Sahara). Hills of sandstone rising like towers, walls and fortified castles and offering great analogy to quadersandstein, bound the American desert towards the west, on the south of Arkansas.) The basin of the steppes is itself the bottom of a sea destitute of islands; it is only on the south of the Apure, between that river and the Meta, near the western bank of the Sierra, that a few hills appear, as Monte Parure, la Galera de Sinaruco and the Cerritos de San Vicente. With the exception of the fragments of tertiary strata above mentioned there is, from the equator to the parallel of 10 degrees north (between the meridian of Sierra Nevada de Merida and the coast of Guiana), if not an absence, at least a scarcity of those petrifactions, which strikes an observer recently arrived from Europe.

The maxima of the height of the different formations diminish regularly in the country we are describing with their relative ages. These maxima, for gneiss-granite (Peak of Duida in the group of Parime, Silla de Caracas in the coast chain) are from 1300 to 1350 toises; for the limestone of Cumanacoa (summit or Cucurucho of Turimiquiri), 1050 toises; for the limestone of Caripe (mountains surrounding the table-land of the Guarda de San Augustin), 750 toises; for the sandstone alternating with the limestone of Cumanacoa (Cuchilla de Guanaguana), 550 toises; for the tertiary strata (Punta Araya), 200 toises.

The tract of country of which I am here describing the geological constitution is distinguished by the astonishing regularity observed in the direction of the strata of which the rocks of different eras are composed. I have already often pointed the attention of my readers to a geognostic law, one of the few that can be verified by precise measurements. Occupied since the year 1792 by the parallelism, or rather the loxodromism of the strata, examining the direction and inclination of the primitive and transition beds, from the coast of Genoa across the chain of the Bochetta, the plains of Lombardy, the Alps of Saint Gothard, the table-land of Swabia, the mountains of Bareuth, and the plains of Northern Germany, I was struck with the extreme frequency, if not the uniformity, of the horary directions 3 and 4 of the compass of Freiberg (direction from south-west to north-east). This research, which I thought might lead to important discoveries relating to the structure of the globe, had then such attractions for me that it was one of the most powerful incentives of my voyage to the equator. My own observations, together with those of many able geologists, convince me that there exists in no hemisphere a general and absolute uniformity of direction; but that in regions of very considerable extent, sometimes over several thousand square leagues, we observe that the direction and (though more rarely) the inclination have been determined by a system of particular forces. We discover at great distances a parallelism (loxodromism) of the strata, a direction of which the type is manifest amidst partial perturbations and which often remains the same in primitive and transition strata. A fact which must have struck Palasson and Saussure is that in general the direction of the strata, even in those which are far distant from the principal ridges, is identical with the direction of mountain chains; that is to say, with their longitudinal axis.

Venezuela is one of the countries in which the parallelism of the strata of gneiss-granite, mica-slate and clay-slate, is most strongly marked. The general direction of these strata is north 50 degrees east, and the general inclination from 60 to 70 degrees north-west. Thus I observed them on a length of more than a hundred leagues, in the littoral chain of Venezuela; in the stratified granite of Las Trincheras at Porto Cabello; in the gneiss of the islands of the lake of Valencia, and in the vicinity of the Villa de Cura; in the transition-slate and greenstone on the north of Parapara; in the road from La Guayra to the town of Caracas, and through all the Sierra de Avila in Cape Codera; and in the mica-slate and clay-slate of the peninsula of Araya. The same direction from north-east to south-west, and this inclination to north-west, are also manifest, although less decidedly, in the limestones of Cumanacoa at Cuchivano and between Guanaguana and Caripe. The exceptions to this general law are extremely rare in the gneiss-granite of the littoral Cordillera; it may even be affirmed that the inverse direction (from south-east to north-west) often bears with it the inclination towards south-west.

As that part of the group of the Sierra Parime over which I passed contains much more granite* than gneiss (* Only the granite of the Baragon is stratified, as well as crossed by veins of granite: the direction of the beds is north 20 degrees west), and other rocks distinctly stratified, the direction of the layers could be observed in this group only on a small number of points; but I was often struck in this region with the continuity of the phenomenon of loxodromism. The amphibolic slates of Angostura run north 45 degrees east, like the gneiss of Guapasoso which forms the bed of the Atabapo, and like the mica-slate of the peninsula of Araya, though there is a distance of 160 leagues between the limits of those rocks.

The direction of the strata, of which we have just noticed the wonderful uniformity, is not entirely parallel with the longitudinal axes of the two coast chains, and the chain of Parime. The strata generally cut the former of those chains at an angle of 35 degrees, and their inclination towards the north-west becomes one of the most powerful causes of the aridity which prevails on the southern declivity* of the mountains of the coast. (* This southern declivity is however less rapid than the northern.) May we conclude that the direction of the eastern Cordillera of New Grenada, which is nearly north 45 degrees east from Santa Fe de Bogota, to beyond the Sierra Nevada de Merida, and of which the littoral chain is but a continuation, has had an influence on the direction (hor. 3 to 4) of the strata in Venezuela? That region presents a very remarkable loxodromism with the strata of mica-slate, grauwacke, and the orthoceratite limestone of the Alleghenies, and that vast extent of country (latitude 56 to 68 degrees) lately visited by Captain Franklin. The direction north-east to south-west prevails in every part of North America, as in Europe in the Fitchtelgebirge of Franconia, in Taunus, Westerwald, and Eifel; in the Ardennes, the Vosges, in Cotentin, in Scotland and in the Tarentaise at the south-west extremity of the Alps. If the strata of rocks in Venezuela do not exactly follow the direction of the nearest Cordillera, that of the shore, the parallelism between the axis of one chain, and the strata of the formations that compose it, are manifest in the Brazil group.* (* The strata of the primitive and intermediary rocks of Brazil run very regularly, like the Cordillera of Villarica (Serra do Espinhaco) hor. 1.4 or hor. 2 of the compass of Freiberg (north 28 degrees east.))

SECTION 3.
NATURE OF THE ROCKS. RELATIVE AGE AND SUPERPOSITION OF THE FORMATIONS. PRIMITIVE, TRANSITION, SECONDARY, TERTIARY, AND VOLCANIC STRATA.

The preceding section has developed the geographical limits of the formations, the extent of the direction of the zones of gneiss-granite, mica-slate-gneiss, clay-slate, sandstone and intermediary limestone, which come successively to light. We will now indicate succinctly the nature and relative age of these formations. To avoid confounding facts with geologic opinions I shall describe these formations, without dividing them, according to the method generally followed, into five groups—primitive, transition, secondary, tertiary and volcanic rocks. I was fortunate enough to discover the types of each group in a region where, before I visited it, no rock had been named. The great inconvenience of the old classification is that of obliging the geologist to establish fixed demarcations, while he is in doubt, if not respecting the spot or the immediate superposition, at least respecting the number of the formations which are not developed. How can we in many circumstances determine the analogy existing between a limestone with but few petrifactions and an intermediary limestone and zechstein, or between a sandstone superposed on a primitive rock and a variegated sandstone and quadersandstein, or finally, between muriatiferous clay and the red marl of England, or the gem-salt of the tertiary strata of Italy? When we reflect on the immense progress made within twenty-five years in the knowledge of the superposition of rocks, it will not appear surprising that my present opinion on the relative age of the formations of Equinoctial America is not identically the same with what I advanced in 1800. To boast of a stability of opinion in geology is to boast of an extreme indolence of mind; it is to remain stationary amidst those who go forward. What we observe in any one part of the earth on the composition of rocks, their subordinate strata and the order of their position are facts immutably true, and independent of the progress of positive geology in other countries; while the systematic names applied to any particular formation of America are founded only on the supposed analogies between the formations of America and those of Europe. Now those names cannot remain the same if, after further examination, the objects of comparison have not retained the same place in the geologic series; if the most able geologists now take for transition-limestone and green sandstone, what they took formerly for zechstein and variegated sandstone. I believe the surest means by which geologic descriptions may be made to survive the change which the science undergoes in proportion to its progress, will be to substitute provisionally in the description of formations, for the systematic names of red sandstone, variegated sandstone, zechstein and Jura limestone, names derived from American localities, as sandstone of the Llanos, limestone of Cumanacoa and Caripe, and to separate the enumeration of facts relative to the superposition of soils, from the discussion on the analogy of those soils with those of the Old World.*

(* Positive geography being nothing but a question of the series or succession (either simple or periodical) of certain terms represented by the formations, it may be necessary, in order to understand the discussions contained in the third section of this memoir, to enumerate succinctly the table of formations considered in the most general point of view.

1. Strata commonly called Primitive; granite, gneiss and mica-slate (or gneiss oscillating between granite and mica-slate); very little primitive clay-slate; weisstein with serpentine; granite with disseminated amphibole; amphibolic slate; veins and small layers of greenstone.

2. Transition strata, composed of fragmentary rocks (grauwacke), calcareous slate and greenstone, earliest remains of organized existence: bamboos, madrepores, producta, trilobites, orthoceratites, evamphalites). Complex and parallel formations; (a) Alternate beds of grey and stratified limestone, anthracitic mica-slate, anhydrous gypsum and grauwacke; (b) clay-slate, black limestone, grauwacke with greenstone, syenite, transition-granite and porphyries with a base of compact felspar; (c) Euphotides, sometimes pure and covered with jasper, sometimes mixed with amphibole, hyperstein and grey limestone; (d) Pyroxenic porphyries with amygdaloides and zirconian syenites.

3. Secondary strata, presenting a much smaller number of monocotyledonous plants; (a) Co-ordinate and almost contemporary formations with red sandstone (rothe todtes liegende), quartz-porphyry and fern-coal. These strata are less connected by alternation than by opposition. The porphyries issue (like the trachytes of the Andes) in domes from the bosom of intermediary rocks. Porphyritic breccias which envelope the quartzose porphyries. (b) Zechstein or Alpine limestone with marly, bituminous slate, fetid limestone and variegated gypsum (Productus aculeatus). (c) Variegated sandstone (bunter sandstein) with frequent beds of limestone; false oolites; the upper beds are of variegated marl, often muriatiferous (red marl, salzthon) with hydrated gypsum and fetid limestone. The gem-salt oscillates from zechstein to muschelkalk. (d) Limestone of Gottingen or muschelkalk alternating towards the top with white sandstone or brittle sandstein. (Ammonitis nodosus, encrinites, Mytilus socialis): clayey marl is found at the two extremities of muschelkalk. (e) White sandstone, brittle sandstein, alternating with lias, or limestone with graphites; a quantity of dicotyledonous mixed with monocotyledonous plants. (f) Jura limestone of complex formation; a quantity of sandy intercalated marl. We most frequently observe, counting from below upwards; lias (marly limestone with gryphites), oolites, limestone with polypi, slaty limestone with fish, crustacea, and globules of oxide of iron (Amonites planulatus, Gryphaea arcuata). (g) Secondary sandstone with lignites; iron sand; Wealden clay; greensand or green sandstone; (h) Chlorite; tufted and white chalk; (planerkalk, limestone of Verona.)

4. Tertiary strata, showing a much smaller number of dicotyledonous plants. (a) Clay and tertiary sandstone with lignites; plastic clay; mollasse and nagelfluhe, sometimes alternating where chalk is wanting, with the last beds of Jura limestone; amber. (b) Limestone of Paris or coarse limestone, limestone with circles, limestone of Bolca, limestone of London, sandy limestone of Bognor; lignites. (c) Silicious limestone and gypsum with fossil bones alternating with marl. (d) Sandstone of Fontainebleau. (e) Lacustrine soil with porous millstone grit. (e) Alluvial deposits.)

1. CO-ORDINATE FORMATIONS OF GRANITE, GNEISS AND MICA-SLATE.

There are countries (in France, the vicinity of Lyons; in Germany, Freiberg, Naundorf) where the formations of granite and gneiss are extremely distinct; there are others, on the contrary, where the geologic limits between those formations are slightly marked, and where granite, gneiss and mica-slate appear to alternate by layers or pass often from one to the other. These alternations and transitions appeared to me less common in the littoral Cordillera of Venezuela than in the Sierra Parime. We recognise successively, in the former of these two systems of mountains, above all in the chain nearest the coast, as predominating rocks from west to east, granite (longitude 70 to 71 degrees), gneiss (longitude 68 1/2 to 70 degrees), and mica-slate (longitude 65 3/4 to 66 1/2 degrees); but considering altogether the geologic constitution of the coast and the Sierra Parime, we prefer to treat of granite, gneiss and mica-slate, if not as one formation, at least as three co-ordinate formations closely linked together. The primitive clay-slate (urthonschiefer) is subordinate to mica-slate, of which it is only a modification. It no more forms an independent stratum in the New Continent, than in the Pyrenees and the Alps.

(a) GRANITE which does not pass to gneiss is most common in the western part of the coast-chain between Turmero, Valencia and Porto Cabello, as well as in the circle of the Sierra Parime, near the Encaramada, and at the Peak of Duida. At the Rincon del Diablo, between Mariara and Hacienda de Cura, and at Chuao, it is coarse-grained, and contains fine crystals of felspar, 1 1/2 inches long. It is divided in prisms by perpendicular vents, or stratified regularly like secondary limestone, at Las Trincheras, the strait of Baraguan in the valley of the Orinoco, and near Guapasoso, on the banks of the Atabapo. The stratified granite of Las Trincheras, giving birth to very hot springs (from 90.5 degrees centigrade), appears from the inclination of its layers to be superposed on gneiss which is seen further southward in the islands of the lake of Valencia; but conjectures of superposition founded only on the hypothesis of an indefinite prolongation of the strata are doubtful; and possibly the granite masses which form a small particular zone in the northern range of the littoral Cordillera, between 70 degrees 3 minutes and 70 degrees 50 minutes longitude, were upheaved in piercing the gneiss. The latter rock is prevalent, both in descending from the Rincon del Diablo southward to the hot-springs of Mariara, and towards the banks of the lake of Valencia, and in advancing on the east towards the group of Buenavista, the Silla of Caracas and Cape Codera. In the region of the littoral chain of Venezuela, where granite seems to constitute an independent formation from 15 to 16 leagues in length, I saw no foreign or subordinate layers of gneiss, mica-slate or primitive limestone.* (* Primitive limestone, everywhere so common in mica-slate and gneiss, is found in the granite of the Pyrenees, at Port d'Oo, and in the mountains of Labourd.)

The Sierra Parime is one of the most extensive granitic strata existing on the globe;* but the granite, which is seen alike bare on the flanks of the mountains and in the plains by which they are joined, often passes into gneiss. (* To prove the extent of the continuity of this granitic stratum, it will suffice to observe that M. Leschenault de la Tour collected in the bars of the river Mana, in French Guiana, the same gneiss-granites (with a little amphibole) which I observed three hundred leagues more to the west, near the confluence of the Orinoco and the Guaviare.) Granite is most commonly found in its granular composition and independent formation, near Encaramada, at the strait of Baraguan, and in the vicinity of the mission of the Esmeralda. It often contains, like the granites of the Rocky Mountains (latitude 38 to 40 degrees), the Pyrenees and Southern Tyrol, amphibolic crystals,* disseminated in the mass, but without passing to syenite. (* I did not observe this mixture of amphibole in the granite of the littoral chain of Venezuela except at the summit of the Silla of Caracas.) Those modifications are observed on the banks of the Orinoco, the Cassiquiare, the Atabapo, and the Tuamini. The blocks heaped together, which are found in Europe on the ridge of granitic mountains (the Riesengebirge in Silesia, the Ochsenkopf in Franconia), are especially remarkable in the north-west part of the Sierra Parime, between Caycara, the Encaramada and Uruana, in the cataracts of the Maypures and at the mouth of the Rio Vichada. It is doubtful whether these masses, which are of cylindrical form, parallelopipedons rounded on the edge, or balls of 40 to 50 feet in diameter, are the effect of a slow decomposition, or of a violent and instantaneous upheaving. The granite of the south-eastern part of Sierra Parime sometimes passes to pegmatite,* composed of laminary felspar, enclosed in curved masses of crystalline quartz. (* Schrift-granit. It is a simple modification of the composition and texture of granite, and not a subordinate layer. It must not be confounded with the real pegmatite, generally destitute of mica, or with the geographic stones (piedras mapajas) of the Orinoco, which contain streaks of dark green mica irregularly disposed.) I saw gneiss only in subordinate layers;* (* The magnetic sands of the rivers that furrow the granitic chain of the Encaramada seem to denote the proximity of amphibolic or chloritic slate (hornblende or chloritschiefer), either in layers in the granite, or superposed on that rock.); but, between Javita, San Carlos del Rio Negro, and the Peak of

Duida, the granite is traversed by numerous veins of different ages, abounding with rock-crystal, black tourmalin and pyrites. It appears that these open veins become more common on the east of the Peak of Duida, in the Sierra Pacaraina, especially between Xurumu and Rupunuri (tributaries of the Rio Branco and the Essequibo), where Hortsmann discovered, instead of diamonds* and emeralds, a mine (four) of rock-crystal. (* These legends of diamonds are very ancient on the coast of Paria. Petrus Martyr relates that, at the beginning of the sixteenth century, a Spaniard named Andres Morales bought of a young Indian of the coast of Paria admantem mire pretiosum, duos infantis digiti articulos longum, magni autem pollicis articulum aequantem crassitudine, acutum utrobique et costis octo pulchre formatis constantem. [A diamond of marvellous value, as long as two joints of an infant's finger, and as thick as one of the joints of its thumb, sharp on both sides, and of a beautiful octagonal shape.] This pretended adamas juvenis pariensis resisted the action of lime. Petrus Martyr distinguishes it from topaz by adding offenderunt et topazios in littore, [they pay no heed to topazes on the coast] that is of Paria, Saint Marta and Veragua. See Oceanica Dec. 3 lib. 4 page 53.)

(b) GNEISS predominates along the littoral Cordillera of Venezuela, with the appearance of an independent formation, in the northern chain from Cerro del Chuao, and the meridian of Choroni, as far as Cape Codera; and in the southern chain, from the meridian of Guigne to the mouth of the Rio Tuy. Cape Codera, the great mass of the Silla of Galipano, and the land between Guayra and Caracas, the table-land of Buenavista, the islands of the lake of Valencia, the mountains between Guigne, Maria Magdalena and the Cerro do Chacao are composed of gneiss;* (* I have been assured that the islands Orchila and Los Frailes are also composed of gneiss; Curacao and Bonaire are calcareous. Is the island of Oruba (in which nuggets of native gold of considerable size have been found) primitive?); yet amidst this soil of gneiss, inclosed mica-slate re-appears, often talcous in the Valle de Caurimare, and in the ancient Provincia de Los Mariches; at Cabo Blanco, west of La Guayra; near Caracas and Antimano, and above all, between the tableland of Buenavista and the valleys of Aragua, in the Montana de las Cocuyzas, and at Hacienda del Tuy. Between the limits here assigned to gneiss, as a predominant rock (longitude 68 1/2 to 70 1/2 degrees), gneiss passes sometimes to mica-slate, while the appearance of a transition to granite is only found on the summit of the Silla of Caracas.* (* The Silla is a mountain of gneiss like Adams Peak in the island of Ceylon, and of nearly the same height.) It would require a more careful examination than I was able to devote to the subject, to ascertain whether the granite of the peak of St. Gothard, and of the Silla of Caracas, really lies over mica-slate and gneiss, or if it has merely pierced those rocks, rising in the form of needles or domes. The gneiss of the littoral Cordillera, in the province of Caracas, contains almost exclusively garnets, rutile titanite and graphite, disseminated in the whole mass of the rock, shelves of granular limestone, and some metalliferous veins. I shall not decide whether the granitiferous serpentine of the table-land of Buenavista is inclosed in gneiss, or whether, superposed upon that rock, it does not rather belong to a formation of weisstein (heptinite) similar to that of Penig and Mittweyde in Saxony.

In that part of the Sierra Parime which M. Bonpland and myself visited, gneiss forms a less marked zone, and oscillates more frequently towards granite than mica-slate. I found no garnets in the gneiss of Parime. There is no doubt that the gneiss-granite of the Orinoco is slightly auriferous on some points.

(c) MICA-SLATE, with clay-slate (thonschiefer), forms a continuous stratum in the northern chain of the littoral Cordillera, from the point of Araya, beyond the meridian of Cariaco, as well as in the island of Marguerita. It contains, in the peninsula of Araya, garnets disseminated in the mass, cyanite and, when it passes to clayey-slate, small layers of native alum. Mica-slate constituting an independent formation must be distinguished from mica-slate subordinate to a stratum of gneiss, on the east of Cape Codera. The mica-slate subordinate to gneiss presents, in the valley of Tuy, shelves of primitive limestone and small strata of graphic ampelite (zeicheschiefer); between Cabo Blanco and Catia layers of chloritic, granitiferous slate, and slaty amphibole; and between Caracas and Antimano, the more remarkable phenomenon of veins of gneiss inclosing balls of granitiferous diorite (grunstein).

188

In the Sierra Parime, mica-slate predominates only in the most eastern part, where its lustre has led to strange errors.

The amphibolic slate of Angostura, and masses of diorite in balls, with concentric layers, near Muitaco, appear to be superposed, not on mica-slate, but immediately on gneiss-granite. I could not, however, distinctly ascertain whether a part of this pyritous diorite was not enclosed on the banks of the Orinoco, as it is at the bottom of the sea near Cabo Blanco, and at the Montana de Avila, in the rock which it covers. Very large veins, with an irregular direction, often assume the aspect of short layers; and the balls of diorite heaped together in hillocks may, like many cones of basalt, issue from the crevices.

Mica-slate, chloritic slate and the rocks of slaty amphibole contain magnetic sand in the tropical regions of Venezuela, as in the most northern regions of Europe. The gannets are there almost equally disseminated in the gneiss (Caracas), the mica-slate (peninsula of Araya), the serpentine (Buenavista), the chloritic slate (Cabo Blanco), and the diorite or greenstone (Antimano). These garnets re-appear in the trachytic porphyries that crown the celebrated metalliferous mountain of Potosi, and in the black and pyroxenic masses of the small volcano of Yana-Urca, at the back of Chimborazo.

Petroleum (and this phenomenon is well worthy of attention) issues from a soil of mica-slate in the gulf of Cariaco. Further east, on the banks of the Arco, and near Cariaco, it seems to gush from secondary limestone formations, but probably that happens only because those formations repose on mica-slate. The hot springs of Venezuela have also their origin in, or rather below, the primitive rocks. They issue from granite (Las Trincheras), gneiss (Mariara and Onoto) and the calcareous and arenaceous rocks that cover the primitive rocks (Morros de San Juan, Bergantin, Cariaco). The earthquakes and subterraneous detonations of which the seat has been erroneously sought in the calcareous mountains of Cumana have been felt with most violence in the granitic soils of Caracas and the Orinoco. Igneous phenomena (if their existence be really well certified) are attributed by the people to the granitic peaks of Duida and Guaraco, and also to the calcareous mountain of Cuchivano.

From these observations it results that gneiss-granite predominates in the immense group of the mountains of the Parime, as mica-slate-gneiss prevails in the Cordillera of the coast; that in the two systems the granitic soil, unmixed with gneiss and mica-slate, occupies but a very small extent of country; and that in the coast-chain the formations of clayey slate (thonschiefer), mica-slate, gneiss and granite succeed each other in such a manner on the same line from east to west (presenting a very uniform and regular inclination of their strata towards the north-west), that, according to the hypothesis of a subterraneous prolongation of the strata, the granite of Las Trincheras and the Rincon del Diablo may be superposed on the gneiss of the Villa de Cura, of Buenavista and Caracas; and the gneiss superposed in its turn on the mica-slate and clay-slate of Maniquarez and Chuparuparu in the peninsula of Araya. This hypothesis of a prolongation of every rock, in some sort indefinite, founded on the angle of inclination presented by the strata appearing at the surface, is not admissible; and according to similar equally vague reasoning we should be forced to consider the primitive rocks of the Alps of Switzerland as superposed on the formation of the compact limestone of Achsenberg, and that [transition, or identical with zechstein?] in turn, as being superposed on the molassus of the tertiary strata.

2. FORMATION OF THE CLAY-SLATE (THONSCHIEFER) OF MALPASSO.

If, in the sketch of the formations of Venezuela, I had followed the received division into primitive, intermediary, secondary and tertiary strata, I might be doubtful what place the last stratum of mica-slate in the peninsula of Araya should occupy. This stratum, in the ravine (aroyo) of Robalo, passes insensibly in a carburetted and shining slate, into a real ampelite. The direction and inclination of the stratum remain the same, and the thonschiefer, which takes the look of a transition-rock, is but a modification of the primitive mica-slate of Maniquarez, containing garnets, cyanite, and rutile titanite. These insensible passages from primitive to transition strata by clay-slate, which becomes carburetted at the same time that it presents a concordant position with mica-slate and gneiss, have also been observed several times in Europe by celebrated geologists. The existence of an independent formation of primitive slate (urthonschiefer) may even be doubted, that is, of a formation which is not joined below by strata containing some vestiges of monocotyledonous plants.

The small thonschiefer bed of Malpasso (in the southern chain of the littoral Cordillera) is separated from mica-slate-gneiss by a co-ordinate formation of serpentine and diorite. It is divided into two shelves, of which the upper presents green steatitous slate mixed with amphibole, and the lower, dark-blue slate, extremely fissile, and traversed by numerous veins of quartz. I could discover no fragmentary stratum (grauwacke) nor kieselschiefer nor chiastolite. The kieselschiefer belongs in those countries to a limestone formation. I have seen fine specimens of the chiastolite (macle) which the Indians wore as amulets and which came from the Sierra Nevada de Merida. This substance is probably found in transition-slate, for MM. Rivero and Boussingault observed rocks of clay-slate at the height of 2120 toises, in the Paramo of Mucuchies, on going from Truxillo to Merida.* (* In Galicia, in Spain, I saw the thonschiefer containing chiastolite alternate with grauwacke; but the chiastolite unquestionably belongs also to rocks which all geologists have hitherto called primitive rocks, to mica-schists intercalated like layers in granite, and to an independent stratum of mica-slate.)

3. FORMATION OF SERPENTINE AND DIORITE (GREEN-STONE OF JUNCALITO.)

We have indicated above a layer of granitiferous serpentine inclosed in the gneiss of Buenavista, or perhaps superposed on that rock; we here find a real stratum of serpentine alternating with diorite, and extending from the ravine of Tucutunemo as far as Juncalito. Diorite forms the great mass of this stratum; it is of a dark green colour, granular, with small grains, and destitute of quartz; its mass is formed of small crystals of felspar intermixed with crystals of amphibole. This rock of diorite is covered at its surface, by the effect of decomposition, with a yellowish crust, like that of basalts and dolerites. Serpentine, of a dull olive-green and smooth fracture, mixed with bluish steatite and amphibole, presents, like almost all the co-ordinate formations of diorite and serpentine (in Silesia, at Fichtelgebirge, in the valley of Baigorry, in the Pyrenees, in the island of Cyprus and in the Copper Mountains of circumpolar America),* traces of copper. (* Franklin's Journey to the Polar Sea page 529.) Where the diorite, partly globular, approaches the green slate of Malpasso, real beds of green slate are found inclosed in diorite. The fine saussurite which we saw in the Upper Orinoco in the hands of the Indians, seems to indicate the existence of a soil of euphotide, superposed on gneiss-granite, or amphibolic slate, in the eastern part of the Sierra Parime.

4. GRANULAR AND MICACEOUS LIMESTONE OF THE MORROS OF SAN JUAN.

The Morros of San Juan rise like ruinous towers in a soil of diorite. They are formed of a cavernous greyish green limestone of crystalline texture, mixed with some spangles of mica, and are destitute of shells. We see in them masses of hardened clay, black, fissile, charged with iron, and covered with a crust, yellow from decomposition, like basalts and amphiboles. A compact limestone containing vestiges of shells adjoins this granular limestone of the Morros of San Juan which is hollow within. Probably on a further examination of the extraordinary strata between Villa de Cura and Ortiz, of which I had time only to collect some few specimens, many phenomena may be discovered analogous to those which Leopold von Buch has lately described in South Tyrol. M. Boussingault, in a memoir which he has recently addressed to me, calls the rock of the Morros a problematic calcariferous gneiss. This expression seems to prove that the plates of mica take in some parts a uniform direction, as in the greenish dolomite of Val Toccia.

5. FELSPATHIC SANDSTONE OF THE ORINOCO.

The gneiss-granite of the Sierra Parime is covered in some few places (between the Encaramada and the strait of Baraguan and in the island of Guachaco) in its western part with an olive-brown sandstone, containing grains of quartz and fragments of felspar, joined by an extremely compact clayey cement. This cement, where it abounds, has a conchoidal fracture and passes to jasper. It is crossed by small veins of brown iron-ore, which separate into very thin plates or scales. The presence of felspar seems to indicate that this small formation of sandstone (the sole secondary formation hitherto known in the Sierra Parime) belongs to red sandstone or coal.* (* Broken and intact crystals of feldspar are found in the todte liegende coal-sandstone of Thuringia. I observed in Mexico a very singular agglomerated felspar formation superposed upon (perhaps inclosed in) red sandstone, near Guanaxuato.) I hesitate to class it with the sandstone of the Llanos, the relative antiquity of which appears to me to be less satisfactorily verified.

6. FORMATION OF THE SANDSTONE OF THE LLANOS OF CALABOZO.

I arrange the various formations in the order which I fancied I could discern on the spot. The carburetted slate (thonschiefer) of the peninsula of Araya connects the primitive rocks of gneiss-granite and mica-slate-gneiss with the transition strata (blue and green slate, diorite, serpentine mixed with amphibole and granular greenish-grey limestone) of Malpasso, Tucutunemo and San Juan. On the south the sandstone of the Llanos rests on this transition strata; it is destitute of shells and composed, like the savannahs of Calabozo, of rounded fragments of quartz,* kieselschiefer and Lydian stone, cemented by a ferruginous olive-brown clay. (* In Germany sandstones which belong unquestionably to red sandstone contain also (near Weiderstadt, in Thuringia) nodules, and rounded fragments. I shall not cite the pudding-stone subordinate to the red sandstone of the Pyrenees because the age of that sandstone destitute of coal may be disputed. Layers of very large rounded nodules of quartz are inclosed in the coal sandstone of Thuringia, and in Upper Silesia.) We there find fragments of wood, in great part monocotyledonous, and masses of brown iron-ore. Some strata, as in the Mesa de Paja, present grains of very fine quartz; I saw no fragments of porphyry or limestone. Those immense beds of sandstone that cover the Llanos of the Lower Orinoco and the Amazon well deserve the attention of travellers. In appearance they approximate to the pudding-stones of the molassus stratum, in which calcareous vestiges are also often wanting, as at Schottwyl and Diesbach in Switzerland; but they appeared to me by their position to have more relation to red sandstone. Nowhere can they be confounded with the grauwackes (fragmentary transition-rocks) which MM. Boussingault and Rivero found along the Cordilleras of New Grenada, bordering the steppes on the west. Does the want of fragments of granite, gneiss and porphyry, and the frequency of petrified wood,* (* The people of the country attribute those woods to the Alcornoco, Bowdichia virgilioides (See Nova Gen. et Spec. Plant. volume 3 page 377), and to the Chaparro bovo, Rhopala complicata. It is believed in Venezuela as in Egypt that petrified wood is formed in our times. I found this dicotyledonous petrified wood only at the surface of the soil and not inclosed in the sandstone of the Llanos. M. Caillaud made the same observation on going to the Oasis of Siwa. The trunks of trees, ninety feet long, inclosed in the red sandstone of Kifhauser (in Saxony), are, according to the recent researches of Von Buch, divided into joints, and are certainly monocotyledonous.) sometimes dicotyledonous, indicate that those sandstones belong to the more recent formations which fill the plains between the Cordillera of the Parime and the coast Cordillera, as the molassus of Switzerland fills the space between the Jura and the Alps? It is not easy, when several formations are not perfectly developed, to determine the age of arenaceous rocks. The most able geologists do not concur in opinion respecting the sandstone of the Black Forest and of the whole country south-west of the Thuringer Waldgebirge. M. Boussingault, who passed through a part of the steppes of Venezuela long after me, is of opinion that the sandstone of the Llanos of San Carlos, that of the valley of San Antonio de Cucuta and the table-lands of Barquisimeto, Tocuyo, Merida and Truxillo belong to a formation of old red sandstone or coal. There is in fact real coal near Carache, south-west of the Paramo de las Rosas.

Before a part of the immense plains of America was geologically examined, it might have been supposed that their uniform and continued horizontality was caused by alluvial soils, or at least by arenaceous tertiary strata. The sands which in the Baltic provinces and in all the north of Germany, cover coarse limestone and chalk, seem to justify these systematic ideas, which have been extended to the Sahara and the steppes of Asia. But the observations which we have been able to collect sufficiently prove that both in the Old and the New World, both plains, steppes, and deserts contain numerous formations of different eras, and that these formations often appear without being covered by alluvial deposits. Jura limestone, gem-salt (plains of the Meta and Patagonia) and coal-sandstone are found in the

Llanos of South America; quadersandstein,* (* The forms of these rocks in walls and pyramids, or divided in rhomboid blocks, seems no doubt to indicate quadersandstein; but the sandstone of the eastern declivity of the Rocky Mountains in which the learned traveller Mr. James found salt-springs (licks), strata of gypsum and no coal, appear rather to belong to variegated sandstone (buntersandstein).) a saliferous soil, beds of coal,* (* This coal immediately covers, as in Belgium, the grauwacke, or transition-sandstone.) and limestone with trilobites,* (* In the plains of the Upper Missouri the limestone is immediately covered by a secondary limestone with turritulites, believed to be Jurassic, while a limestone with grypheae, rich in lead-ore and which I should have believed to be still more ancient than oolitic limestone, and analogous to lias, is described by Mr. James as lying above the most recent formation of sandstone. Has this superposition been well ascertained?) fill the vast plains of Louisiana and Canada. In examining the specimens collected by the indefatigable Caillaud in the Lybian desert and the Oasis of Siwa, we recognize sandstone similar to that of Thebes; fragments of petrified dicotyledonous wood (from thirty to forty feet long), with rudiments of branches and medullary concentric layers, coming perhaps from tertiary sandstone with lignites;* (* Formation of molassus.); chalk with spatangi and anachytes, Jura limestone with nummulites partly agatized; another fine-grained limestone* employed in the construction of the temple of Jupiter Ammon (Omm-Beydah) (* M. von Buch very reasonably inquires whether this statuary limestone, which resembles Parian marble, and limestone become granular by contact with the systematic granite of Predazzo, is a modification of the limestone with nummulites, of Siwa. The primitive rocks from which the fine-grained marble was believed to be extracted, if there be no deception in its granular appearance, are far distant from the Oasis of Siwa.); and gem-salt with sulphur and bitumen. These examples sufficiently prove that the plains (llanos), steppes and deserts have not that uniform tertiary formation which has been too generally supposed. Do the fine pieces of riband-jasper, or Egyptian pebbles, which M. Bonpland picked up in the savannahs of Barcelona (near Curataquiche), belong to the sandstone of the Llanos of Calabozo or to a stratum superposed on that sandstone? The former of these suppositions would approach, according to the analogy of the observations made by M. Roziere in Egypt, the sandstone of Calabozo, or tertiary nagelfluhe.

7. FORMATION OF THE COMPACT LIMESTONE OF CUMANACOA.

A bluish-grey compact limestone, almost destitute of petrifactions, and frequently intersected by small veins of carburetted lime, forms mountains with very abrupt ridges. These layers have the same direction and the same inclination as the mica-slate of Araya. Where the flank of the limestone mountains of New Andalusia is very steep we observe, as at Achsenberg, near Altdorf in Switzerland, layers that are singularly arched or turned. The tints of the limestone of Cumanacoa vary from darkish grey to bluish white and sometimes pass from compact to granular. It contains, as substances accidentally disseminated in the mass, brown iron-ore, spathic iron, even rock-crystal. As subordinate layers it contains (1) numerous strata of carburetted and slaty marl with pyrites; (2) quartzose sandstone, alternating with very thin strata of clayey slate; (3) gypsum with sulphur near Guire in the Golfo Triste on the coast of Paria. As I did not examine on the spot the position of this yellowish-white fine-grained gypsum I cannot determine with any certainty its relative age.

([Footnote not indicated:] This sandstone contains springs. In general it only covers the limestone of Cumanacoa, but it appeared to me to be sometimes enclosed.)

The only petrifactions of shells which I found in this limestone formation consist of a heap of turbinites and trochites, on the flank of Turimiquiri, at more than 680 toises high, and an ammonite seven inches in diameter, in the Montana de Santa Maria, north-north-west of Caripe. I nowhere saw the limestone of Cumanacoa (of which I treat specially in this article) resting on the sandstone of the Llanos; if there be any such superposition it must be found on descending the table-land of Cocollar towards the Mesa de Amana. On the southern coast of the gulf of Cariaco the limestone formation probably covers, without the interposition of another rock, a mica-slate which passes to carburetted clay-slate. In the northern part of the gulf I distinctly saw this clayey formation at the depth of two or three fathoms in the sea. The submarine hot springs appeared to me to gush from mica-slate like the petroleum of Maniquarez. If any doubts remain as to the rock on which the limestone of

Cumanacoa is immediately superposed, there is none respecting the rocks which cover it, such as (1) the tertiary limestone of Cumana near Punta Delgada and at Cerro de Meapire; (2) the sandstone of Quetepe and Turimiquiri, which, forming layers also in the limestone of Cumanacoa, belongs properly to the latter soil; the limestone of Caripe which we have often identified in the course of this work with Jura limestone, and of which we shall speak in the following article.

8. FORMATION OF THE COMPACT LIMESTONE OF CARIPE.

Descending the Cuchillo de Guanaguana towards the convent of Caripe, we find another more recent formation, white, with a smooth or slightly conchoidal fracture, and divided in very thin layers, which succeeds to the bluish grey limestone formation of Cumanacoa. I call this in the first instance the limestone formation of Caripe, on account of the cavern of that name, inhabited by thousands of nocturnal birds. This limestone appeared to me identical (1) with the limestone of the Morro de Barcelona and the Chimanas Islands, which contains small layers of black kieselschiefer (slaty jasper) without veins of quartz, and breaking into fragments of parallelopiped form; (2) with the whitish grey limestone with smooth fracture of Tisnao, which seems to cover the sandstone of the Llanos. We find the formation of Caripe in the island of Cuba (between the Havannah and Batabano and between the port of Trinidad and Rio Guaurabo), as well in the small Cayman Islands.

I have hitherto described the secondary limestone formations of the littoral chain without giving them the systematic names which may connect them with the formations of Europe. During my stay in America I took the limestone of Cumanacoa for zechstein or Alpine limestone, and that of Caripe for Jura limestone. The carburetted and slightly bituminous marl of Cumanacoa, analogous to the strata of bituminous slate, which are very numerous* in the Alps of southern Bavaria (* I found them also in the Peruvian Andes near Montau, at the height of 1600 toises.), appeared to me to characterize the former of these formations; while the dazzling whiteness of the cavernous stratum of Caripe, and the form of those shelves of rocks rising in walls and cornices, forcibly reminded me of the Jura limestone of Streitberg in Franconia, or of Oitzow and Krzessowic in Upper Silesia. There is in Venezuela a suppression of the different strata which, in the old continent, separate zechstein from Jura limestone. The sandstone of Cocollar, which sometimes covers the limestone of Cumanacoa, may be considered as variegated sandstone; but it is more probable that in alternating by layers with the limestone of Cumanacoa, it is sometimes thrown to the upper limit of the formation to which it belongs. The zechstein of Europe also contains a very quartzose sandstone. The two limestone strata of Cumanacoa and Caripe succeed immediately each other, like Alpine and Jura limestone, on the western declivity of the Mexican table-land, between Sopilote, Mescala and Tehuilotepec. These formations, perhaps, pass from one to the other, so that the latter may be only an upper shelf of zechstein. This immediate covering, this suppression of interposed soils, this simplicity of structure and absence of oolitic strata, have been equally observed in Upper Silesia and in the Pyrenees. On the other hand the immediate superposition of the limestone of Cumanacoa on mica-slate and transition clay-slate—the rarity of the petrifactions which have not yet been sufficiently examined—the strata of silex passing to Lydian stone, may lead to the belief that the soils of Cumanacoa and Caripe are of much more ancient formation than the secondary rocks. We must not be surprised that the doubts which arise in the mind of the geologist when endeavouring to decide on the relative age of the limestone of the high mountains in the Pyrenees, the Apennines (south of the lake of Perugia) and in the Swiss Alps, should extend to the limestone strata of the high mountains of New Andalusia, and everywhere in America where the presence of red sandstone is not distinctly recognized.

9. SANDSTONE OF THE BERGANTIN.

Between Nueva Barcelona and the Cerro del Bergantin a quartzose sandstone covers the Jura limestone of Cumanacoa. Is it an arenaceous rock analogous to green sandstone, or does it belong to the sandstone of Cocollar? In the latter case its presence seems to prove still more clearly that the limestones of Cumanacoa and Caripe are only two parts of the same system, alternating with sandstone, sometimes quartzose, sometimes slaty.

10. GYPSUM OF THE LLANOS OF VENEZUELA.

Deposits of lamellar gypsum, containing numerous strata of marl, are found in patches on the steppes of Caracas and Barcelona; for instance, in the table-land of San Diego, between Ortiz and the Mesa de Paja; and near the mission of Cachipo. They appeared to me to cover the Jura limestone of Tisnao, which is analogous to that of Caripe, where we find it mixed with masses of fibrous gypsum. I have not given the name formation either to the sandstone of the Orinoco, of Cocollar, of Bergantin or to the gypsum of the Llanos, because nothing as yet proves the independence of those arenaceous and gypsous soils. I think it will one day be ascertained that the gypsum of the Llanos covers not only the Jura limestone of the Llanos, but that it is sometimes enclosed in it like the gypsum of the Golfo Triste on the east of the Alpine limestone of Cumanacoa. The great masses of sulphur found in the layers, almost entirely clayey, of the steppes (at Guayuta, valley of San Bonifacio, Buen Pastor, confluence of the Rio Pao with the Orinoco) may possibly belong to the marl of the gypsum of Ortiz. These clayey beds are more worthy of attention since the interesting observations of Von Buch and several other celebrated geologists respecting the cavernosity of gypsum, the irregularity of the inclination of its strata and its parallel position with the two declivities of the Hartz and the upheaved chain of the Alps; while the simultaneous presence of sulphur, oligist iron and the sulphurous acid vapours which precede the formation of sulphuric acid, seem to manifest the action of forces placed at a great depth in the interior of the globe.

11. FORMATION OF MURIATIFEROUS CLAY (WITH BITUMEN AND LAMELLAR GYPSUM) OF THE PENINSULA OF ARAYA.

This soil presents a striking analogy with salzthon or leberstein (muriatiferous clay) which I have found accompanying gem-salt in every zone. In the salt-pits of Araya (Haraia) it attracted the attention of Peter Martyr d'Anghiera at the beginning of the sixteenth century. It probably facilitated the rupture of the earth and the formation of the gulf of Cariaco. This clay is of a smoky colour, impregnated with petroleum, mingled with lamellar and lenticular gypsum and sometimes traversed by small veins of fibrous gypsum. It incloses angular and less friable masses of dark brown clay with a slaty and sometimes conchoidal fracture. Muriate of soda is found in particles invisible to the naked eye. The relations of position or superposition between this soil and the tertiary rocks does not appear sufficiently clear to enable me to pronounce with certainty on this element, the most important of positive geology. The co-ordinate layers of gem-salt, muriatiferous clay and gypsum present the same difficulties in both hemispheres; these masses, the forms of which are very irregular, everywhere exhibit traces of great commotions. They are scarcely ever covered by independent formations; and after having been long believed, in Europe, that gem-salt was exclusively peculiar to Alpine and transition limestone, it is now still more generally admitted, either from reasoning founded on analogy or from suppositions on the prolongation of the strata, that the true location of gem-salt is found in variegated sandstone (buntersandstein). Sometimes gem-salt appears to oscillate between variegated sandstone and muschelkalk.

I made two excursions on the peninsula of Araya. In the first I was inclined to consider the muriatiferous clay as subordinate to the conglomerate (evidently of tertiary formation) of the Barigon and of the mountain of the castle of Cumana, because a little to the north of that castle I had found shelves of hardened clay containing lamellar gypsum inclosed in the tertiary strata. I believed that the muriatiferous clay might alternate with the calcareous conglomerate of Barigon; and near the fishermen's huts situated opposite Macanao, conglomerate rocks appeared to me to pierce through the strata of clay. During a second excursion to Maniquarez and the aluminiferous slates of Chaparuparu, the connexion between tertiary strata and bituminous clay seemed to me somewhat problematical. I examined more particularly the Penas Negras near the Cerro de la Vela, east-south-east of the ruined castle of Araya. The limestone of the Penas is compact, bluish grey and almost destitute of petrifactions. It appeared to me to be much more ancient than the tertiary conglomerate of Barigon, and I saw it covering, in concordant position, a slaty clay, somewhat analogous to muriatiferous clay. I was greatly interested in comparing this latter formation with the strata of carburetted marl contained in the Alpine limestone of Cumanacoa. According to the opinions now most generally received, the rock of the Penas

Negras may be considered as representing muschelkalk (limestone of Gottingen); and the saliferous and bituminous clay of Araya, as representing variegated sandstone; but these problems can only be solved when the mines of those countries are worked. Those geologists who are of opinion that the gem-salt of Italy penetrates into a stratum above the Jura limestone, and even the chalk, may be led to mistake the limestone of the Penas Negras for one of the strata of compact limestone without grains of quartz and petrifactions, which are frequently found amidst the tertiary conglomerate of Barigon and of the Castillo de Cumana; the saliferous clay of Araya would appear to them analogous to the plastic clay of Paris,* (* Tertiary sandstone with lignites, or molassus of Argovia.) or to the clayey shelves (dief et tourtia) of secondary sandstone with lignites, containing salt-springs, in Belgium and Westphalia. However difficult it may be to distinguish separately the strata of marl and clay belonging to variegated sandstone, muschelkalk, quadersandstein, Jura limestone, secondary sandstone with lignites (green and iron sand) and the tertiary strata lying above chalk, I believe that the bitumen which everywhere accompanies gem-salt, and most frequently salt-springs, characterizes the muriatiferous clay of the peninsula of Araya and the island of Marguerita, as linked with formations lying below the tertiary strata. I do not say that they are anterior to that formation, for since the publication of M. von Buch's observations on the Tyrol, we must no longer consider what is below, in space, as necessarily anterior, relatively to the epoch of its formation.

Bitumen and petroleum still issue from the mica-slate; these substances are ejected whenever the soil is shaken by a subterranean force (between Cumana, Cariaco and the Golfo Triste). Now, in the peninsula of Araya, and in the island of Marguerita, saliferous clay impregnated with bitumen is met with in connexion with this early formation, nearly as gem-salt appears in Calabria in flakes, in basins inclosed in strata of granite and gneiss. Do these circumstances serve to support that ingenious system, according to which all the co-ordinate formations of gypsum, sulphur, bitumen and gem-salt (constantly anhydrous) result from floods passing across the crevices which have traversed the oxidated crust of our planet, and penetrating to the seat of volcanic action. The enormous masses of muriate of soda recently thrown up by Vesuvius,* (* The ejected masses in 1822 were so considerable that the inhabitants of some villages round Vesuvius collected them for domestic purposes.) the small veins of that salt which I have often seen traverse the most recently ejected lavas, and of which the origin (by sublimation) appears similar to that of oligist iron deposited in the same vents,* (* Gay-Lussac on the action of volcanoes in the Annales de Chimie volume 22 page 418.) the layers of gem-salt and saliferous clay of the trachytic soil in the plains of Peru and around the volcano of the Andes of Quito are well worthy the attention of geologists who would discuss the origin of formations. In the present sketch I confine myself to the mere enumeration of the phenomena of position, indicating, at the same time, some theoretic views, by which observers in more advantageous circumstances than I was myself may direct their researches.

12. AGGLOMERATE LIMESTONE OF THE BARIGON, OF THE CASTLE OF CUMANA, AND OF THE VICINITY OF PORTO CABELLO.

This is a very complex formation, presenting that mixture and that periodical return of compact limestone, quartzose sandstone and conglomerates (limestone breccia) which in every zone peculiarly characterises the tertiary strata. It forms the mountain of the castle of San Antonio near the town of Cumana, the south-west extremity of the peninsula of Araya, the Cerro Meapire, south of Caraco and the vicinity of Porto Cabello. It contains (1) a compact limestone, generally of a whitish grey, or yellowish white (Cerro del Barigon), some very thin layers of which are entirely destitute of petrifactions, while others are filled with cardites, ostracites, pectens and vestiges of lithophyte polypi: (2) a breccia in which an innumerable number of pelagic shells are found mixed with grains of quartz agglutinated by a cement of carbonate of lime: (3) a calcareous sandstone with very fine rounded grains of quartz (Punta Arenas, west of the village of Maniquarez) and containing masses of brown iron ore: (4) banks of marl and slaty clay, containing no spangles of mica, but enclosing selenite and lamellar gypsum. These banks of clay appeared to me constantly to form the lower strata. There also belongs to this tertiary stratum the limestone tufa (fresh-water formation) of the valleys of Aragua near Vittoria, and the fragmentary rock of Cabo Blanco, westward of the port of La Guayra. I must not designate the latter by the name of nagelfluhe, because that term indicates rounded fragments, while the fragments of Cabo Blanco are generally angular, and composed of gneiss, hyaline quartz and chloritic slate, joined by a limestone cement. This cement contains magnetic sand,* (* This magnetic sand no doubt owes its origin to chloritous slate, which, in these latitudes, forms the bed of the sea.) madrepores, and vestiges of bivalve sea shells. The different fragments of tertiary strata which I found in the littoral Cordillera of Venezuela, on the two slopes of the northern chain, seem to be superposed near Cumana (between Bordones and Punta Delgada); in the Cerro of Meapire; on the [Alpine] limestone of Cumanacoa; between Porto Cabello and the Rio Guayguaza; as well as in the valleys of Aragua; on granite; on the western declivity of the hill formed by Cabo Blanco, on gneiss; and in the peninsula of Araya, on saliferous clay. But this is perhaps merely the effect of apposition.* (* An-nicht Auflagerung, according to the precise language of the geologists of my country.) If we would range the different members of the tertiary series according to the age of their formation we ought, I believe, to regard the breccia of Cabo Blanco with fragments of primitive rocks as the most ancient, and make it be succeeded by the arenaceous limestone of the castle of Cumana, without horned silex, yet somewhat analogous to the coarse limestone of Paris, and the fresh-water soil of Victoria. The clayey gypsum, mixed with calcareous breccia with madrepores, cardites and oysters, which I found between Carthagena and the Cerro de la Popa, and the equally recent limestones of Guadalope and Barbadoes (limestones filled with seashells resembling those now existing in the Caribbean Sea) prove that the latest deposited strata of the tertiary formation extend far towards the west and north.

These recent formations, so rich in vestiges of organized bodies, furnish a vast field of observation to those who are familiar with the zoological character of rocks. To examine these vestiges in strata superposed as by steps, one above another, is to study the Fauna of different ages and to compare them together. The geography of animals marks out limits in space, according to the diversity of climates, which determine the actual state of vegetation on our planet. The geology of organized bodies, on the contrary, is a fragment of the history of nature, taking the word history in its proper acceptation: it describes the inhabitants of the earth according to succession of time. We may study genera and species in museums, but the Fauna of different ages, the predominance of certain shells, the numerical relations which characterize the animal kingdom and the vegetation of a place or of a period, should be studied in sight of those formations. It has long appeared to me that in the tropics as well as in the temperate zone the species of univalve shells are much more numerous than bivalves. From this superiority in number the organic fossil world furnishes, in every latitude, a further analogy with the intertropical shells that now live at the bottom of the ocean. In fact, M. Defrance, in a work* full of new and ingenious ideas, not only recognizes this preponderance of the univalves in the number of the species, but also observes that out of 5500 fossil univalve, bivalve and multivalve shells, contained in his rich collections, there are 3066 univalve, 2108 bivalve, and 326 multivalve; the univalve fossils are therefore to the bivalve as three to two. (* Table of Organized Fossil Bodies, 1824.)

13. FORMATION OF PYROXENIC AMYGDALOID AND PHONOLITE, BETWEEN ORTIZ AND CERRO DE FLORES.

I place pyroxenic amygdaloid and phonolite (porphyrschiefer) at the end of the formations of Venezuela, not as being the only rocks which I consider as pyrogenous, but as those of which the volcanic origin is probably posterior to the tertiary strata. This conclusion is not deduced from the observations I made at the southern declivity of the littoral Cordillera, between the Morros of San Juan, Parapara and the Llanos of Calabozo. In that region local circumstances would possibly lead us to regard the amygdaloids of Ortiz as linked to a system of transition rocks (amphibolic serpentine, diorite, and carburetted slate of Malpasso); but the eruption of the trachytes across rocks posterior to the chalk (in the Euganean Mountains and other parts of Europe) joined to the phenomenon of total absence of fragments of pyroxenic porphyry, trachyte, basalt and phonolite (The fragments of these rocks appear only in tufas or conglomerates which belong essentially to basaltic formations or surround the most recent volcanoes. Every volcanic formation is enveloped in breccia, which is the effect of the eruption itself.), in the conglomerates or fragmentary rocks anterior to the recent tertiary strata, renders it probable that the appearance of trap rocks at the surface of the earth is the effect of one of the last revolutions of our planet, even where the eruption has taken place by crevices (veins) which cross gneiss-granite, or the transition rocks not covered by secondary and tertiary formations.

The small volcanic stratum of Ortiz (latitude 9 degrees 28 minutes to 9 degrees 36 minutes) formed the ancient shore of the vast basin of the Llanos of Venezuela: it is composed on the points where I could examine it of only two kinds of rocks, namely, amygdaloid and phonolite. The greyish blue amygdaloid contains fendilated crystals of pyroxene and mesotype. It forms balls with concentric layers of which the flattened centre is nearly as hard as basalt. Neither olivine nor amphibole can be distinguished. Before it shows itself as a separate stratum, rising in small conic hills, the amygdaloid seems to alternate by layers with the diorite, which we have mentioned above as mixed with carburetted slate and amphibolic serpentine. These close relations of rocks so different in appearance and so likely to embarrass the observer give great interest to the vicinity of Ortiz. If the masses of diorite and amygdaloid, which appear to us to be layers, are very large veins, they may be supposed to have been formed and upheaved simultaneously. We are now acquainted with two formations of amygdaloid; one, the most common, is subordinate to the basalt: the other, much more rare,* (* We find examples of the latter in Norway (Vardekullen, near Skeen), in the mountains of the Thuringerwald; in South Tyrol; at Hefeld in the Hartz, at Bolanos in Mexico etc.) belongs to the pyroxenic porphyry.* (* Black porphyries of M. von Buch.) The amygdaloid of Ortiz approaches, by its oryctognostic characters, to the former of those formations, and we are almost surprised to find it joining, not basalt, but phonolite,* an eminently felspathic rock, in which we find some crystals of amphibole, but pyroxene very rarely, and never any olivine. (* There are phonolites of basaltic strata (the most anciently known) and phonolites of trachytic strata (Andes of Mexico). The former are generally above the basalts; and the extraordinary development of felspar in that union, and the want of pyroxene, have always appeared to me very remarkable phenomena.) The Cerro de Flores is a hill covered with tabulary blocks of greenish grey phonolite, enclosing long crystals (not fendillated) of vitreous felspar, altogether analogous to the phonolite of Mittelgebirge. It is surrounded by pyroxenic amygdaloid; it would no doubt be seen below, issuing immediately from gneiss-granite, like the phonolite of Biliner Stein, in Bohemia, which contains fragments of gneiss embedded in its mass.

Does there exist in South America another group of rocks, which may be preferably designated by the name of volcanic rocks, and which are as distinct from the chain of the Andes, and advance as far towards the east as the group that bounds the steppes of Calabozo? Of this I doubt, at least in that part of the continent situated north of the Amazon. I have often directed attention to the absence of pyroxenic porphyry, trachyte, basalt and lavas (I range these formations according to their relative age) in the whole of America eastward of the Cordilleras. The existence even of trachyte has not yet been verified in the Sierra Nevada de Merida which links the Andes and the littoral chain of Venezuela. It would seem as if volcanic fire, after the formation of primitive rocks, could not pierce into eastern America. Possibly the scarcity of argentiferous veins observed in those countries may be owing to the absence of more recent volcanic phenomena. M. Eschwege saw at Brazil some layers (veins?) of diorite, but neither trachyte, basalt, dolerite, nor amygdaloid; and he was therefore much surprised to see, in the vicinity of Rio Janeiro, an insulated mass of phonolite, exactly similar to that of Bohemia, piercing through gneiss. I am inclined to believe that America, on the east of the Andes, would have burning volcanoes if, near the shore of Venezuela, Guiana and Brazil, the series of primitive rocks were broken by trachytes, for these, by their fendillation and open crevices, seem to establish that permanent communication between the surface of the soil and the interior of the globe, which is the indispensable condition of the existence of a volcano. If we direct our course from the coast of Paria by the gneiss-granite of the Silla of Caracas, the red sandstone of Barquisimeto and Tocuyo, the slaty mountains of the Sierra Nevada de Merida, and the eastern Cordillera of Cundinamarca to Popayan and Pasto, taking the direction of west-south-west, we find in the vicinity of those towns the first volcanic vents of the Andes still burning, those which are the most northerly of all South America; and it may be remarked that those craters are found where the Cordilleras begin to present trachytes, at a distance of eighteen or twenty-five leagues from the present coast of the Pacific Ocean.* (* I believe the first hypotheses respecting the relation between the burning of volcanoes and the proximity of the sea are contained in Aetna Dialogus, a very eloquent though little-known work by Cardinal Bembo.) Permanent communications, or at least communications frequently renewed, between the atmosphere and the interior of the globe, have been preserved only along that immense crevice on which the Cordilleras have been upheaved; but subterranean volcanic forces are not less active in eastern America, shaking the soil of the littoral Cordillera of Venezuela and of the Parime group. In describing the phenomena which accompanied the great earthquake of Caracas,* on the 26th March, 1812, I mentioned the detonations heard at different periods in the mountains (altogether granitic) of the Orinoco. (* I stated in another place the influence of that great catastrophe on the counter-revolution which the royalist party succeeded in bringing about at that time in Venezuela. It is impossible to conceive anything more curious than the negociation opened on the 5th of April, by the republican government, established at Valencia in the valleys of Aragua, with Archbishop Prat (Don Narciso Coll y Prat), to engage him to publish a pastoral letter calculated to tranquilize the people respecting the wrath of the deity. The Archbishop was permitted to say that this wrath was merited on account of the disorder of morals; but he was enjoined to declare positively that politics and systematic opinions on the new social order had nothing in common with it. Archbishop Prat lost his liberty after this singular correspondence.) The elastic forces which agitate the ground, the still-burning volcanoes, the hot sulphurous springs, sometimes containing fluoric acid, the presence of asphaltum and naphtha in primitive strata, all point to the interior of our planet, the high temperature of which is perceived even in mines of little depth, and which, from the times of Heraclitus of Ephesus, and Anaxagoras of Clazomenae, to the Plutonic theory of modern days, has been considered as the seat of all great disturbances of the globe.

The sketch I have just traced contains all the formations known in that part of Europe which has served as the type of positive geology. It is the fruit of sixteen months' labour, often interrupted by other occupations. Formations of quartzose porphyry, pyroxenic porphyry and trachyte, of grauwacke, muschelkalk and quadersandstein, which are frequent towards the west, have not yet been seen in Venezuela; but it may be also observed that in the system of secondary rocks of the old continent muschelkalk and quadersandstein are not always clearly developed, and are often, by the frequency of their marls, confounded with the lower layers of Jura limestone. The muschelkalk is almost a lias with encrinites; and quadersandstein (for there are doubtless many above the lias or limestone with gryphites) seems to me to represent the arenaceous layers of the lower shelves of Jura limestone.

I have thought it right to give at some length this geologic description of South America, not only on account of the novel interest which the study of the formations in the equinoctial regions is calculated to excite, but also on account of the honourable efforts which have recently been made in Europe to verify and extend the working of the mines in the Cordilleras of Columbia, Mexico, Chile and Buenos Ayres. Vast sums of money have been invested for the attainment of this useful end. In proportion as public confidence has enlarged and consolidated those enterprises, from which both continents may derive solid advantage, it becomes the duty of persons who have acquired a local knowledge of these countries to publish information calculated to create a just appreciation of the relative wealth and position of the mines in different parts of Spanish America. The success of a company for the working of mines, and that of works undertaken by the order of free governments, is far from depending solely on the improvement of the machines employed for draining off the water, and extracting the mineral, on the regular and economical distribution of the subterraneous works, or the improvements in preparation, amalgamation, and melting: success depends also on a thorough knowledge of the different superposed strata. The practice of the science of mining is closely linked with the progress of geology; and it would be easy to prove that many millions of piastres have been rashly expended in South America from complete ignorance of the nature of the formations, and the position of the rocks, in directing the preliminary researches. At the present time it is not precious metals solely which should fix the attention of new mining companies; the multiplication of steam-engines renders it indispensable, wherever wood is not abundant or easy of transport, to seek at the same time to discover coal and lignites. In this point of view the precise knowledge of the red sandstone, coal-sandstone, quadersandstein and molassus (tertiary formation of lignites), often covered with basalt and dolerite, is of great practical importance. It is difficult for a European miner, recently arrived, to judge of a country presenting so novel an aspect, and when the same formations cover an immense extent. I hope that the present work, as well as my Political Essay on New Spain, and my work on the Position of Rocks in the Two Hemispheres, will contribute to diminish those obstacles. They may be said to contain the earliest geologic information respecting places whose subterraneous wealth attracts the attention of commercial nations; and they will assist in the classification of the more precise notions which later researches may add to my labours.

The republic of Colombia, in its present limits, furnishes a vast field for the enterprising spirit of the miner. Gold, platinum, silver, mercury, copper, gem-salt, sulphur and alum may become objects of important workings. The production of gold alone amounted, before the outbreak of the political dissensions, on the average, to 4700 kilogrammes (20,500 marks of Castile) per annum. This is nearly half the quantity furnished by all Spanish America, a quantity which has an influence the more powerful on the variable proportions between the value of gold and silver, as the extraction of the former metal has diminished at Brazil, for forty years past, with surprising rapidity. The quint (a tax which the government raises on gold-washings) which in the Capitana of Minas Geraes was, in 1756, 1761 and 1767, from 118, 102 and 85 arobas of gold (of 14 3/5 kilogrammes), has fallen, during 1800, 1813 and 1818, to 30, 20 and 9 arobas; an arob of gold having, at Rio Janeiro, the value of 15,000 cruzados. According to these estimates the produce of gold in Brazil, making deductions for fraudulent exportation, was, in the middle of the eighteenth century, the years of the greatest prosperity of the gold-washings, 6600 kilogrammes, and in our days, from 1817 to 1820, 600 kilogrammes less. In the province of San Paulo the extraction of gold has entirely ceased; in the province of Goyaz, it was 803 kilogrammes in 1793 and in 1819 scarcely 75. In the province of Mato Grosso it is almost nothing; and M. Eschwege is of opinion that the whole produce of gold in Brazil does not amount at present to more than 600,000 cruzados (scarcely 440 kilogrammes). I dwell on these particulars because, in confounding the different periods of the riches and poverty of the gold-washings of Brazil, it is still affirmed in works treating of the commerce of the precious metals, that a quantity of gold equivalent to four millions of piastres (5800 kilogrammes of gold*) flows into Europe annually from Portuguese America. (* This error is twofold: it is probable that Brazilian gold, paying the quint, has not, during the last forty years, risen to 5500 kilogrammes. I heretofore shared this error in common with writers on political economy, in admitting that the quint in 1810 was still (instead of 26 arrobas or 379 kilogrammes) 51,200 Portuguese ounces, or 1433 kilogrammes; which supposed a product of 7165 kilogrammes. The very correct information afforded by two Portuguese manuscripts on the gold-washings of Minas Geraes, Minas Novas and Goyaz, in the Bullion Report for the House of Commons, 1810, acc. page 29, goes as far only as 1794, when the quinto do ouro of Brazil was 53 arrobas, which indicates a produce of more than 3900 kilogrammes paying the quint. In Mr. Tooke's important work, On High and Low Prices part 2 page 2) this produce is still estimated (mean year 1810 to 1821) at 1,736,000 piastres; while, according to official documents in my possession, the average of the quint of those ten years amounted only to 15 arrobas, or a product quint of 1095 kilogrammes, or 755,000 piastres. Mr. John Allen reminded the Committee of the Bullion Report, in his Critical Notes on the table of M. Brongniart, that the decrease of the produce of the gold-washings of Brazil had been extremely rapid since 1794; and the notions given by M. Auguste de Saint Hilaire indicate the same desertion of the gold-mines of Brazil. Those who were miners have become cultivators. The value of an arroba of gold is 15,000 Brazilian cruzados (each cruzado being 50 sous). According to M. Franzini the Portuguese onca is equal to 0.028 of a kilogramme, and 8 oncas make 1 mark; 2 marks make 1 arratel, and 32 arratels 1 arroba.) If, in commercial value, gold in grains prevails, in the republic of Columbia, over the value of other metals, the latter are not on that account less worthy to fix the attention of government and of individuals. The argentiferous mines of Santa Anna, Manta, Santo Christo de las Laxas, Pamplona, Sapo and La Vega de Sapia afford great hope.

The facility of the communications between the coast of Columbia and that of Europe imparts the same interest to the copper-mines of Venezuela and New Grenada. Metals are a merchandize purchased at the price of labour and an advance of capital; thus forming in the countries where they are produced a portion of commercial wealth; while their extraction gives an impetus to industry in the most barren and mountainous districts.